Catalogue of Indian Insects

THE EDITOR

Thomas Bainbrigge Fletcher (1878-1950) was an English entomologist. He worked in the Royal Navy; he became an expert on the "microlepidoptera" and was appointed as the second Imperial Entomologist in India, succeeding Harold Maxwell-Lefroy at the Imperial Agricultural Research Institute at Pusa. He was a meticulous naturalist and very careful on matters of systematics and taxonomic nomenclature.

He produced a list of publications on India Entomology and Catalogue of Indian Insects. He also wrote general works on entomology including Some South Indian Insects, and a Veterinary Entomology for India and hints on collecting and preserving insects. He was Fellow of the Linnean Society, the Royal Entomological Society, Zoological Society of London and a President of the Cotteswold naturalists field club.

Catalogue of Indian Insects

Volume 4

–Editor–
Thomas Bainbrigge Fletcher

2016
Daya Publishing House®
A Division of
Astral International Pvt. Ltd.
New Delhi – 110 002

Catalogue of Indian Insects (5 Volumes Set)

Publisher's note:

The book "***Catalogue of Indian Insects***" by Thomas Bainbrigge Fletcher was published in 1921. But presently complete set not available anywhere. We have tried to reproduced the complete valuable set from different sources in a form as close to the original possible. This historical reproduction fulfill this long felt needs of students, teachers and researchers engaged in the field of entomology, zoology and other related discipline.

Cataloging in Publication Data—DK
Courtesy: D.K. Agencies (P) Ltd. <docinfo@dkagencies.com>

Catalogue of Indian insects / editor, Thomas Bainbrigge Fletcher.

volumes cm
First published in 1921.
ISBN: 978-93-86071-09-5 (Vol. 4)
ISBN: 978-93-5130-922-2 (Set) (International Edition)

1. Insects—Catalogs and collections—India. I. Fletcher, T. B. (Thomas Bainbrigge), 1878-1950, editor.

QL468.2.C38 2016 DDC 595.70954 23

Published by : **Daya Publishing House®**
A Division of
Astral International Pvt. Ltd.
– ISO 9001:2008 Certified Company –
4760-61/23, Ansari Road, Darya Ganj
New Delhi-110 002
Ph. 011-43549197, 23278134
E-mail: info@astralint.com
Website: www.astralint.com

Printed at : **Replika Press Pvt. Ltd.**

डा. एस. अय्यप्पन
सचिव एवं महानिदेशक
Dr. S. AYYAPPAN
SECRETARY & DIRECTOR GENERAL

भारत सरकार
कृषि अनुसंधान और शिक्षा विभाग एवं
भारतीय कृषि अनुसंधान परिषद
कृषि एवं किसान कल्याण मंत्रालय, कृषि भवन, नई दिल्ली 110 001

GOVERNMENT OF INDIA
DEPARTMENT OF AGRICULTURAL RESEARCH & EDUCATION
AND
INDIAN COUNCIL OF AGRICULTURAL RESEARCH
MINISTRY OF AGRICULTURE AND FARMERS WELFARE
KRISHI BHAVAN, NEW DELHI 110 001
Tel.: 23382629; 23386711 Fax: 91-11-23384773
E-mail: dg.icar@nic.in

Foreword

With rapidly increasing human population that hardly matched by the rate of increase in agriculture production, the per capita availability of food and other products is bound to fall short. Therefore, increasing production per unit area is the most plausible alternative to sustain self sufficiency in food, fodder and fibre. Insects have got both positive and negative impacts in agriculture. Insects become pests when they interfere with our welfare, aesthetics or profits. A reduction or loss in yield and quality of crops is partly due to result of insect feeding activities and as vector/host to pathogen transmissions. On the other hand these small creatures serve us a lot as pollinating agents of many important agricultural crops, as biological controlling agents of pests and weeds as well as by producing honey, silk and many other materials essential to us.

This book 'Catalogue of Indian Insects' is divided into 5 volumes and covers 29 parts. This historical compilation is a dedicated effort to present the different aspects of the insects in a very concise and relevant manner. I believe that the students and scientist working in the area of entomology, agriculture, fisheries, zoology, botany and veterinary science will find this book as a source of useful and authentic information.

I appreciate the efforts of Daya Publishing House, A Division of Astral International (P) Ltd. New Delhi.

(Dr. S. Ayyappan)
Secretary (DARE) & Director General (ICAR)
Krishi Bhawan, New Delhi – 110001

Preface

There is a immense number of insect species on the planet. The class Insecta is a large and diverse as well as important ecologically. When compared to other groups, the number of insect species is much larger and represents more than three quarters of all known animal species. Most insects are either beneficial or harmless. Yet there are large numbers which are economically important as pests, pollinators, natural enemies and other beneficial insects like honey bees, silkworms, etc. Insects are fascinating to investigate, their classification is complex. It important to learn the main differences among insects so you can distinguish one group from another. Knowledge of insect classification is of primary importance in carrying out proper control procedures and quality pest management. Proper insect classification and identification is extremely important. If a beneficial or harmless insect is improperly identified, the resultant unnecessary and undesirable pesticide application usually disrupts natural control agents.

Insects deliver a host of ecological services fundamental to the survival of humankind. Insects play an important role in plant reproduction. Thousands of pollinator species have been identified and almost all of these are insects. Over 90 percent of the flowering plant species depend on pollinators. Insects play an equally vital role in waste biodegradation. They clean up dead plant matter, breaking down organic matter until it is fit to be consumed by fungi and bacteria. Insects also play a significant role in decomposing animal carcasses and manure. Insects have a large array of natural enemies, predators and parasitoids, keeping them under economic threshold levels. Due to large number of insects species it is necessary that these catalogued so as to enable locating the species into correct classification. One shall also know what are the species knows from a specific group. Especially in case of Indian Insects where large number of species are known. These are important Catalogue of Indian Insects as a compilation serves this purpose.

'Catalogue of Indian Insects" is divided into 5 volumes and covers 29 parts. Volume 1 deals with Part I to Part IX. It covers Acrydidae, Culicidae, Bombyliidae, Trypetidae, Nitidulidae, Stapylinidae, Lasiocampidae, Amatidae and Zygaenidae. Volume 2 gives a detailed account of Part X to Part XVII. This volume covers Stephanidae, Brenthidae, Tabanidae,

Cicindelidae, Palpicornia, Cecidomyidae, Cosmopterygidae and Yponomeutidae. Volume 3 is devoted to Part XVIII i.e. Carabidae. Volume 4 covers in detail of Part XIX to Part XXV. This volume deals with Gyrinoidea, Alucitidae, Lycidae, Phalonidae and Chlidanotidae, Chalcidoidea, Evanidae and Thysanoptera. Volume 5 of the book describes the Part XXVI to Part XXIX. This volume deals with Serphoidea, Isoptera, Anthribidae and Asilidae.

This historical reproduction fulfill the long felt needs of students, teachers, researchers engaged in the field of entomology, agriculture, ecology, fisheries, forestry, paleontology, zoology, botany, plant science, medical and veterinary science, towards knowing insects.

We are extremely grateful to Dr. S. Ayyapppan, Secretary and Director General, Indian Council of Agricultural Reserarch, New Delhi for providing a foreword. The cooperation and support of Dr. V.V. Ramamurthy, Principal Scientist (Retd.), Division of Entomology, IARI, New Delhi is duly acknowledged.

All suggestions for future improvement for this book will be highly solicited.

Publisher

Contents

VOLUME 3

VOLUME 4

VOLUME 5

PART XIX - GYRINOIDEA

Superfamily: GYRINOIDEA

Gyrinoidea, Leng, Cat. Col. N. America 26 (1920); Tillyard, Ins. of Austral. and New Zeal. 187 (1926).

Family: GYRINIDAE.

Dytiscua, Lin., Syst. Nat. ed. X 412 (1758) [*partim*].

Gyrinus, Geoffr., Hist. Ins. Paris I 193 (1762); Linn., Syst. Nat. ed. XII 567 (1767); Fab., Syst. Ent. 234 (1775); Modeer, Physiogr. Sällsk. Handl. I 155 (1776); Oliv., Ent. III (41) I (1795); Fab., Syst. Eleuth. I 274 (1801).

Hydrocanthares, Latr., Hist. Nat. Crust. et Ins. VIII 149 (1803) [*partim*].

Otiophori, Latr., Gen. Crust. et. Ins. II 57 (1807) [*partim*].

Gyrinus, Schönh., Syn. Ins. 36 (1808).

Gyrini, Ahrens, Neue Schrift. Nat. Ges. Halle II 41 (1812); Forsb., Nov. Act. Upsal. VIII 297 (1821).

Gyrinii, Say, Trans. Amer. Phil. Soc. II 2 (1825).

Gyrinidœ, Macleay, Annul. Javan. ed. I 30 (1825), ed. II 132 (1833).

Gyriniarii, Perty, Delect. Anim. Artic. 15 (1830-34).

Hydrocanthares, Dejean, Cat. Col. ed. II 58 (1833) [*partim*].

Gyriniens, Brullé, Hist. Nat. Ins. V 226 (1835); Aubé, Icon. V 375 (1836-38), Spec. Col. VI 649 (1838).

Gyrinites, Lap., Etud. Ent. I 107 (1835), Hist. Nat. I 169 (1840).

Gyronecha, Kirby, Fn. bor. Amer. IV 78 (1837).

Gyrinen, Erichs., Käfer Mark Brandenburg I 190 (1837).

Gyrinida, (Leach) Heer, Fn. Helv. 165 (1839).

Gyriniteos, Sol., in Gay Hist. Fis. Chile Zool. IV 290 (1849-51).

Gyrinia, Mannerh., Bull. Soc. Imp. Moscou XXVI (II) 113 (1853).

Gyrinides, Lacord., Gen. Col. I 433 (1854).

Gyrinidœ, Rég., Ann. S. E. Fr. (6) II 379 Pl. 10-12 (1882), (6) III 121 Pl. 6, 381 Pl. 11-14 (1883), (6) VI 247 Pl. 4 (1886), LX p. 663 Pl. 18-19 (1891), LXXVI 137 (1907) [monograph]; Seidl., Verh. Naturf. Ver. Brünn XXV 26 (1887); Fowl., Col. Brit. Islds. I 209 (1887); Sévérin, Ann. S. E. Belg. XXXIII 149 (1889) [catalogue]; Ganglb., Käfer Mitteleurop. I 518 (1892); Sharp, Cambr. Nat. Hist. VI 215 (1899); Lameere, Ann. S. E. Belg. XLIV 376 (1900); Rég., in Wytsm Gen. Ins. I 1 (1902); Jacobs., Käfer Russl. I 436 (1905); Ahlw., in Junk-Schenkling Col. Cat. XXI 3 (1910) [catalogue]; Fowl., Fn. Brit. Ind. Col. 65 (1912); Sharp and Muir, Trans. E.S. London 493 Pl. XLIX f. 42, 43, 43a (1912).

Biremi, Gozis, Misc. Ent. XXIII (3) 1 (1920).

Gyrinidœ, Kerhervé, Bull. Soc. Sc. et Méd. Ouest XXXI 1 [sep.] (1922); Hatch, Bull. Brookl. E.S. XX 101 (1925); Handl., in Schröders Handb. d. Ent. III 554 (1925); Hatch, Pap. Mich. Acad. V 429 (1925), VII 311 Pl. XXXXIV (1926); Ochs, Ent. Zeitschr. Frankf. XXXX 61 (1926); Hatch, Bull. Brookl. E.S. XXII 89 (1927) [fossil]; Ochs, Rec. Ind. Mus. XXXI 248 (1929) [faunal distribution in India].

Subfamily I: GYRININAE, Rég.

Gyrinini, Rég., ann. S. E. Fr. (6) II 391 (1882), (6) III 124 (1883), in Wytsm. Gen. Ins. I 3 (1902).

Gyriniœ, Ahlw., in Junk–Schenkling Co., Cat. XXI 12 (1910); Ochs, Ent. Blätter XX 229 (1924).

Gyrinini, Hatch, Pap. Mich. Aead. V 443, 456 (1925).

Gyrininœ, Ochs., Ent. Zeitschr. Frankf. XXXX 62 (1926).

1. **Aulonogyrus** Motsch. (1853).

Type: *concinnus,* Klug (=*strigipennis,* Suffr.=*strictus* Motsch. 1853) [palaearctic].

Aulonogyrus, Motsch., Hydroc. Russ. 9 (1853); Rég., Ann. S. E. Fr. (6) III 124 (1883), in Wytsm. Gen. Ins. I 6 (1902); Hatch, Pap. Mich. Acad. V 456 (1925).

obliquus, Walker

Gyrinus obliquus, Walk., Ann. Mag. Nat. Hist. (3) II 205 (1858)[1]; Motsch., Bull. Mosc. XXXIV 109 (1861)[2]

Aulonogyrus obliquus, Rég., Ann. S. E. Fr. (6) III 137 (1883)[3], LXII 104 (1893)[4], LXXI 470 (1902)[5], LXXII 334 (1903)[6]; Zimmerm., Ent. Mitt. VI 142 (1917)[7].

Ind. merid.[3]
Ceylon: [1,2,3,7]—
Nuwara Eliya.[4]
Bandarawela.[5]
Pattipola.*
Maskeliya.*
Halgranoya.*
Diyatalawa.*
Kandy.*
Palni Hills:
Kukha.*
Kodaikanal.*
Nilgiris:
Ootacamund.*
Coonor R.[6],*
Benhope Bungalow*
Neduvattam.*
Salem Dt.: Shevaroys, Yercaud.*
Pondicherry.*
Madras.*

2. **Paragyrinus** Ochs (1924)

Type: *sinensis,* Ochs (China).

Paragyrinus, Ochs, Ent. Blätter XX 231 (1924).

arrowi, Rég.

Aulonogyrus arrowi, Rég., Ann. S. E. Fr. LXXVI 161 (1907)[1].

Paragyrinus arrowi, Ochs, Ent. Blätter XX 231 (1924)[2].

N.W. India.[1]
U.P., Chakrata, Bodyar.*
Nepal, Soondrijal.*
Assam, Shillong.*
Leos, Luang Prabang, Pak-Ten.[1]

N.W. India?*

vitalísi, Paschet.

Aulonogyrus vitalisi, Peschet, Opusc. Inst. Sc. Saigon I 5 (1923)[1]

Paragyrinus vitalisi, Ochs, Ent. Blätter XX 231 (1924).[2]

3. Gyrinus, Geoffroy (1762).

Type: *natator* L. (Palæarctic).

Dytiscus, Lin., Syst. Nat. ed. X 412 (1758).

Gyrinus, Geoffr., Hist. Ins. Paris I 193 (1762); Mod., Vet. Acad. Handl. XXXI 324 (1770); Degeer, Mém. Hist. Ins. Iv 361 Pl. 13 f. 16-20 (1774); Lacord., Fn. Ent. Paris I 340 (1835); Brullé, Hist. Nat. Ins. (V) II 235 (1835); Aubé, Icon. V 380 (1836-38), Spec. Col. VI 655 (1838); Suffr., Ent. Zeit. Stettin III 219 (1842); Lacord., Gen. Col. I 438 (1854); Rég., Ann. S. E. Fr. (6) III 141 (1883), in Wytsm. Gen. Ins. I 7 (1902); Zaitz., Rev. Russe d'Ent. VII 238 (1907); Zimmerm., Ent. Mitt. VI 143 (1917); Perfil., Russ. Hydrobiol. Zeit. IV 139 (1925) [larva]; Hatch. Pap. Mich. Acad. V 457 (1925); Eggers, Zool. Anzeiger LXVIII 184 (1926), LXXI 136 (1927) [biol.]; Bott, Zeitschr. f. Morphol. u. Oekol. X 207 Pl. 8-9 (1928) [biol., physiol.].

Subgenus: **Gyrinus** s. str (Geoffr.) Hatch. Subgenus *Gyrinus,* Hatch, Pap. Mich. Acad. V 458 (1925).

caspius, Mén., Cat. rais. Caucas. 142 (1832); Aubé, Icon. V 386 Pl. 44 f. 1 (1836-38), Spec. Col. VI 679 (1838).

———————

Central Europe, Mediterranean, W. Asia to Chinese Turkestan.

Ceylon.[1]

Gyrinus angustatus, Aubé, Icon. V 387 Pl. 44 f. 2 (1836-38), Spec. Col. VI 681 (1838).

Gyrinus elongatus, Aubé, Icon. V 384 Pl. 43 f. 4 (1836-38), Spec. Col. VI 676 (1838).

Gyrinus celox, Schiödte, Danm. Eleuth. 565 Pl. 23 f. a (1841).

Gyrinus caspius, Suffr., Stett. Ent. Z. III 244 (1842).

Gyrinus distinctus, Suffr., Stett. Ent. Z. III 245 (1842).

Gyrinus elongatus, Rég., Ann. S. E. Fr. (6) III 156 Pl. 6 f. 85 (1883); Seidl., Verh. Nat. V. Brünn XXV 114 (1887).

Gyrinus caspius, Seidl., Verh. Nat. V. Brünn XXV 114 (1887); Ganglb., Käf. Mitteleurop. I 528 (1892); Zaitz., Revue Russe d'Ent. VII 239 (1907).

Gyrinus elongatus, Sharp, Ent. Monthl. Mag. (2) XXV 134 Pl. 9 f. 6 (1914).

Gyrinus caspius, Zimmerm., Ent. Mitt. VI 160 Pl. 6 f. 8 (1917).

ceylonicus, Rég., Ann. S. E. Fr. (6) III 164 (1883)[1].

Ind. orient.[1]

convexiusculus, Macleay.

Gyrinus nitidulus, Aubé, Spec. Col. VI 700 (1838) [*partim*];[1] Motsch., Bull. Soc. Imp. Moscou XXXIV 109 (1861)[2].

Gyrinus convexiusculus, Macleay, Trans. E. S. N. S. Wales II 128 (1871)[3].

Gyrinus huttoni, Pascœ, Ann. Mag. Nat. Hist. (4) XIX 141 (1877)[4]; Broun, Man. N. Zeal Col. 76 (1880)[5].

Gyrinus convexiusculus, Rég., Ann. S. E. Fr. (6) III 166 Pl. 6 f. 88 (1883)[6]; Fauvel, Rev. d'Ent. II 349 (1883)[7]; Rég., Ann. S. E. Fr. (6) VI 257 (1886)[8]; Séverin, Bull. S. E. Belg. XXXIV 194 (1890)[9]; Fauvel, Rev. d'Ent. XXII 255 (1903)[10]; Rég., Ann. S. E. Fr. LXXVI 171 (1907)[11]; Zaitzev, Rev. Russe d'Ent. VII 242 (1907)[12]; Peschet, Opusc. Inst. Sc. Saigon I 7 (1923)[13]; Ochs, Ent. Blätter XX 229 (1924)[14].

Gyrinus huttoni, Tilly., Ins. of Austr. and N. Zeal. 193 (1926)[15].

Gyrinus convexiusculus, Ochs, Suppl. Ent. XV 115, 121 (1927)[16], Ent. Blätter XXIV 42 (1928), Rec. Ind. Mus. XXXI 241 (1929)[17].

Ceylon:[2, 16, 17]
Diyatalawa.[14]
Maskeliya.[14]
S. India:
Madras.[6, 14, 17]
Madras-Town.[17]
Kodaikanal.[14]
Palni Hills, Poubat-Havangi.[17]
Nilgiri-Hills, Coonoor-River.[17]
Mysore.[14, 17]
Bangalore.[14, 17]
Centr. India:
Bombay Pres., Satara dist., Medha, Yenna-Valley.[17]
Rawa-State, Amarkantak.[17]
Rawa-State, Dhud Dhara Falls.[17]
Rewa-State, Baratnala.[17]
Rewa-State, Sarai,[17]
Rewa-State, Umaria, Jaithari Range.[17]

convexiusculus, subsp. ?, Ochs, Suppl. Ent. XV 115 (1927)[1].

Himalayas.[13, 17]
Eastern Himalayas.[17]
Nepal, Soondrijal.[14, 17]
Bengal: Tetara.[9]
Kunbir-Nowatoli.[9]
Burma, S. Shan States: Hsamonghkam.[17]

Kalaw.[17]
Annam:[14, 16, 17]
Hué.[13]
Tibet.[?][6, 7, 10, 12, 13, 14, 16, 17]
China.[6, 7, 10, 12, 13, 14, 16, 17]
Sumatra.[14, 16, 17]
Java.[14, 16, 17]
N. Caledonia.[6, 7, 10, 12, 13, 14, 16, 17]
Australia[3, 6, 7, 9, 10, 12, 13, 14, 16, 17]
N. Zealand.[4, 5, 8, 11, 14, 15, 16, 17]
Liu-Kiu-Islands, Kagoshima.[1]

distinctus, Aubé, Icon. V. 383 Pl. 43 f. 3 (1836-38), Spec. Col. VI 666 (1838).

Gyrinus niloticus, Waltl., Isis VI 453 (1838).

Gyrinus colymbus, Erichs., Käf. Mark. Brandenburg I 191 (1837); Suffr., Stett. Ent. Z. III 247 (1842).

Gyrinus caspius, Sharp, Ent. Monthl. Mag. V 57 (1868); Rég., Ann. S. E. Fr. (6) III 158 (1883), (6) VI 251, 253 (1886).

Gyrinus œgyptiacus, Rég., Ann. S. E. Fr. (6) III 161 (1883).

Gyrinus niloticus, Rég., Ann. S. E. Fr. (6) III 161 (1883).

Gyrinus distinctus, Scidl., Verh. Nat. V. Brünn XXV 115 (1887).

Gyrinus colymbus, Edw., Entomologist XXIII 108 (1890); Zaitz., Rev. Russe d'Ent. VII 241 (1907); Sharp, Ent. Monthl. Mag. (2) XXV 135 Pl. 9 f. 7, Pl. 10 f. 14 (1914).

Kashmir: Chinar Bagh Nullah.*

Europe.
Egypt, Sennaar.
W. Asia to Chinese Turkestan and Mongolia.

Gyrinus distinctus, Zimmerm., Ent. Mitt. VI 158 Pl. 6 f. 10 (1917).	
orientalis, Rég., Ann. S. E. Fr. (6) III 167 (1883)[1].	India[?].[5]
Gyrinus dejeani, Rég., And. S. E. Fr. (6) III 169 (1883) [*partim*][2].	Yünnan.[8]
Gyrinus orientalis, Rég., Ann. S. E. Fr. LX 678 (1891)[3], LXXVI 170 (1907)[4]; Zaitzev, Rev. Russe d'Ent. VII 242 (1907)[5].	China.[1, 2, 3, 4, 5, 6, 7, 8]
Gyrinus dejeani, Zaitzev, Rev. Russe d'Ent. VII 242 (1907) [*partim*][6].	
Gyrinus orientalis, Zimmerm., Ent. Mitt. VI 163 (1917)[7]; Ochs, Proc. U.S. Nat. Mus. LXXV (III) 2 (1929)[8].	
orientalis politus, Ochs [female form], Proc. U. S. Nat. Mus. LXXV (III) 2 (1929).[1]	Szechuan.[1]
sericeolimbatus, Rég., Ann. S. E. Fr. (6) III 185 (1883)[1], LXXVI 181 (1907)[2]; Zimmerm., Ent. Mitt. VI 164 (1917)[3]; Ochs, Phil. Journ. of Sc. XXIV 82 (1924)[4]; Peschet, Treubia VII 6 (1925)[5]; Ochs, Suppl. Ent. XV 115 (1927)[6], Journ F. M. S. Mus. XIII 242 (1927)[7], Ent. Blätter XXIV 43 (1928)[8].	Penins. Siam.[6, 7, 8] Sumatra.[2, 4, 5, 6, 7, 8] Java.[1, 3, 4, 5, 6, 7, 8] Celebes.[1, 4, 5, 6, 7, 8] Buru.[5, 7, 8] N. Guinea.[2, 4, 5, 6, 7, 8] Philipp. Islands.[1, 4, 5, 6, 7, 8]
smaragdinus, Rég., Ann. Mus. Civ. Genova (2) X 547 (1891)[1], Ann. S. E. Fr. LX 682 (1891)[2]; Zimmerm., Ent. Mitt. VI 164 (1917)[3]; Peschet, Opusc. Inst. Sc. Saigon I 7 (1923)[4]; Ochs, Ent. Blätter XX 230 (1924)[5].	Assam:[5] Khasi-Hills.[1, 2, 5] Shilong.* Dumpep.* Burma: Taunggyi.* Carin Ghecu.[1, 2, 3, 5] Tenasserim.[5] Yünnan.[4] Upper Tonkin: Chapo.[4]

Subfamily II: ENHYDRINAE, Rég.

Enhydrini, Rég., Ann. S. E. Fr. (6) II 391, 392 (1882), in Wytsm. Gen. I 2 (1902).

Enhydrinæ, Ahlw., in Junk-Schenkling Col. Cat. XXI 4 (1910); Ochs, Ent. Blätter XX 233 (1924).

Enhydrini, Hatch, Pap. Mich. Acad. V 443, 444 (1925).

Enhydrinæ, Ochs, Ent. Zeitschr. Frankf. XXXX 62 (1926).

Tribe: DINEUTINI, Ochs.

Dineutini, Ochs, Ent. Zeitschr. Frankf. XXXX 63 (1926).

4. Dineutus, Macleay (1825). — Type: *politus* Macleay (Java).

Type: *politus,* Macleay (Java).

Dineutus, Macleay, Ann. Jav. ed. I 30 (1825), ed. II 133 (1833). — Type: *australis* Fabr. (Malayo-Australian).

Cyclous, Eschsch., in Dejean Cat. ed. II 58 (1833).

Dyneutes, Lap., Etud. Ent. 109 (1835).

Dineutes, Brullé, Hist. Ins. (V) II 234, 240 (1835).

Cyclinus, Kirby, in Richards. Fauna bor. Amer. IV 78 (1837). — Type: *assimilis* Kirby (Nearctic).

Dineutes, Aubé, Icon. V 407 (1836-38), Spec. Col. VI 761 (1838).

Dyneutes, Lap. Hist. Nat. I 170 (1840).

Dineutus, Lacord., Gen. Col. I 439 (1854).

Dineutes, Rég., Ann. S. E. Fr. (6) II 394 (1882); Lec. and Horn. Classif. 69 (1883); Rég., in Wytsm. Gen. Ins. I 3 (1902); Zimmerm., Ent. Blätter XVI 231 (1920).

Dineutus, Ochs, Ent. Blätter XX 233 (1924); Hatch, Pap. Mich. Acad. V 444, 447 (1925); Ochs. Ent. Zeitschr. Frankf. XXXX 64. 136 (1926).

Subg. **Spinosodineutes** Hatch (1925).

Spinosodineutes, Hatch, Pap. Mich. Acad. V 447 (1925). — Type: *spinosus* Fabr.

Cyclous [*Partim*], Hatch, Pap. Mich. Acad. V 447 (1925).

Gyrinodineutus, Ochs, Ent. Zeitschr. Frankf. XXXX 65, 195(1926). — Type: *unidentatus* Aubé.

Spinosidineutes, Hatch, pap. Mich. Acad. VII 343 (1926).

orientalis. Modeer.

Gyrinus orientalis, Mod., Physiogr. Sällsk. Handl. I (3) 160 (1776).

Dineutes marginatus, Sharp, Trans. E.S. London 56 (1873).

Dineutes quadrispina, Fairm., Ann. S.E. Fr. (5) VIII 88 (1878).

Ind. orient.[4, 7]
Yünnan:
Tengyueh to Talifu.*
Djo-Kou-La.*
Tche-Ping-Tcheou.*
Kouang-Si-Hien.*
Sze-Tsong.*

Dineutes marginatus, Rég., Ann. S.E. Fr. (6) II 425 Pl. 12f. 47, 47*a* (1882); Schönf., Ent. Nachr. 170 (1890); Rég., Ann. S. E. Fr. LXXVI 152 (1907); Peschet, Opusc. Inst. Sc. Saigon I 3 (1923).

Dineutus marginatus, Ochs, Ent. Blätter XX 236 (1924).

Dineutus (Gyrinodineutus) Orientalis, Ochs, Ent. Zeitschr. Frankf. XXXX 136, 190 (1926).

Yi-Leang.*
Laos:
Tranninh.[7], Tonkin:
Laokay, Chapo.*
Hoabinh.[7]
Mts. Mauson.*
Hanoi.[7], Hadong.[7]
Annam:
Phuc-Son.*
Cochin-China.[7]

China.[1, 3, 4, 7, 9]
Manchuria.[2, 7]
Korea.[7, 8] Formosa.*
Liu-Kiu-Islands.[5]
Japan.[2, 4, 7]
Java.[7][?]

spinosus, **Fabr.**

Gyrinus spinosus, Fabr. Spec. Ins. 298 (1781)[1], Mant. Ins. I 194 (1787)[2], Ent. Syst. I 203 (1792)[3]; Oliv., Ent. III 41, 13 Pl. 1 f. 7 (1795)[4]; Fabr., Syst. Eleuth. I 275 (1801)[5]; Forsb., Nov. Act. Upsal. VIII 300 (1821)[6]; Perty, Obs. Col. Ind. 12 (1831)[7].

Dineutes spinosus, Aubé, Spec. Col. VI 789 (1838)[8]; Motsch., Bull. S. Imp. Moscou XXXIV 109 (1861)[9]; Rég., Ann. S.E. Fr. (6) II 425 Pl. 12 f. 48, 48*a* (1882)[10], Ann. Mus. Civ. Genova (2) VI 619 (1888)[11], (2) X 547 (1891)[12], Mission Pavie III 81 (1904)[13]; Nowrojee, Mem. Dept. Agric. Ind. Ent. Ser. II, 9 177 (1912).[14]

Dineutus (Gyrinodineutus) spinosus, Ochs, Ent. Zeitschr. XXXX 136 (1926)[15], Kol. Rundsch. XIII 40 (1927)[16], Journ. F. M. S. Mus. XIII 245 (1927)[17], Ent. Blätter XXIV 43 (1928)[18], Journ. F. M.S. Mus. XIV 401 (1929)[19], Rec. Ind. Mus. XXXI 247 (1929)[20].

Ind. orient.[8, 10, 18, 19, 20]
Bengal:[7], Calcutta.*
Murshidabad.*
Sherpurtown,
Mymen. singh.*
Orissa Coast: Puri.*
Madras:
Coromandel.[1,3,4,5,6,7,16]
Pondicherry.*
Tranquebar.*
Ceylon.[6, 9]
U.P.: Lucknow.*
Naini Tal Dt.,
Chuharwala.*
Bihar:, Pusa.[14]
Kierpur.*
E. Himalayas:
Siliguri.*, Assam:[19]
Mangaldai.*, Tezpur.*
Selai-Kusi.*
Goalundo-Gauhati.*
Kowpati.*
Chittagong Hills:
Maini-Mukh.*
Burma:[19], Bhamo.[11, 12]
Teinzo.[11, 12]

spinosus, **Fabr.**— Katha.[11, 12] Rangoon.[11, 12] Pegu.[12], Palon.[12] Toungoo.[12], Bassein.* Tenasserim. Kawkareik.[11, 12] Peninsular Siam: Lampan-Petalung.[17] Malay Peninsula: Kedah, Jitra.[19] Perak, Bukit Merah.[19] Batang Padang.[19] Singapore.* Siam:[10, 19] Bangkok.*, Laos.* Tonkin: Than-Moi.* Annam: Phuc-Son.* Pha-Rang.*

Borneo.[10, 18, 19] Sumatra.*[?]

spinosus nepalensis, Ochs, Rec. Ind. Mus. XXXI 247 (1929)[1]. — Nepal: Soondrijal.[1] Katmandu.*

unidentatus, Aubé. — Ind. orient.[2, 6, 10, 12] Bengal:, Calcutta.[11] Berhampore Court.[11], Konbir.[11] Bowring.[11] Sherpur Town, Mymensingh.[11] Chota Nagpur: Manbhum.[11] Rewa State: Koilari.[11] Orissa Coast: Puri[11] Chilka Lake, Sta. 139[11] Ganjam dist., Rambha.[11], Madras: Pondicherry.[11] Tranquebar.[11] Madura.[11]

Dineutes unidentatus, Aubé, Spec. Col. VI 788 (1838)[1]; Rég., Ann. S.E. Fr. (^) II 424 Pl. 12 f. 46 (1882)[2], (6) IX 156 (1889)[3]; Sévérin, Bull. S.E. Belg. XXXIV 194 (1890)[4]; Nowrojee, Mem. Dept. Agric. Ind., Ent. Ser. II 9 177 [habits] Pl, 26 f. 7, 5 [larva], 6 [pupa], 4 [cocoon] (1912)[5]; Peschet, Opusc. Inst. Sc. Saigon I 4 (1923)[6].

Dineutus (Cyclous) unidentatus, Hatch, Bull. Brookl. E. S. XX 110 (1925) [larva][7] Dineutus (Gyrinodineutus) unidentatus, Ochs. Ent. Zeitschr. Frankf. XXXX 112, 136 (1926)[8], Journ. F. M. S Mus. XIII 245 (1927)[9], Ent. Blätter XXIV 43 (1928)[10], Rec. Ind. Mus. XXXI 242, 245 [larva] (1929)[11], Journ. F.M. S. Mus. XIV 401 (1929).[12]

unidentatus, Aubé.—

Ceylon:[10, 11, 12] Kandy.* Ambalangoda.* Karachi[?].[11] U.P., Lucknow.[11] Bihar: Pusa.[5, 11] Kierpur.[11] Penins. Siam:[11, 12] Lampan-Patalung.[9] Perak:[11] Kuala Kangsar.[12] Burma.[11, 12], Siam.[11] Tonkin:[11, 12] Hanoi,[6] Dap-Cau,[6] Kep.[6] Annam: Song-Dinh.* Cochin China: Saigon.[2], Mytho.[3] Pnomh-Penh.[3] Indo-china.[4, 6] Brasilia (erron.!)[1] China.[2, 6, 11] Java.[6, 10, 11, 12]

unidentatus curtulus, Rég.

Dineutes curtulus, Rég., Ann. S.E. Fr. LXXVI 151 (1907)[1]; Ochs, Phil. Journ. of Sc. XXIV 82 (1924)[2].

Dineutus (Gyrinodineutus) curtulus, Ochs, Ent. Zeitschr. Frankf. XXXX 136 (1926)[3].

Dineutus (*Gyrinodineutus*) *unidentatus* subsp. *curtulus,* Ochs. Ent. Blätter XXIV 43 (1928)[4], Rec. Ind. Mus. XXXI 242 (1929)[5].

Philippine Islds.[1, 2, 3, 4, 5]

Subg. **Protodineutus,** Ochs.

Protodineutus, Ochs, Ent. Zeitschr. Frankf. XXXX 66 (1926); Hatch, Pap. Mich. Acad. VII 344 (1926).

Type: *æreus Klug* (Africa, Arabia).

indicus, Aubé.

Dineutes indicus, Aubé, Spec. Col. VI 772 (1838)[1]; Rég., Ann. S.E. Fr. (6) II 405 (1882)[2]; Sévérin, Bull. S.E. Belg. XXXIV 194 (1890)[3]; Sharp, Scient. Res. 2nd

Ind. orient.[1, 2] Bengal: Calcutta.* Berhampore Court.* Murshidabad.* Burju.*, Raniganj.*

Yark. Miss. 39 (1890)[4]; Rég., Ann. S.E. Fr. LX 666 (1891)[5], LXII 104 (1892)[6]; Lesne, Bull. S.E. Fr. 86 [larva, probably of this species] (1902)[7]; Rég., Ann. S.E. Fr. LXXI 470 (1902)[8], LXXII 334 (1903)[9]; Fowl., Faun. Brit. Ind. f. 30 (1912)[10]; Zimmerm, Ent. Mitt. VI 135(1917)[11]; Kempers, Ent. Mitt. XII 75 f. 62 [wing] (1923)[12].

Dineutus indicus, Ochs, Ent. Blätter XX 234 (1924)[13].

Dineutus (*Protodineutus*) *indicus,* Ochs, Ent. Zeitschr. Frankf. XXXX 137 (1926)[14], Rec. Ind. Mus. XXXI 242, 245 [larva] (1929)[15].

Tetara.[3], Konbir.[3, 5] Parisnath.*, Bihar: Sahibganj.* Rajmahal.*, Tinpahar.* Birbhum.* Chota Nagpur.* Singhbhum dist. Ghadkuri-Manharpur.* Orissa: N. Simlipal Range.*

Subg. Dineutus [s. str.] (Macleay) Hatch.

Subg. *Dineutus,* Hatch, Pap. Mich. Acad. V 447 (1925); Ochs, Ent. Zeitschr. Frankf. XXXX 65 (1926); Hatch, Pap. Mich. Acad. VII 344 (1926).

Type: *politus* Macleay (Java).

mellyi, Rég.

Dineutes fulgidus, Rég. Not. Leid. Mus. II 214 (1880) [*partim*][1].

Dineutes mellyi, Rég., Ann. S.E. Fr. (6) II 399 (1882)[2].

Dineutes sauteri, Uyttenboogaart, Zool. Mededeel. 140 (1915)[3].

Dineutes mellyi, Zimmerm. Arch. f. Naturgesch. LXXXII (A, 5) 122 -1917)[4].

Dineutus mellyi, Ochs, Ent. Blätter XXI 174 (1925)[5], Ent. Zeitschr. Frankf. XXXX 139 (1926)[6].

Yünnan: Yünnan-fou.* Yünnan-sen.* Kut-Sin-Fou.* Kouang-Si-Hien.* Mit-Sao.* Tonkin: Laokay, Chapo.*

China.[1, 2] Formosa.[3, 4]

mellyi insularis, Rég.,

Dineutes insularis, Rég., Ann. S.E. Fr. LXXVII 140 (1907)[1].

Dineutus mellyi insularis, Ochs, Ent. Zeitschr. Frankf. XXXX 139, 193 (1926)[2],

Liu-Kiu-Islands.[1]

Subg. **Porrorhynchus** (Lap.)

Porrorhynchus, Lap., Etud. Ent. 108 (1835); Brullé, Hist. Ins. V (II) 239 (1835); Aubé, Icon. V 404 (1836-38), Spec. Col. VI 759 (1838); Lap., Hist. Nat. I. 170

Kerhervé. Type: *marginatus* Lap.

(1840); Lacord., Gen. Col. I 440 (1854); Rég., Ann. S.E. Fr. (6) II 427 (1882), in Wytsm. Gen. Ins. I 4 (1902); Hatch, Pap. Mich. Acad. V 444 (1925).

Subg. *Porrorhynchus,* Kerhervé, Soc. Sc. Méd. Ouest XXXI 2 [sep.] (1922); Ochs, Ent. Zeitschr. Frankf. XXXX 64 (1926); Hatch, Pap. Mich. Acad. VII 344 (1926).

barthelemyi, Rég.

Annam: Arroyos des Mois.[1]

Porrorhynchus barthelemyi, Reg., Ann. S. E. Fr. LXXVI 153 (1907)[1].

Dineutus (Porrorhynchus) barthelemyi, Ochs, Ent. Zeitschr. Frankf. XXXX 139 (1926)[2].

indicans, Walker.

Ceylon:[1, 2, 3, 4] Maskeliya.*

Java.[3, 4] (erron.!)

Dineutes indicans, Walker, Ann. Mag. Nat. Hist. (3) II 205 (1858)[1].

Porrorhynchus indicans, Motsch., Bull. Soc. Imp. Moscou XXXIV (I) 109 (1861)[2].

Porrorhynchus brevirostris, Rég., Ann. S. E. Fr. (5) VII 113 Pl. 6 f. 5 (1877)[3], (6) II 429 Pl. 12 f. 50 (1882)[4].

Porrorhynchus indicans Rég., Ann. S. E. Fr. (6) VI 250 (1886)[5].

Dineutus (Porrorhynchus) indicans, Ochs, Ent. Zeitschr. Frankf. XXXX 139 (1926)[6].

landaisi, Rég.

Tonkin: Cao-Bang, Ban-Khau.[1, 4] Thuyen-Quan.[3, 4] Vietri.[4], Thatké.[4] Backan.*, Laos.*

S.W. China.[3, 4]

Porrorhynchus landaisi, Rég., Ann. S. E. Fr. LX 667 Pl. 18 f. 1 (1891)[1], in Wytsm Gen. Ins. I f. 12 (1902)[2], LXXVI 152 (1907)[3]; Peschet, Opusc. Inst. Sc. Saigon I 5 (1923)[4]; Kempers, Ent. Mitt. XII 75 f. 63 [wings] (1923)[5]; Hatch, Journ. N. Y. E. S. XXXIV Pl. 26 f. 8 [eyes] (1926)[6], Pap. Mich. Acad. VII 311 [anatomy, numerous fig.] (1926)[7].

Dineutus (Porrorhynchus) landaisi, Ochs, Ent. Zeitschr. Frankf. XXXX 139 (1926)[8].

landaisi latilimbus, Ochs.

Hainan.[1, 2, 3]

Porrorhynchus landaisi [partim], Rég., Ann. S. E. Fr. LXXVI 153 (1907)[1]; Peschet, Opusc. Inst. Sc. Saigon I 5 (1923)[2].

Dineutus (Porrorhynchus) landaisi latilimbus, Ochs, Ent. Zeitschr. Frankf. XXXX 139, 193 (1926)[3].

marginatus, Lap.

Porrorhynchus marginatus, Lap., Etud. Ent. 108 (1835)[1]; Brullé, Hist. Ins. (V) II 239 Pl. 10 f. 4 (1835)[2]; Aubé, Icon. V 406 Pl. 46 f. 3 (1836-38)[3], Spec. Col. VI 759 (1838)[4]; Lap., Hist. Nat. I 170 (1840)[5]; Lacord., Gen. Col. I 440 Pl. 13 f. 4 (1854)[6]; Rég., Ann. S. E. Fr. (5) VII 110 Pl. 6 f. 3 (1877)[7], (6) II 427 Pl. 12 f. 49, 49a (1882)[8], Mission Pavie III 81 (1904)[9], Ann. S. E. Fr. LXXVI 152 (1907)[10]; Zimmerm., Ent. Mitt. VI 139 (1917)[11]; Peschet, Opusc. Inst. Sc. Saigon I 4 (1923)[12]; Kempers, Ent. Mitt. XII 76 f. 87 [wings] (1923)[13].

Dineutus (Porrorhynchus) marginatus, Ochs, Ent. Zeitschr. Frankf. XXXX 139 (1926)[14], Suppl. Ent. XV 116, 121 (1927)[15], Journ. F. M. S. Mus. XIII 242 (1927)[16], Ent. Blätter XXIV 44 (1928)[17].

Lower Burma: Tavoy.[15]
Tenasserim.[15, 16, 17]
Mergui.*, Perak:[16]
Kuala Kangsar.[10, 16]
Temangoh.[16]
Lenggong.[16]
Sungei Gemas, Lubok Kedondong.[16]
Pahang:[16]
Sungei Tahan.[16]
Kuala Tahan.[16]
Kuala Lipis.[16]
Kuala Teku.[16]
Lower Siam: Trong.*
Luang-Prabang: Theng.[9]
Upper Laos.[19]
Upper Mekong.[12]
Sunda-Archipelago:[7]
Java [?][1, 2, 3, 4, 5, 8, 12, 15, 16, 17]
Borneo.[7, 8, 12, 15, 17]

marginatus mjöbergi, Ochs, Ent. Zeitschr. Frankf. XXXX 139, 193 (1926)[1], Ent. Blätter XXIV 44 (1928)[2].

Sumatra.[11, 12, 15, 16, 17]
Banka.[15, 16, 17]
Nias. [?]*, Borneo.[1, 2]

marginatus tenuirostris, Rég.

Porrorhynchus tenuirostris, Rég., Ann. S. E. Fr. (5) VII 111 Pl. 6 f. 4 (1877)[1].

Porrorhynchus marginatus tenuirostris, Rég., Ann. S. E. Fr. (6) II 428 (1882)[2], LXXII 152 (1907)[3]; Peschet, Opusc. Inst. Sc. Saigon I 4 (1923)[4].

Dineutus (Porrorhynchus) marginatus tenuirostris, Ochs, Ent. Zeitschr. Frankf. XXXX 139 (1926)[5], Ent. Blätter XXIV 44 (1928)[6].

Siam.[2, 4]
Indo-china: Lakhon Hills.[2]
Annam:[4, 6]
Phuc-Son.[3, 6]
Song-Dinh.*
Pha-Rang.[6]
Cochin China:[4]
Phu-Quoc.

Subfamily III: ORECTOCHILINAE, Rég.

Orectochilini, Rég., Ann. S. E. Fr. (6) II 391 (1882), (6) III 381 (1883), in Wytsm. Gen. Ins. I 3 (1902).

Orectochilinæ, Ahlw., in Junk-Schenkling Col. Cat. XXI 26 (1910); Ochs, Ent. Blätter XX 238 (1924).

Orectochilini, Hatch, Pap. Mich. Acad. V 443, 451 (1925).

Orectochilinæ, Ochs, Ent. Zeitschr. Frankf. XXXX 62 (1926).

5. **Orectochilus** Eschsch. (1833).

Type: *villosus* O. Müller (Palaearctic).

Orectochilus, Eschsch., in Dejean Cat. Col. ed. II 59 (1833); Lacord., Fn. Ent. Paris I 344 (1835); Brullé, Hist. Nat. V., II 234, 238 (1835); Aubé, Icon. Col. V 399 (1836-38), Spec. Col. VI 726 (1838); Lacord., Gen. Col. I 441 (1854); Rég., Ann. S. E. Fr. (6) III 385, 410 (1883), in Wytsm. Gen. Ins. I 9 (1902); Hatch, Pap. Mich. Acad. V 452 (1925). — Type: *villosus* O. Müller (Palæarctic).

Patrus, Aubé Icon. Col. V. 397 (1836-38), Spec. Col. VI 724 (1838); Lacord., Gen. Col. I 441 (1854) — Type: *javanus* Aubé (Java).

Subgenus: **Orectochilus** s. str. Hatch.

Subgenus *Orectochilus* (s. str.), Hatch, Pap. Mich. Acad. V 453 (1925). — Type: *villosus* O. Müller (Palæarctic).

argenteolimbatus, Peschet, Opusc. Inst. Sc. Saigon I 17 (1923)[1]. — Upper Tonkin: Chapa.[1] Upper Mekong: Nam Long.[1]

castetsi. Rég., Ann. S. E. Fr. LX 709 (1891)[1]. — Kodaikanal.[1] Trichinopoly.* Palni Hills.* "Madras."*

murinus, Rég., Ann. S. E. Fr. LX 709 (1891)[1] LXXVI 215 (1907)[2]; Zimmerm., Ent. Mitt. VI 168 (1917)[3]; Peschet, Opusc. Inst. Sc. Saigon I 17 (1923)[4]. — E. Himalayas: Padong.[1, 4] Brit. Bhutan.[2, 4] Singla.* Kalimpong.* Pashok.* Kurseong.[4]

Ghumti.*
Upper Tonkin: Chapa.[4]
W. Himalayas: Kumaon, Malva Tal.* Valley of Sutlej, below Simla* Dharampur.* Mussoorie. Arni Gad.* Chakrata, Binal Gad.* Simla Hills, Ghuma.* Dehra Dun, Rikhikesh.*

tomentosus, Rég., Ann. Mus. Civ. Genova (2) X 553 (1891)[1], Ann. S. E. Fr. LX 711 Pl. 18 f. 15 (1891)[2]; Zimmerm., Ent. Mitt. VI 168 (1917)[3].
Burma: Carin Ghecu.[1, 2, 3]

villosovittatus, Rég., Ann. Mus. Civ. Genova (2) X 552 (1891)[1], Ann. S. E. Fr. LX 710 Pl. 18 f. 14 (1891)[2]; Zimmerm., Ent. Mitt. VI 168 (1917)[3]; Peschet, Opusc. Inst. Sc. Saigon I 17 (1923)[4].
Burma:[4] Carin Ghecu.[1, 2] Upper Mekong: Nam-Tiene.[4]

Subgenus: **Patrus** (Aubé) Hatch.

Subgenus *Patrus*, Hatch, Pap. Mic. Acad. V 452 (1925).
Type: *O. javanus* Aubé (Java).

aeneipennis, Rég., Ann. S. E. Fr. LXXVI, 217 (1907)[1]; Nowrojee, Mem. Dept. Agric. Ind. Ent. Ser. II 180 (1912)[2]; Zimmermann, Ent. Mitt. VI 168 (1917)[3].
Kanara.[1, 3] Pusa.[2] [?]+).

andamanarum, Ochs.

Orectochilus scalaris (ex parte), Rég., Ann. S. E. Fr. (6) III 430 Pl. 12 f. 134a (1883)[1], LX 705 (1891)[2]; Zimmermann, Ent. Mitt. VI 168 (1917)[3];
Andaman Islds.[1, 2, 3, 4]

Orectochilus scalaris subsp. *andamanarum*, Ochs, Rec. Ind. Mus. XXVII 200 (1925)[4];

Orectochilus andamanarum, Ochs, Journ. F. M. S. Mus. XIII 244 (1927)[5], Suppl. Emt. XV 118 (1927)[6].

+) This citation is certainly erroneous. As I was informed by Prof. T. B. Fletcher the only specimen of *O. œneipennis* in the collection of the Agricultural Research Institute at Pusa is a Paratype from Kanara. The species was not collected at Pusa, and Nowrojee meant probably *O. cylindricus* Rég., which was abundantly taken there together with *O. gangeticus* Wied.

andamanicus, Rég., Ann. S. E. Fr. (6) III 435 Pl. 12 f. 138 (1883)[1], (6) VI 265 Pl. 4 f. 14 (1886)[2]; Ochs, Ent. Blätter XX 238 (1924)[3]; Journ. F. M. S. Mus. XIV 403 (1929)[4].	Andaman Islad.[1, 3, 4] Calcutta.[2, 4] Burma: Mergui.[4] Perak.[4]
angusticinctus, Rég., Ann. S. E. Fr. LXXVI 213 (1907)[1].	Eastern Cochin China Reu Hai.[1] Tonkin: Lak-Tho.*
angusticinctus, subsp.?	Burma: Mergui.*
annadalei, Ochs, Rec. Ind. Mus. XXVII 203 (1925)[1].	Travancore: Tenmalai.[1] Madras: Nilambur.* Base of Nilgiris: Mettupalaiyam.*
apicalis, Rég., Ann. Mus. Civ. Genova (2) X 548 (1891)[1], Ann. S. E. Fr. LX 691 Pl. 18 f. 4 (1891)[2].	Burma: Carin Hills Asciuii Ghecu.[1, 2]
apicalis subapicalis, Ochs, nov. subsp.**) *Orectochilus apicalis,* Peschet, Opusc. Inst. Sc. Saigon I 8 (1923)[1].	Upper Laos: Nam Mat, Pou Lan, Nam Tiene.[1]
bírmanicus, Rég., Ann. Mus. Civ. Genova (2) X 550 (1891)[1]; Ann. S. E. Fr. LX 694 Pl♂18 f. 7. (1891)[2]; Zimmerm. Ent. Mitt. VI 167 (1917)[3]; Peschet, Opusc. Inst. Sc. Saigon I 10 (1923)[4], Bull. Soc. Ent. Fr. 254 (1925)[5].	Burma: Carin Ghecu.[1, 2, 3] French Ind-china: Bab-Bab, Nam-Long, Nam-Teine, Nam-Mat, Chapa.[4]
cameroni, Ochs, Rec. Ind. Mus. XXVII 195 (1925)[1].	Western-Himalayas: Dehra Dun, Kaligad.[1] Mussoorie, Arni Gad.[1]
cameroni subsp.?	Kumaon, Bhowali.[1] Valley of Sutlej, below Simla.*

** I have seen the type specimen of *O. apicalis* Rég., a single ♀ in the Museum at Genova, and am indebted to Mr. Peschet for some of his examples from Upper Laos. Compared with the type, the females of the latter differ by the smooth portion of elytra being less strongly sinuate posteriorly, so the apical terminations of this smooth portion are less prominent and less acute. Moreover, in the specimens from Laos, the exterior apical angles of elytra are less prominent, and to exterior apical angles of the anterior tibiæ are obtuse, not produced. It certainly represents a well distinguished varietal form, for which I propose the name mentioned above. Holotype in my collection, Allotype ♀in the Museum at Genova; type locality: Nam Mat.

cardiophorus, Rég., Ann. Mus. Civ. Genova (2) VI 620 (1888)[1], (2) X 548 (1891)[2], Ann. S. E. Fr. LX 693 Pl. 18 f. 6 (1891)[3]; Zimmerm., Ent. Mitt. VI 166 (1917)[4].	Burma: Catcin.[1, 2, 3] Carin Ghechu.[2, 3, 4] Carin Asciuii Ghecu.[2, 3] Tenasserim: Plapoo, Mt. Moleyit.[1, 2, 3]
cardiophorus peguensis, Peschet	Pegu.
cardoni, Rég., Ann. S. E. Fr. LX 714 Pl. 18 f. 17 (1891)[1], Ann. S. E. Belg. XXXVI 121 (1892)[2].	Bengal: Tetara.[1, 2] Nilgiris.*
cavernicola, Ochs, Rec. Ind. Mus. XXVII 196 (1925)[1]; cf. Chopra, Rec. Ind. Mus. XXVI 14, 18, 19 (1924)[2].	Assam: Garo-Hills, Siju Cave.[1, 2]
ceylonicus, Redtenb.	
Gyretes ceylonicus, Redtenb., Reise Novara Zool. II first section A[1] 24 Pl. 1 f. 10 (1868)[1]. *Orectochilus conspicuus,* Rég., Not. Leyd. Mus. IV 64 (1882)[2]. *Orectochilus ceylonicus,* Rég., Ann. S. E. Fr. (6) III 416 Pl. 12 f. 125, 125a (1883)[3]; Ochs, Ent. Blätter XX 238 (1924)[4].	Ceylon.[1, 3] Ceylon: Rambodde.[2] Lindula.[4], Maskeliya.* Madras.9(?)[3]
choprai, Ochs.	
Orestochilus chopræ, Ochs, Rec. Ind. Mus. XXVII 197 (1925)[1]; cf. Chopra, Rec. Ind. Mus. XXVI 14, 18, 19 (1924)[2].	Assam: Garo-Hills, Siju Cave.[1, 2]
coimbatorensis, Ochs, Rec. Ind. Mus. XXVII 194 (1925)[1].	Tinnevelly-Hills.[1]
coomani, Peschet, Bull. Soc. Ent. Fr. 253 (1925).	Tonkin: Hoah Binh,[1] Lac Tho.*
cardatus, Rég., Ann. Mus. Civ. Genova VI (2) 621 (1888),[1] X (2) 551 (1891)[2], Ann. S. E. Fr. LX 696 Pl. 18 f. 8 (1891)[3].	Tenasserim: Thagata, Mt. Moleyit.[1, 2, 3] Tenasserim.* Burma: Tavoy, Nwalabo.* Tonkin: Lac Tho.*
corniger, Zaitzev, Rev. Russe d'Ent. X 224 fig. (1910)[1]; Zimmerm., Ent. Mitt. VI 168 (1917)[2]; Peschet, Opusc. Inst. Sc. Saigon I 11, 12 (1923)[3]; Ochs, Suppl.	Tonkin: Hoah Binh.[3, 4, 5] Annam: Song Dinh.*

Ent. XV 117, 122 (1927)[4], Ent. Blätter XXIV 101 (1928)[5].	——— Japan. [?][1, 3, 4, 5] Borneo.[4, 5] Java.* Sumatra.[1, 5]
coronatus, Ochs, Rec. Ind. Mus. XXVII 203 (1925)[1].	N. E. Assam: Sadiya. Dikrang River.[1]
corporaali, Peschet, Bull. Soc. Ent. Fr. 293 (1922); Ochs, Suppl. Ent. XV. 116, 122 (1927)[2], Journ. F. M. S. Mus. XIII 244 (1927)[3], XIV 401 (1929)[4].	Malacca.[2] Selangor.[3, 4] Perak.[4], Penang.[4] Singapore.[4] ——— Sumatra.[1, 2, 3], Java.*
corpulentus, Rég., Ann. S. E. Fr. (6) III 411 Pl. 12 f. 121 (1883)[1], Ann. Mus. Civ. Genova VI (2) 619 (1888)[2], X (2) 548 (1891)[3], Ann. S. E. Fr. LX 689 (1891)[4]; Zimmerm., Ent. Mitt. VI 165 (1917)[5]; Ochs, Journ. F. M. S. Mus. XIII 245 (1927)[6], Ent. Blätter XXIV 79 (1928)[7].	Tenasserim: Thagata, Mt. Moleyit.[2, 3, 4, 5] Burma: Carin Hills, Cheba.[3, 4, 5] Burma.[7] Lower Burma: Amherst Dt., Dawna-Hills.* Upper Tenasserim.* Tenasserim.*, Tavoy.* Tavoy, Nwalabo.* Salanga.[6] Siam: Renong.[4] ——— Borneo[?][1, 2, 7] Java?[1]
cribratellus, Rég., Ann. Mus. Civ. Genova (2) X 551 (1891)[1], Ann. S. E. Fr. LX 700 Pl. 18 f. 12, 12a (1891)[2]; Zimmerm., Ent. Mitt, VI 168 (1917)[3].	Burma: Carin Ghecu.[1, 2] Upper Tenasserim.*
cribratellus metallescens, Rég., Ann. S. E. Fr. LXXVI 206 (1907)[1]; Zimmerm., Ent. Mitt. VI 168 (1917)[2].	Sikkim: Padong.[1] Kurseong.[1] Brit. Bhutan.[1] Kalimpong.* Sitong near Mungpu.*, Pashok.* Ghumti.*
cribratellus subsp.?	Assam: Garo Hills, above Tura.*

euneatus, Rég., Ann. S. E. Fr. LX 702 (1891)[1], LXXVI 202 (1907) [female][2]; Zimmerm., Ent. Mitt. VI 166 (1917)[3].

Sikkim:[1], Padong.[2]
Sukna.*, Singla.*
Kurseong.*
Kalimpong.*
Pashok.*, Ghumti.*
W. Himalayas:
Dharampur.*
Kathgodam,
Kumaon.*
Kali Gad.
Dehra Dun.*
Dwaharath,
Almora.*

cylindricus, Rég., Ann. S. E. Fr. LX 713 (1891)[1]; Ochs, Rec. Ind. Mus. XXVII 202 (1925) [female][2].

Murshidabad.[1, 2]
Pusa.[2]

desgodinsi, Rég., Ann. S. E. Fr. (6) VI 260 Pl. 4 f. 10 (1886)[1], LX 691 (1891)[2]; Zimmerm., Ent. Mitt. VI 166 (1917)[3]; Ochs, Rec. Ind. Mus. XXVII 194 (1925)[4].

Sikkim:
Padong.[1, 2, 4]
Brit. Bhutan.*
Kalimpong.[4]
Mungpu.[4], Sureil.[4]
Pashok.*, Ghumti.*
Bengal: Buxar.[4]

desgodinsi assamensis, Ochs, Rec. Ind. Mus. XXVII 194 (1925)[1]l cf. Rég., Ann. S. E. Fr. LXXVI 193 (1907)[2].

Assam: Garo Hills above Tura.[1]
Assam.[1]
Khasi Hills.[1, 2]

dilatatus, Redtenbacher.

Patrus dilatatus, Redtenb., Reise Novara, Zool. II, first section A, 1 25 Pl. 1 f. 12 (1868)[1].

Orectochilus dilatatus, Rég., Ann. S. E. Fr. (6) III 419 Pl. 12 f. 126 (1883)[2], LX 695 (1891)[3], LXXVI 203 (1907)[4].

Ceylon.[1, 2, 3, 4]
Madras [?][2]

discifer, Walker.

Gyrinus discifer, Walker, Ann. Mag. Nat. Hist. (3) III 51 (1859)[1].

Gyretes discifer, Motsch., Bull. Soc. Imp. Moscou XXXIV 109 (1861)[2]; Redtenb., Reise Novara, Zool. II, first section A, 1 24 Pl. 1 f. 11 (1868)[3].

Orectochilus discifer, Rég., Ann. S. E. Fr. (6) III 416 (1883)[4], (6) VI 262 (1886)[5], LXII 104 (1893)[6], LXXI 470 (1902)[7], LXXII 334 (1903)[8]; Zimmerm., Ent.

Ceylon:[1, 2, 3, 4, 10, 11]
Peradeniya.*
Kandy.[8], Halgamar.*
Trincomali.*
Nawalapitiya.[6]
Bandarawela.[2]
Nalanda.[2]
South India.[10]
Travancore

Mitt. VI 168 (1917)[9]; Ochs, Journ. F. M. S. Mus. XIII 243 (1927)[10], Rec. Ind. Mus. XXXI 245 (1929)[11].	Wallardi.[11] Tenmalai, Courtallum.[11] S. Malabar, Dhoni Forest.[11], Cochin.* Cochin State. Parambikulam to Kavalai.[11], Nilgiris. Benhope, Bungalow.[11] Madras, Nilambur.[11] Kodaikanal.[5] Central India: Umaria., Rewa. Jaithari Range.[11] Rewa State, Amarkantak.[11] Satpura Hills. Pachmarhi.[11] Bombay Press.: Poona Dist. Khandala.[11] Bengalen?[11] Calcutta?[11] Malacca.[4, 10] [?]
fairmairei, Rég., Ann. S. E. Fr. (6) III 428 Pl. 12f. 135 (1883)[1].	Ceylon.[1] Maskeliya.*
fallax, Peschet, Opusc. Inst. Sc. Saigon I 16 (1923)[1].	Luang Prabang: Pak-Ten.[1]
figuratus, Rég., Ann. S. E. Fr. LX 698 Pl. 18 f. 10 (1891)[1]. *Orectochilus oblongiusculus* var. *figuratus,* Rég., Ann. S. E. Fr. LXXVI 199 (1907)[2]. *Orectochilus figuratus,* Zimmerm., Ent. Mitt. VI 166 (1917)[3]; Peschet, Opusc. Inst. Sc. Saigon I 11 (1923)[4].	Padong.[1, 2] Kurseong.[2] Solan near Simla.[2] [?] Singla.*[1] Pashok.*
fletcheri, Ochs, Rec. Ind. Mus. XXVII 201 (1925)[1], XXXI 243 (1929)[2].	S. Malabar: Kollengode.[1, 2] Rewa State: Koilari.[2] Chota Nagpur: Singbhum Dt. Chaibassa.[2] Chakradharpur, Sanjai River.[2]

fraternus, Rég., Ann. S. E. Fr. (6) III 417 Pl. 12f. 142 (1883)[1].

Ceylon.[1]

frustorferi, Rég., Ann. S. E. Fr. LXXVI 194 (1907)[1]; Zimmerm., Ent. Mitt. VI 165 (1917)[2].

Tonkin: Mts. Mauson.[1, 2]

gengeticus, Wiedemann.

Gyrinus gangeticus, Wied., in Germar, Megazin d. Ent. IV 119 (1821)[1]; Perty, Obs. Col. Ind. 12 (1831)[6].

Orectochilus gangeticus, Aubé, Spec. Col. VI 740 (1838)[2]; Rég., Ann. S. E. Fr. (6) III 434 Pl. 12f. 137, 137a, 137b (1883)[3], LX 714 (1891)[4]; Nowrojee, Mem. Dept. Agric. Ind., Ent. Ser. II, 179 1912[5] [habits].

Bengal:[6], Ganges.[1]
India orient.[2, 3]
Pusa.[5]
Murshidabad.[4]
Berhampore Court.*
Narail near Jessore.[4]
Birbhum.*
Base of E. Himalayas:
Jalpaiguri, Tista R.*
Sikkim.[4]
Darjiling Dt.*
Assam: Kairchana.*

Java.[3][?]

haemorrhous, Rég., Ann. S. E. Fr. LX 706 Pl. 18f. 13, 13a (1891)[1]; Zimmerm., Ent. Mitt. VI 168 (1917)[2]; Ochs, Rec. Ind. Mus. XXXI 242 (1929)[3].

Madras.[1, 3]
Travancore:
Allepey.*[3]
Bangalore.*[3]
Central Prov.:
Rewa State, Koilari.[3]
Rewa State, Khetgaon.[3]
Mandla Dt., Sakka, Kharmer Nullah.[3]
Mohgaon, Burhner River.[3]
Nerbudda River: Padmi.[3]
between Dupta and Gorakhpur.[3]
Barham Kalan.[3]
Chota Nagpur:
Singbhum Dt., Chaibasa.*[3]
Chakradharpur, Sanjai River.*[3]

haemorrhous lobifer, Rég., Ann. S. E. Fr. LXXVI 214 (1907)[1].

Base of Himalayas.[1, 3]
Bihar: Pusa.[3]

indicus, Rég., Ann. S. E. Fr. (6) III 435 Pl. 12 f. 141 (1883)[1], LX 714 (1891)[2], Ann. S. E. Belg. XXXVI 121 (1892)[3], Ann. S. E. Fr. LXXI 470 (1902)[4], LXXVI 217 (1907)[5]; Zimmerm., Ent. Mitt. VI 168 (1917)[6] [male]; Ochs, Rec. Ind. Mus. XXXI 246 (1929)[7].

India.[1, 6]
Bengal. [?][2, 3, 7]
Bangalore: 3000 ft.[7]
Chikkangalur.[5]
Central Prov.: Barham Kalan.[7]
Ceylon:[2, 6, 7]
Bandarawela.[4]
Nalanda. [4, 6]
S.O.*
E.P., Niro-Munai.*
Diyatalawa.*

intermedius, Ochs.

Orectochilus frushtorferi subsp. *intermedius,* Ochs, Rec. Ind. Mus. XXVII 193 (1925)[1].

Dawna-Hills: Sukli.[1]
Upper Tenasserim.*
Tavoy: Nwalabo.*

kempi, Ochs, Rec. Ind. Mus. XXVII 198 (1925).[1]

East Assam: Yembung.[1]

klynstrai, Ochs, Suppl. Ent. XV 117, 122 (1927)[1], Journ. F. M. S. Mus. XIV 402 (1929)[2].

Singapore.[2]
Pahang.[2]

Sumatra.[1,2]

lancaisi, Rég., Ann. S. E. Fr. LX 692 Pl. 18 f. 5 (1891)[1], LXXVI 194 (1907)[2]; Zaitzev, Rev. Russe d'Ent. X 225, 2 fig. (1910)[3] [male]; Peschet, Opusc. Inst. Sc. Saigon I 8 (1923)[4].

Upper Tonkin.[1,3,4]
N. Tonkin: Laokay,[3, 4]
Laokay: Chapa.*, Laos.*
Tonkin: Hoa-Binh.*

S.W. Chima.[2, 3, 4]

limbatus, Rég., Ann. S. E. Fr. (6) III 424 (1883)[1], LXXVI 211 (1907)[2] [male].

Orectochilus gravelyi, Ochs, Rec. Ind. Mus. XXVII 199 (1925)[3].

Orectochilus limbatus, Ochs, Rec. Ind. Mus. XXXI 243 (1929)[4].

Ind. Orient.[1,4]
Barway.[2, 4]
Khandesh.[2, 4]
Chota Nagpur: Chaibassa.[3]
Chakradharpur, Sanjai-River.[4]
Bombay, Matheran,

limbatus,—

Charlotte Lake.[4] Ratnagiri Dt., Vashishty Valley, Pophli.[3]

Satara Dr.: Yenna Valley, Medha[3]. Koyna Valley, Helvak.[3] Central India.: Umaria, Rewa, Jaithari Range.[4] Rewa State, Kiolari.[4] Rewa State. Pongatola, Kornu Nala.[4] Mandla[4] Mandla Dt., Sakka, Deo Nullah.[4] Kharmer Nullah.[4] Mandla Dt., Mohgaon, Burhner River.[4] Rewa State, Amarkantak.[4]

Nerbudda River: Between Dupta and Gorakpur.[4] Padmi.[4], Sandia.[4]

Nilgiris: Mettupalaiyam, Bhavani River.[4] Ceylon.[4]

lucidus, Rég., Not. Leyd. Mus. IV 66 (1882)[1], Ann. S. E. Fr. (6) III 425 Pl. 12 f. 131 (1883)[2], LXXVI 210 (1907)[3]; Ochs, Suppl. Ent. XV 116, 122 (1927)[4].

Java.[1, 2, 3, 4] Nœsa Kambangan.[4] Sumatra.[3, 4]

lucidus subsp.?

Orectochilus lucidus, Rég., Ann. Mus. Civ. Genova VI (2) 623 (1888)[1], X (2) 552 (1891)[2], Ann. S. E. Fr. LX

Tenasserim: Thagata.[1, 2, 3, 4, 5]

Tavoy: Nwalabo.[5]

Burma: Mergui.*

704 (1891)[3], LXXVI 210 (1907)[4]; Ochs, Suppl. Ent. XV 116, 122 (1927)[5].

Ind. orient. post.*

marginepennis, Aubé, Spec. Col. VI 744 (1838)[1].

Orectochilus subsulcatus, Rég., Not. Leyd. Mus. II 215 (1880)[2].

Orectochilus marginipennis, Rég., Ann. S. E. Fr. (6) III 429 Pl. 12 f. 133, 133a, 133b (1883)[3].

Orectochilus subsulcatus, Rég., Midd. Sumatra IV, 612 Pl. 1 f. 10 (1887)[4].

Orectochilus marginipennis, Rég., Ann. Mus. Civ. Genova X (2) 552 (1891)[5], Ann. S. E. Fr. LX 704 (1891)[6], LXXVI 214 (1907)[2]; Zimmerm., Ent. Mitt. VI 168 (1917)[8]; Kempers, Ent. Mitt. XII 76 f. 92 (1923); Ochs, Rec. Ind. Mus. XXVII 201 (1925)[9], Suppl. Ent. XV 118, 122 (1927)[10], Journ. F. M. S. Mus. XIII 243, 245 (1927)[11], XIV 402 (1929)[12].

Burma: Carin Ghecu.[5, 6, 9, 10, 11]
Assam: Cherrapunji. [9, 10, 11]
Perak, [7, 9, 10, 11]
Selangor.[2]
Pahang.[10, 11]
Upper Tonkin: Cao-Bang.[7, 10, 11]
Cochinchina oritent: Reu-Hai.[7, 10, 11]

Java.[1, 3, 9, 10, 11]
Sumatra.[2, 3, 4, 8, 9, 10, 11]

marginipennis angustilimbus, Ochs, Rec. Ind. Mus. XXVII 202 (1925)[1].

W. Himalayas: Dehra Dun, Lachiwala.[1]

marginipennis parvilimbus, Ochs, Rec. Ind. Mus. XXVII 202 (1925)[1].

China.
Canton Prov.[1]

metallicus, Rég., Ann. S. E. Fr. (6) III 418 Pl. 12 f. 143 (1883)[1], (6) VI 262 (1886)[2] [male], LX Pl. 18 f. 11 (1891)[3], LXXVI 206 (1907)[4]; Zimmerm., Ent. Mitt. VI 168 (1917)[5]; Peschet, Opusc. Inst. Sc. Saigon I 13 (1923)[6].

India.[1]
E. Himalayas: Padong.[2, 4]
Kurseong.[4]
Kurseong, Dow Hill.*, Gantok*
Sureil*, Between Ghoom & Sonada.*
Ghom.*, Mangpoo*
Pashok.*, Ghumti.*

Upper Laos: Nam Mat.[6]

W. Himalayas: Dehra Dun.*
Kumaon, Bhowali.*
Chakrata, Bodyar.*

neglectus, Ochs, Rec. Ind. Mus. XXVII 199 (1925)[1].	W. Himalayas: Naini Tal dt. Bareni. (type loc.) Kiari.[1] Umballa Dt., Kalka.[1] Kumaon, Kathgodam.[1] Patiala State, Dharampur Koa.[1] Punjab, Kasauli.*
neglectus var.?, Ochs, Rec. Ind. Mus. XXVII 199 (1925)[1].	Almora Dt., Dwaharat.[1]
neglectus subsp.?	E. Himalayas: Mariambari.*
nítens, Peschet, Opusc. Inst. Sc. Saigon I 13 (1923)[1].	Luang Prabang: Pak-Ten.[1]
oblongusculus, Rég., Ann. S. E. Fr. (6) VI 262 Pl. 4 f. 11 (1886)[1], LX 698 (1891)[2], LXXVI 199 (1907)[3]; Zimmerm., Ent. Mitt. VI 166 (1917)[4]; Peschet, Opusc. Inst. Sc. Saigon I 11 (1923)[5].	Padong, [1, 2, 3] Kurseong.[3] Kalimpong.* Singla.*, Sukna.* Ghumti.*, Pashok.* Solan near Simla.* Kumaon, Bhim Tal.* Simla Hills, Barogh.* Loas: Sala Nam Pik.* Tonkin: Hoah-Binh.*
oblongiusculus feai, Reg. *Orectochilus feœ,* Rég., Ann. Mus. Civ. Genova (2) VI 621 (1888)[1], (2) X 551 (1891)[2]. *Orectochilus* var, *Feai,* Rég., Ann. S. E. Fr. LX 697 (1891), LXXVI 199 (1907); Zimmerm., Ent. Mitt. VI 166 (1917).	Tenasserim: Thagata.[1, 2, 3] Tenasserim.[4] W. Side of Dawna Hills, Misty Hollow.* Tavoy: Nwalabo.* Carin Ghecu.[2, 3]
oblongiusculus subsp.?, Ochs, Ent. Blätter XX 238 (1924)[1].	Khasi Hills.[1]
oblongiusculus parkeri, Ochs. *Orectochilus parkeri,* Ochs, Rec. Ind. Mus. XXVII 196 (1925)[1].	W. Himalayas: Almora, Gonai.[1]

oxygonus, Rég., Ann. S. E. Fr. LXXVI 203 (1907)[1]; Ochs, Journ. F. M. S. Mus. XIII 243 (1927)[2]. — Singapore,[1, 2]

patellimanus, Rég., Ann. S. E. Fr. LXXVI 202 (1907)[1]; Ochs, Ent. Blätter XX 238 (1924)[2], Journ. F. M. S. Mus. XIII 243 (1927)[3], XIV 401 (1929)[4]. — Dindings Islds.[1, 3] Penang.[1, 3, 4] Selangor.[3] Perak, Taiping.* Siamese Malay States: Nawngchik, Bukit Besar.[2, 3]

procerus, Rég., Ann. S. E. Fr. (6) III 415 (1883)[1]. — Cochinchina: Lakhon.[1]

productus, Rég., Ann. S. E. Fr. (6) III 422 Pl. 12 f. 130 (1883)[1], Ann. Mus. Civ. Genova (2) VI 623 (1888)[2], Ann. S. E. Fr. (6) IX 156 (1889)[3], Ann. Mus. Civ. Genova (2) X 552 (1891)[4], Ann. S. E. Fr. LXII 104 (1893)[5], LXXII 334 (1903)[6]; Zimmerm., Ent. Mitt. VI 167 (1917)[7]; Peschet, Opusc. Inst. Sc. Saigon I 12 (1923)[8]; Ochs, Journ. F. M. S. Mus. XIII 244 (1927)[9], Ent. Blätter XXIV 100 (1928)[10], Rec. Ind. Mus. XXXI 245 (1929)[11], Journ. F. M. S. Mus. XIV 402 (1929)[12]. — India. [1, 7, 8, 12] Ceylon: Colombo.[5] Wackwelle.[5] Mahé.[6], Calicut.[6] Madras.[11] Cochin State, Chalakudi.[11] Travancore, Keruma-dadi.[11] Rewa State, Amarkantak.[11] Umaria, Rewa, Jaithari Range.[11] C.P., Mandla Dt., Dindori.[11] C.P., between Dupta and Gorakpur.[11] Chota Nagpur, Singhbum Dt., Manbhum.[11] Bihar, Pusa.[11] Bankipur.[11] Bengal.[11] Kunbir.[11], Calcutta.[11] Garia near Calcutta.[11] Sunderbans.[11] Base of E. Himalyas: Siliguri.[11] N.E. Assam, Dibrugarh.[11] Assam:[11, 12]

productus,—

Mangaldai.[11]
Garo Hills,
Damalgiri.[11]
Seali Kirai.[11]
Burma:[11, 12]
Rangoon, [2, 4, 11]
Bassein.[11]
Lower Burma:
Moulmein,[11]
Kawkareik.[11]
Malay Penisula.[11]
Selangor.[9], Perak.[12]
Siam.[11]
Indo-china.[10, 12]
Cambodia:[11]
Pnom Penh.[3]
Cochinchina:[1, 8]
Mytho.[3], Annam:[11]
Phuc Son.*
Thuan An.[3]
Tonkin:[11], Than Moi.*
Dap Cau.[8]

S. China.[1, 7, 8, 10, 11]
Hongkong.*
Borneo,[1, 7, 8, 10, 11, 12]

punctilabris, Rég., Ann. S. E. Fr. LXXVI 212 (1907)[1]. — Upper Tonkin: Song-Bo.[1, 2]

punctilabris [subsp.?], Ochs, Journ. F. M. S. Mus. XIV 402 (1929)[2]. — Burma: Mergui.[2] Malay Penins, West Coast Langkawi Island.[2]

punctulatus, Rég., Ann. S. E. Fr. (6) VI 261 (1886)[1]. — Kodaikanal.[1] Tenmalai, Courtallum.*

ritsemai, Rég.

Orectochilus ritsemæ, Rég., Not Leyd. Mus. IV 67 (1882)[1], Ann. S. E. Fr. (6) III 431 Pl. 12 f. 136 (1883)[2].

Orectochilus ritsemai, Peschet, Opusc. Inst. Sc. Saigon I 13 (1923)[3]; Ochs, Rec. Ins. Mus. XXXI 243 (1929)[4].

Centr. Prov.:
Rewa State, Koilari.[4]
Barham Kalan.[4]
Mandla.[4], Sandia.
Burma: Mergui.[4]
Assam: Kairchana.*

ritsemai, Rég.—

Tonkin:
Hat-Lon-Man.[3, 4]
Hoa-Binh.[4]

Java.{?}[1, 2, 3, 4]

rivularis, Rég., Ann. S. E. Fr. (6) III 427 (1883)[1].

Cochinchina:
Phu-Quoc.[1]
Tokin: Lak-Tho.*

rivularis, subsp.?

Assam:
Cherrapunji.*

salvazai, Peschet.

Orectochilus cardiophorus, Peschet, Opusc. Inst. Sc. Saigon I 7 (1923)[1].

French Indochina:
Muong-You.[1]
Sala-Nam-Pik.[1]
Pou-Hai-Katoui.[1]
Ban-Samang.[1]
Pak-Sa.[1]
Siam:
Pah Meeung Mt.*

scalaris, Rég., Not. Leyd. Mus. II 215 (1880)[1], Ann. S. E. Fr. (6) III 430 Pl. 12 f. 134 (1883)[2], Midd. Sumatra IV, 6, 12 Pl. 1 f. 11 (1887)[3], Ann. S. E. Fr. LX 705 (1891)[4]; Zimmerm., Ent. Mitt. VI 168 (1917)[5]; Peschet, Opusc. Inst. Sc. Saigon I 11 (1923)[6]; Ochs, Rec. Ind. Mus. XXVII 200 (1925)[7], Suppl. Ent. XV 118, 122 (1927)[8], Journ. F. M. S. Mus. XIII 244, 245 (1927)[9], XIV 402 (1929)[10].

Selangor.[9, 10]
Perak.[10], Tonkin:[8, 9]
Lak-Tho.*
Hoa-Binh.[6]
Cambodia.[2.]

Sumatra.[1, 2, 3, 4, 7, 8, 9]
Java.[2, 4, 7, 8, 9]

scalaris parvulus, Ochs.

Orectochilus scalaris, Rég., Ann. Mus. Civ. Genova (2) VI 623 (1888)[1], (2) X 552 (1891)[2], Ann. S. E. Fr. LX 705 (1891)[3]; Zimmerm., Ent. Mitt. VI 168 (1917)[4].

Orectochilus scalaris subsp. *parvulus,* Ochs, Rec. Ind. Mus. XXVII 200 (1925)[5]; Suppl. Ent. XV 118 (1927)[6]; Journ. F. M. S. Mus. XIII 244, 245 (1927)[7].

Tenasserim:[6, 7]
Kawkareik.[1, 2, 3, 5]
Thagata.[1, 2, 3, 5]

semivestitus, Guérin, Rev. Zool. III 38 (1840)[1]; Rég., Ann. S. E. Fr. (6) II Pl. 10 f. 4 (1882)[2], (6) III 413 Pl. 12 f. 123, 123a (1883)[3], (6) VI 262 (1886)[4], in Wytsm. Gen. Trs. f. 29 (1902), LXXII 334 (1903)[5]; Zimmerm., Ent. Mitt. VI 166 (1917)[6].

Nilgiris: [1, 6]
Ootacamund.*
Coonoor River.[5]
Benhope Bungalow.*
Kotagiri.*, Mahe.[5]
Cochin.*

semivestitus,—	Palni Hills: Neutral Saddle.* Kodaikanal.[4,6] Shembaganur.[6] Dindigul.*, Madura.[6] Trichinopoli.[6] Pondicherry.[2] Madras.*
similis, Ochs, Rec. Ind. Mus. XXXI 251 (1929)[1].	Assam: Shillong.[1]
spiniger, Rég., Not. Leyd. Mus. II 214 (1880)[1], Ann. S. E. Fr. (6) III 436 Pl. 12 f. 139 (1883)[2], Midd. Sumatra IV, 6 12 Pl. 1 f. 9 (1887)[3], Ann. S. E. Fr. LX 715 (1891)[4], Miss. Pavie III 81 (1904)[5]; Peschet, Opusc. Inst. Sc. Saigon I 9 (1923)[6]; Ochs, Suppl. Ent. XV 120, 122 (1927)[7], Journ. F. M. S. Mus. XIII 244 (1927)[8].	India.[2,4] Malacca.[4,8] Cambodia.[4,5] Cochinchina.[2,6] Laos.[2,4] French Indochina.[8] Pak-Lay.[6] Xieng-Kuang.[6] Keng-Luong.[6] Lak-Bona.[6], Pak-Pa.[6] ―――――― Sumatra.[1,2,3,4,7,8]
spinfger jacobsoni, Ochs, Suppl. Ent. XV 120 (1927)[1].	―――――― Sumatra.[1]
striolifer, Rég., Ann. S. E. Fr. LXXVI 205 (1907)[1]; Peschet, Opusc. Inst. Sc. Saigon I 10 (1923)[2].	Cochinchina: Reu-Hai.[1,2] French Indochina: Son-Hai.[2]
sublineatus, Rég., Ann. S. E. Fr. LX 695 (1891)[1], Miss. Pavie III 81 (1904)[2]; Ann. S. E. Fr. LXXVI 195 (1907)[3]; Peschet, Opusc. Inst. Sc. Saigon I 10 (1923)[4].	Laos.[1,4] Luang-Prabang: Theng.[2] French-Indochina: Nam-Tiena.[4] Ma-Om.[4], Van-Nam.[4] Ban-Ban.[4] Sala-Nam-Tien.[4] Eastern Cochinchina: Reu-Hai.[3,4] Assam.[1,4][?}
sulcipennis, Rég., Ann. S. E. Fr. LX 705 (1891)[1].	Tonkin: Cao-Bang.[1] Burma: Mergui.*

tonkinensis, Rég., Ann. S. E. Fr. LX 701 (1891)[1]. — Tonkin: Cao-Bang.[1]

trianguliger, Rég., Ann. Mus. Civ. Genova (2) VI 622 (1888)[1], (2) X 552 (1891)[2], Ann. S. E. Fr. LX 696 Pl. 18 f. 9 (1891)[3]. — Tenasserim: Thagata.[1, 2, 3] Burma: Mergui.*

undulans, Rég., Ann. S. E. Fr. LX 711 Pl. 18 f. 16 (1891)[1]. — Tonkin: Cao-Bang.[1]

vitalisí, Peschet, Opusc. Inst. Sc. Saigon I 14 (1923)[1]; Ochs, Journ. F. M. S. Mus. XIV 402 (1929)[2]. — Luang-Prabang: Pak-Ten.[1, 2] Upper Mekong. Ban-Mone.[1, 2]

vitalisi smedleyi, Ochs, Journ. F. M. S. Mus. XIV 403 (1929)[1]. — Singapore.[1]

volubilis, Ochs, Rec. Ind. Mus. XXXI 252 (1929)[1]. — Assam: Shillong.[1]

wehnekei, Rég., Ann. S. E. Fr. (6) III 414 Pl. 12 f. 124 (1883)[1], LXXVI 193 (1907)[2]. — Ceylong.[1]

Total 92 species.

PART XX — ALUCITIDÆ (PTEROPHORIDAE)

Phalaenae Alucitae [part]: Linn., Syst. Nat. (ed. x) I. 542 (1758); Vill., Linn. Ent. Faun. Suec. II. pp. 116 [redescr.], 530-535 [species] (1789).

Phalaena Alucita [part]: Linn., Faun. Suec. p. 370 (1761) [non-descr.]; Linn., Syst. Nat. (ed. xii) I. ii. pp. 809 predescr.], 899 [species] (1767); Gmelin, Linn. Syst. Nat. (ed. xiii)I. v. p. 2615 (1790) [redescr.]

Pterophorus [part]: Geoffroy, Hist. abr. Ins. II pp. 24-25, 90-91 (1762): Schluga, Primae Lineae Cogn. Ins., p. 37(1767) [redescr.]; Fb., Syst. Ent., pp. [XXV], 671 (1775); Fb., Genera Ins. p. 168 (1777) [redescr.]; Fb., Spec. Ins. II 310 (1781) [non-descr.]; Fb., Mantissa Ins. I, p. xviii, II. 258 (1787) [redescr.]; Fb., Ent. Syst. III i. p. [vi] (1783) [redescr.]; ii. 345 (1794) [redescr.]; Lamarck, Syst. Anim. Sans. Vert., p. 228 (1801) [redeacr.].

Phalænæ Pterophori sive Alucitae Linn.: Gmelin, Linn. Syst. Nat. (ed. xiii) I. V.p. 2402 (1790) [redescr.] [part].

Pterophorii: Latreille, Hist. Nat. Crust. Ins. III 418 (1802); Latr., H.N. Crust. Ins. XIV 255-256 (1805) [redescr.].

Peterophoridae: Latrille, Hist. Nat. Crust. Ins. V. 322 (1803) [charact.]; Zeller, Isis, p. 755 (1841) [part]; Bruand, Mem. Soc. Emul. Doubs III 56 (1849); Brd., Cat. Microlep. Doubs, p. 90 (1850) [part]; Stainton, Syst. Cat. Brit. Tin. Pteroph., p. 31 (1849) [part]; de Graaf, Tijids. Ent. II 35-58 (1859) [Dutch spp.]; Wallengren. K. Svensk. Vet. Akad. Handl. III, No. 7 pp. 1-24 (1862) [Swedish spp.; redescr.]; Zeller, Stett. Ent. Ztg. XXVIII 321-339 (1867) [review of Wlgn.]; Wlsm., Pteroph. Calif. Oregon, pp. 3-65 (1880); Meyr., Tr. N.Z Inst. XVII 122-129 (1885) [New Zealand spp.]; Meyr., T.E.S 1886, 1-21 (1886)[part] [revision of genera; new spp.]; Meyr., T.E.S. 1890, 483-490 figs. (1890) [Europ. genera and spp.]; Tutt, Pteroph, Brit., pp. 1-155 (1896) exe. *Chrysocorys*); Meyr., Handb., pp. 429-430 (1895) [redeser.]; Fernald, Pteroph. N. Amer., pp. 1-61, tt. 1-9 (1898) [revision of N. Amer. spp.]; Rebel, Cat. Pal. Lep. II pp. 70-78 (1901) [pal. spp.] ; Fernald, U.S. Nat. Mus. Bull 52, pp. 441-447 (1903) [N. Amer. spp.]; Barrett., Lep. Brit. Isds. IX 340-347, tt. 413-419 (1904) [Brit. spp.]; Fletcher, Spolia Zeylan. VI 1-39, tt. A-F, map (1909)

[Ceylon spp.]; Spuler, Schmett Eur. II 317-318 (1910) [redescr.]; Meyr., Wyts. Gen. Ins., fasc. 100 (1910) [revision]; Meyr., Junk's Cat. Lep., Part 17, pp. 1-39 (1913); Fletcher, Ind. Agric. Ent. Mem. VI 1-31 (1921) [biol. Indian spp.]; Barnes and Lindsey, Contrib. N.H. Lep. N. Amer. IV, No. 4, pp. 281-288 (1921) [redescr.; key to genera]; Lindsey, Dension Univ. Bull. XX 187-192, figs. t. 20 (1924) [neuration; phylogeny; key to genera]; Forbes, Lep. N. York. pp. 639-642, figs. (1924) [redescr.]; Tillyard, Ins. Australia New Zealand, pp. 428-429 (1926) [redescr.]; Meyr., Rev. Handb., pp. 447-448 (1928) [redescr.]; Hudson, New Zealand Lep., p. 207 (1928) [redescr.].

Pterophorae: Hübner, Tentamen, p. 2 (1806).

Pterophorites [part]: Latreille, Consid. gen. Ins., pp. 349, 442 (1810); Duponchel, Cat. meth. Lep Eur., p. 380 (1846).

Alucitides: Leach, Edinb. Encycl. IX 135 (1815) [part]; Tutt, Brit. Lep. V 70-127 (1906).

Alucitadæ: Samouelle, Ent. Comp., p. 255 (1819) [part].

Alucitæ: Hübner, Verz. bek. Schmett., p. 428 (1826) [part].

Alucitidæ [part]: Stephens, Cat. Brit. Ins. II 229 (1829); Steph., III. Brit. Ent. Haust. IV 369 (1825) [redescr.]; Westwood, Mod. Classif. Ins. II 413 (1840); Westw., Brit. Moths II 258 (1845).

Pterophorides: Zetterstedt, Ins. Lapp., p. 1012 (1840).

Pterophoriden: Zeller, Linn. Ent. VI 319-404 (1852) [revision]; Frey, Tin. Pter. Schweiz, pp. 399-420 (1856) [redescr.].

Pterophorina: Stainton, List Brit. Anim. B. M. XVI 172 (1854); Herrich-Schäffer, Schmett. Eur. V 361-385 (1855); Stainton, Manual II 439-440 (1859) [redescr.]; Wocke, Hein. Kleinschm. Deutschl. II. ii 780 (1876) [redescr.]; Snellen, Vlind. Ned. Microlep., pp. 1008-1012 (1882) [redescr.].

Agdistidae and Pterophoridae: Wlsm. E. M. M. XXVII 216 (1891); Wlsm., Novit. Lep. I [sep]., pp. 1-7), t. 12 (1892).

Pterophorinen: Hofmann, Ber. Nat. Ver. Regensb. V 25-219, tt. 1-3 (1896) [revis. German spp.].

Agdistidae and Alucitidæ: Tutt, Ent. Rec. XVII 34-37 (1905) [genotypes].

Agdistides and Alucitides: Tutt, Proc. S. Lond. Ent. Soc. 1905-05, pp, 1-5, t. 1 (1906).

Amcitida (Pterophoridae): Fletcher, T. E.S. 1925, p. 59-637, figs. (1926) [Walker's types].

PLATYPTILIANAE

1. Cosmoclostis, Meyrick 1886.

Cosmoclosits, Meyr., T.E.S. 1886. 7 Type: *agladesma,* M. (1886); Meyr. Wytsm. Gen. Ins., fasc. 100, p. 4 (1910) [syn. of *Diacrotricha*]; Meyr., Cat. Lep. Pteroph., p. 3 (1913) [distinct].

Type: *aglaodesma,* M.

aglaodesma, Meyrick 1886.

Cosmoclostis aglaodesma, Meyr., T.E.S. 1886. 12-13 (1886) ♀[1]; Meyr., T.E.S. 1885 422 (1885) [non-descr.][2]; Meyr., T.E.S. 1887 266 (1887)[3]; Meyr., B.J. XVII 134 (1906)[4]; Meyr., Cat. Pteroph., p. 3 (1913)[5].

Diacrotricha aglaodesma, Fletcher, Spol. Zeylan. VI. 32 (1909)[6]; Meyr., Wyts. Gen. Ins. fasc. 100, p. 4, tab. f. 1 (1910)[7]; Meyr., Ark. Zool XIV, No. 15, p. 1 (1922)[8].

Life-history not known. Has been bred from pupa found on leaf of *Gmelina arborea* in Assam.

Ceylon:[5]
Puttalam.[4, 8]
Anuradhapura.[6]
Matalé.[6]
Kurunegala.[6]
Kegalle.[6], Colombo.[6]
Arawa.[6], Trincomali.[6]
Wellawaya.
India:[6], Karwar.
Assam.

E. Australia:[5, 6]
Queensland.
Malanda.[8]
N.S. Wales:
Sydney.[1, 2]
Solomon Isds[3, 5, 7]

auxileuca, Meyrick 1908.

Cosmoclostis auxilecua, Meyr., T.E.S. 1907. 471-472 (1908) ♂ ♀[1]; Meyr., Cat. Pteroph., p. 3 (1913)[2].

Diacrotricha auxileuca, Meyr., Wyts. Gen. Ins., fasc. 100, p. 4 (1910)[3].

Assam:[2, 3]
Khasis.[1]

pesseuta, Meyrick 1906.

Cosmoclostis pesseuta, Meyr., B.J. XVII 134 (1906) ♂ ♀[1]; Meyr., Cat. Pteroph., p. 3 (1913)[2].

Diacrotricha pesseuta, Fletcher, Spol. Zeylan. VI 32 (1909)[3].

Ceylon:[2]
Puttalam.[1, 3]
Hambankota[3]
Wellawaya.
Assam:[2]

Transvaal:[2]
Barberton.

2. Diacrotricha, Zeller 1852.

Diacrotricha, Zeller, Linn. Ent. VI 399 (1852); Meyr., Wyts. Gen. Ins., fasc. 100, p. 4 (1910) [redescr.]; Meyr., Cat. Pteroph., p. 3 (1913).

Type: *fasciola, Z.*

fasciola, Zeller 1852.

Diacrotricha fasciola, Zell., Linn. Ent. VI 399-400 (1852)? ♀[1]; Meyr., T.E.S. 1907. 471 (1908)[2]; Fletcher, Spol. Zeylan. VI 31-32, t. A. f. 6, t. F. ff. 5, 6, 8 (1909)[3]; Meyr., Wyts. Gen. Ins., fasc. 100, p. 4 (1910)[4]; Meyr., Cat. Pteroph., p. 3 (1913)[5]; Fletcher, Ind. Agric. Ent. Mem. VI 1-2 (1921)[6].

Diacrotricha callimeres, Lefroy, Ind. Ins. Life, p. 528 (1909)[7].

Biology[3, 6] Larva descr.[3, 6], on flowers of *Averrhoa bilimbi,*[3, 6] *A. carambola.*[6, 7] Pupa descr.[3, 6].

Ceylon:[2, 5] Galle.[3, 6], Kandy[3] Badulla.[3] Arawa.[3], India:[3, 5] Pusa.[2, 6, 7] Gauhati.[6]

Khasis.

Java.[1, 3, 5] New Guinea.[5] Kei Isds.[2, 3]

3. Trichoptilus, Walsingham 1880.

Trichoptilus, Wlsm., Pteroph. Calif. Oregon, pp. 62-63 (1880); Meyr., T.E.S. 1886. 7 (1886) [redescr.]; Meyr. T.E.S. 1890. 484 (1890); Tutt, Pteroph. Brit., p. 77 (1895); Meyr., Handb. pp. 430-431 (1895) [redescr.]; Hofmann, Ber. Nat. Ver. Regens. V. 49, 145-146, t. 3 f. 10 [*paludum,* genit.] (1896) [redescr.]; Fernald, Pteroph. N. Am., p. 13 (1989) [redescr.]; Rebel, Cat. Pal. Lep. II 70 (1901); Fornald, U.S. Nat Mus. Bull. 52, p. 441 (1903); Tutt, Ent. Rec. XVII 36 (1905); Fletcher, Spol. Zeyl. V. 20 (1907); Meyr., T.E.S. 1907, 412 (1908); Fletcher, Spol. Zeylan. VI 27 (1909); Spuler, Schmett. Eur. II 325, f. 100 [neur., *paludum*] (1910) [redescr.]; Meyr., Wyts. Gen. Ins. fasc. 100, pp. 3, 4-5 (1910) [redescr.]; Meyr., Cat. Pteroph., p. 3 (1913); Wlsm., Biol. Centr. Am., Het. IV 434 (1915); Barnes and Lindsey, Contrib. N. H. Lep. N. Am. IV 288-289 (1921) [redescr.]; Lindscy, Denison Univ. Bull. XX 187-191, ff. 1, 2, t. 20 f. 1 (1924) [structure]; Forbes, Lep. N. York, p. 645 (1924) [redescr.]; Meyr. Rev. Handb., p. 448 (1928) [redescr.].

Trichoptylus, Barrett, Lep. Brit. Isds. IX 396-397 (1904) [redescr.].

Type: *pygmœus,* Wlsm. [N. America.]

Aciptilia (part), Snollen, Vlind. Ned., Microlep., p. 1054, f. 483 [*paludum*, head] (1882) [redescr.].

Buckleria, Tutt, Brit. Lep. V 491-493 (1906); Tutt, Ent. Rec. XVII 37 (1905) [non-descr.].

Type: *paludum*, Z. [Europe]

Stangeia, Chapman in Tutt, Brit. Lep. V. 492 (1906); Tutt, Ent. Rec. XX 53 (1908).

Type: *siceliota*, Z. [S. Eurpoe]

archaeodes, Meyrick 1913.

Trichoptilus archaeodes, Meyr., Exot. Micr. I 105 (1913) ♂[1]; Meyr., Cat. Pteroph. p. 4 (1913)[2].

N. Kanara.[2]
Bhatkal.[1]

bidens, Meyrick 1930.

Trichoptilus bidens, Meyr., Exot. Micr. III 564 (1930) ♂[1].

Khasis.[1]

defectalis, Walker 1864.

Pterophorus defectalis, Wlk., Cat. XXX 943 (1864) ♂ ♀[1]; Fletcher, T.E.S. 1925 621-622, f. 10 (1926) [type redescr.][2].

Pterophorus congrualis, Wlk., Cat. XXX 943-944 (1864) ♂ ♀[3]; Wlsm., P.Z.S. 1885. 885 (1885)[4]; Cotes and Swinh., Cat. Moths Ind., p. 668, No. 4545 (1889)[5]; Fletcher, T.E.S. 1925. 622 (1926) [type][6].

Pterophorus oxydactylus, Wlk., Cat. XXX 944 (1864) ♂ ♀ [nec *Pterophorus oxydactylus*, Stdgr. 1859][7]; Wlsm., P.Z.S. 1885. 885 (1885)[8]; Moore, Lep. Ceylon III 529, t. 2009 f. 16 (1887)[9]; Cotes and Swinhoe, Cat. Moths Ind., p. 669, No. 4549 (1889)[10]; Fletcher, T.E.S. 1925. 623 (1926) [type][11].

Aciptilia hawaiiensis, Butl., A.M.N.H. (5) VII 408 (1881) ♂ ♀[12].

Trichoptilus ochrodactylus, Fish, Canad. Ent. XIII 142-143 (1881) ♂ ♀[13]; Fernald, Pter. N. Am., p. 15 (1898) [redescr.][14]; Fernald, Dyar's List N. Am. Lep., p. 441, No. 4930 (1903)[15].

Trichoptilus compsochares, Meyr., T.E.S. 1886. 16 (1886) ♀[16].

Trichoptilus centetes, Meyr., T.E.S. 1886. 16-17 (1886) ♀[17]; Meyr., T.E.S. 1887. 226 (1887)[18]; Wlsm., P.Z.S. 1891. 494-495 (1892)[19]; Wlsm., P.Z.S. 1897. 56 (1897)[20].

Ceylon:[7, 29, 11] Throughout low country.[27] Colombo.[7, 11, 26] Galle.[26], Trincomali. Anuradhapura. Kandy. Maskeliya.
India:[29] Throughout Plains Coimbatore.[3, 6] Tranquebar., Poona. Bassein Fort.[26] Pusa.[26] Chapra.[26] Lyallpur.[26] Rawalpindi. Peshawar.[26] Hangu.[26]

Chagos Isds.[22, 25] China: Fuchaufu.[6] Formosa, Anping.[30] New Guinea[29] Port Moresby.[20] Bismarck Isds.[21, 29] Queensland.[18, 29] N.W. Australia,

Trichoptilus ralumensis, Pagenstecher, Zooloica XXIX 239 (1900)[21].

Trichoptilus defectalis, Fletcher, Tr. Linn. Soc. (Zool. ii) XIII. 312 (1910)[22]; Wlsm., Biol. Centr. Am., Het. IV 434-435 (1915)[23]; Barnes and Lindsey, Contrib. N. H. Lep. N. Am. IV 290-291, t. 41 f. 1, t. 49 f. 9 (1921) [redescr.][24].

Buckleria defectalis, Fletcher, Tr. Linn. Soc. (Zool. ii) XIII 398-399 (1910)[25]; Fletcher, Ind. Agr. Ent. Mem. VI 6-8, t. 1 f. 8 (1921)[26].

Trichoptilus congrualis, Fletcher, Spolia Zeylan. VI 28-30, t. A. f. 8, t. F. ff. 2, 3 (1909)[27]; Meyr., Ann. Transv. Mus. II. 218 (1911)[28]; Meyr., Cat. Pteroph., p. 4 (1913)[29]; Meyr., Supplem. Ent. III 45 (1914)[30]; Meyr., Voyage Alluaud Afr. Orient., Lep. p. 37 (1920)[31]; Meyr., Ark. Zool. XIV, No. 15, p. 1 (1922)[32]; Meyr., T.E.S. 1923. 546 (1924)[33].

Trichoptilus oxydactylus, Wlsm., Fauna Hawaii. I 471-472, t. 10 f. 3 (1907)[34]; Rebel, Denkschr. math-naturw. Klasse K. Akad. Wiss. Wien. LXXI ii. 115 (1907)[35].

Biology[27, 26] Larva descr.[26, 27], on seeds of *Boerhaavia repens.*[26, 27]. Pupa descr[26].

Kimberley.[32]
Hawaiian Isds.[12, 29, 34]
Chile, Arica.[23]
Peur,[29], Lima.[23]
Colombia.[23]
Brazil. Bahia.[23]
Panama.[23]
Mexico.[23]
Arizona.[23, 24]
Texas.[13, 14, 15, 23, 29, 24]
Florida.[24, 29]
W. Indies, St. Vincent.[19, 20, 23]
St. Croix.[19, 20, 23]
St. Thomas.[19, 20, 23]
Jamaica.[23, 29]
Bahamas.[23]
AFRICA.
C. Verde Isds.[16, 29]
Sierra Leone[1, 4]
Congo.[1], C. Africa.[29]
S. Africa.[29]
Transvaal.[28]
E. Africa.[31]
Aden.[35], Sokotra.[35]
Mauritius.[22, 25, 29]
Rodriguez.[23]
Seychelles.[25, 28]
Aldabra.[25]
Amirantes.[22, 28]
Coetivy Isd.[22, 15]
Farquhar Isd.[22, 25]

paludicola, Fletcher 1907.

Trichoptilus paludicola, Fletcher, Spolia Zeylan. V. 20-32, figs. (1907) ♂ ♀[1]; Fletcher, Spol. Zeylan. VI 31, t. A. f. 7 (1909)[2].

Trichoptilus paludum [part.], Meyr., Cat. Pteroph. p. 4 (1913)[3].

Buckleria paludicola, Fletcher, Ind. Agric. Ent. Mem. VI 2-5 (1921)[4].

Biiology.[1, 4] Larva descr.[1, 4] on *Drosera burmanni.*[1, 4] Pupa descr[1].

Ceylon,
Diyatalawa.[1, 2, 4]
Madulsima.[2, 4]
Kegalle.[4]
Khasi Hills:[1, 2, 4]
Shillong.[4]
Laitlyngkot.
Dumpep.

pelias, Meyrick 1908.

Trichoptilus pelias, Meyr., T.E.S. 1907. 472 (1908) ♂ ♀[1]; Fletcher, Spolia Zeylan. VI 27 (1909)[2]; Meyr., Wyts. Gen. Ins., fasc. 100, p. 5 (1910)[3]; Meyr., Cat. Pteroph., p. 4 (1917)[4].

Ceylon: Colombo.[2]
Nalanda.[2]
Trincomali.[2]
Kegalle.
India:[4]
Coorg.[1]
N. Kanara.
Khasis.[1, 3]

wahlbergi, Zeller 1852.

Pterophorus wahlbergi, Zell., Linn. Ent. VI 346-348 (1852) ♀[1]; Zell., Micr. Caffr., p. 117 (1852)[2]; Wlsm., T.E.S. 1881. 280 (1881)[3].

Pterophorus rutilalis, Wlk., Cat. XXX 943 (1864) ♂[4]; Fletcher, T.E.S. 1925. 622 (1926) [*wahlbergi*][5].

Oxyptilus rutilans, Wollaston, A.M.N.H. (5) III 441 (1879)[6].

Trichoptilus pyrrhodes, Meyr., P. Linn. Soc. N.S.W. XIV 1113 (1889) ♂[7].

Oxyptilus ? wahlbergi, Wlsm., Swinh. Cat. Het. Mus. Oxon. II 541 (1900)[8].

Trichoptilus wahlbergi, Meyr., B. J. XVII 134 (1906)[9]; Fletcher, Spol. Zeylan. VI 27, t.A.f. 10 (1909)[10]; Meyr., Wyts. Gen. Ins., fasc. 100, p. 5 (1910)[11]; Meyr., Cat. Pteroph., p. 4 (1913)[12]; Meyr., T. E. S. 1923. 546 (1924)[13].

Buckleria wahlberig, Fletcher, Tr. Linn. Soc. (Zool. ii) XIII 399, f. 2 (1910)[14]; Fletcher, Ind. Agr. Ent. Mem. VI 8-9, t.1 f.2 (1921)[15].

Larva, Pupa figured[15], Larva on *Oxalis, Ipomœ butatas, Vitis indica.*

Ceylon:[12, 14]
Kandy distr.[10, 15]
Maskeliya.[10]
Pundaluoya.[8]
Nawalapitiya.[8]
Bandarawala.[10]
Haldummulla.[15]
Badulla.[10]
Madulsima.[10, 16]
India:[12, 14]
Palni Hills.[15]
Kodaikanal.
Coimbatore.[15]
Shevaroys.[15]
Coorg.[15]
Bangalore.[8]
Ganjam.[8]
Pusa.[15], Shillong.[15]
Margherita., Bhim Tal.[15], Dharmsala.[8]
Abbottabad.[15]
Peshawar.[15]
Parachinar.[15]

E. Australia:[12]
Queensland.[14]
Duaringa.[7]
S. Africa:[3, 11, 12, 14]
Caffraria.[1]
Port Natal.[4, 5]
(Durban).
Mauritius.[12]
Rodriguez.[13]
Seychelles.[12, 14]
St. Helena.[6, 12, 14]

xerodes, Meyrick 1886.

Trichoptilus xerodes, Meyr., T. E. S. 1886. 14-15 (1886) ♂ ♀[1]; Meyr., T. E. S. 1885. 422 (1885) [non-descr.][2]; Meyr., T. E. S. 1887. 267 (1887)[3]; Meyr., B. J. XVII 134 (1906)[4]; Fletcher, Spol. Zeylan. VI 30, t. F. f. 4 [larva] (1909)[5]; Meyr., Gen. Ins. fasc. 100, p. 5, tab. f. 2 (1910)[6]; Meyr., Cat. Pteroph., p. 4 (1913)[7]; Meyr., Voyage Alluaud Afr. Orient., Lep. p. 37 (1920)[8]; Fletcher, Ind. Agr. Ent. Mem. VI6 (1921) [Buckleria][9]; Barraud, Entom. LVI 59 (1923)[10]; Meyr., Bull. Hill Mus. II 232 (1928) [may be a geogr. form of *siceliota,* Zell.][11]. Larva descr,[5] on *Gynandropsis s*][5]., on pods of *Cajanus indicus*[9].	Ceylon:[7, 8] Colombo.[5] Peradeniya.[5, 9] Haldummulla.[9] Madulsima.[5] Trincomali.[5, 9] India: S. India.[7] Pollibetta.[8], (Coorg). N. Kanara., Nagpur.[9] Khurda.[9] ——— New Guinea.[7, 8] Australia:[7, 8] Queensland Toowoomba.[1, 2] N.S. Wales, Bathurst.[2] S. Australia. Adelaide.[1, 2] Wirrabara.[1, 2] Port Lincoln.[1, 2] W. Australia, York.[3] S. Africa.[8] E. Africa.[8] Palestine, Haifa.[10]

4. Tetraschalis, Meyrick 1887.

Tetrachalis, Meyr., T. E. S. 1887. 267 (1887); Meyr., Wytsm. Gen. Ins., fasc. 100, p. 5 (1910) [redescr.].	Type: *arachnodes,* M. [N.S. Wales.]

deltozela, Meyrick 1924.

Tetraschalis deltozela, Meyr., Exot. Micr. III. 92-93 (1924)♂[1].	Masuri.[1]

ischnites, Meyrick 1908.

Tetraschalis ischnites, Meyr., T. E. S. 1907. 474-475 (1908) ♂ ♀[1]; Meyr., Cat. Pteroph., p. 5 (1913)[2].	Assam:[2] Khasis.[1]

ochrias, Meyrick 1908.

Tetraschalis ochrias, Meyr., T. E. S. 1907. 475-476 (1908) ♂[1]; Meyr., Cat. Pteroph., p. 5 (1913)[2].	India:[2], Assam. Khasis.[1], N. Kanara.

ochrias, Meyrick 1908.—

New Guinea:[2]
Kei Isds.[1]
Philippines.[1]

5. Sphenarches, Meyrick 1886.

Type: [*caffer*, *Z*=] *synophrys*, M.

Sphenarches, Meyr., T. E. S. 1886. 8 (1886); Meyr., Wytsm. Gen. Ins. fasc. 100, p. 6 (1910) [redescr.].

caffer, Zeller 1852.

Oxyptilus caffer, Zell., Linn. Ent. VI 348-349 (1852) ♂[1]; Zell., Micr. Caffr., p. 118 (1852)[2]; Wlk., Cat. XXX 934 (1864)[3]; Wlsm., T. E. S. 1881. 279 (1881)[4].

Oxyptilus anisodactylus, Wlk., Cat. XXX 934-935 (1864) ♂ ♀[5]; Moore, Lep. Ceylon III 528 (1887)[6]; Cotes and Swinhoe, Cat. Moths India, p. 668, No. 4542 (1889)[7]; Fletcher, T. E. S. 1925. 611 (1926) [type].

Pterophous diffusalis, Wlk., Cat. XXX 945 (1864) ♂[9]; Fletcher, T. E. S. 1925. 623-624 (1926) [type][10].

Oxyptilus direptalis [nec. Wlk.], Wlk., Cat. XXX 934 (1864) [specimen *b*][11]; Fletcher, T. E. S. 1925. 609 (1926) [specimen *b* is *S. caffer*][12].

Oxyptitus walkeri, Wlsm., T. E. S. 1881. 279-280 (1881)[13].

Sphenarches synophrys, Meyr., T. E. S. 1886. 17-18 (1886) ♂[14].

Sphenarches caffer, Meyr., T. E. S. 1887. 268 (1887) [=*anisodactylus*=*walkeri*= *synophrys*][15]; Wlsm., Ind. Mus. Notes II 20, figs. (1891) [biol.][16]; Cotes, Ind. Mus. Notes II 163 (1893)[17]; Wlsm., P. Z. S. 1897. 56-57 (1897)[18]; Meyr., Gardiner's Faun. Geogr. Maldives I ii. 125 (1902)[19]; Hering, Stett. Ent. Zts. LXIV 96 (1903)[19a]; Lefroy, Ind. Agr. Ent. Mem. I 220 (1907)[20]; Fletcher, Spol. Zeylan. VI 21-22, t. E. ff. 8, 10, t. F. f. 1 (1909) [biol.][21]; Lefroy, Ind. Ins. Life, p. 528, fig. (1909)[22]; Fletcher, Tr. Linn. Soc. (Zool. ii) XIII 399 (1910)[23]; Meyr. Wyts. Gen. Ins. fasc. 100 p. 6, tab. f. 3 (1910)[24]; Meyr., Tr. N. Z, Inst. XLII 72 (1910)[25]; Meyr., Cat. Pteroph., p. 5 (1913)[26]; Fletcher, S. Ind. Ins., pp. 443-444, f. 320 (1917)[27]; Fletcher,

Ceylon:[5, 6, 7, 15, 26]
Throughout low-country.[21]
Colombo.[21], Galle.[21]
Weligama.[21]
Hambantota.[21]
Trincomali.[21]
Jaffna.[21], Kegalle.
Peradeniya.[21]
Maskeliya.[21]
Haldummulla.
Diyatalawa.[21]
Madulsima.[21]
Alutnuwara.[21]
Maldives: Hulule.[19]
India:[20, 21, 24, 26. 33]
Anamalais.[31]
Palnis.
Ootacamund.[31]
Coorg.[31]
Bababudins.[31]
Surat.[31], Pusa.[31]
Allahabad.[31]
Bhim Tal.[31]
Sarai Saleh.[31]
(Hazara.)
Peshawar.[31]
Shillong.[31]
Burma:[21, 26]
Maymyo.[31]

Java.[15a, 24, 26, 33]
Sumatra.[26, 31]
Philippines.[24, 26]

Proc. Second. Ent. Meeting, pp. 44, 56, 306 (tab) (1917)[28]; Meyr., Voyage Alluaud Afr. Orient., Lep. p. 37 (1920)[29]; Fletcher, Proc. Third. Ent. Meeting, pp. 136, 336, 338, 340, 344, 346, 351, 838 (1920)[30]; Fletcher, Ind. Agr. Ent. Mem. VI 9-13, t.2 (1921) [biol.][31]; Barraud, Entom. LVI 59 (1923)[32]; Meyr., Ins. Samoa III 68-69 (1927)[33].

Biology[16, 21, 27, 31]. Larva descr.[21, 31] on Cucurbitaceae[16, 20, 31], *Dolichos lablab*[16, 20], *Cajanus indicus*,[20] *Averrhoa bilimbi*[21], *Biophytum sensitivum*...[21].

Japan.[19a, 21, 26, 31, 38]
New Guinea.[26, 33]
Tenimber Isd.
Nw Hebrid es.[14, 26]
E. Australia:[24, 26, 33]
Queensland.
Moreton Bay.[9, 10]
N.S. Wales.
Sydney.[15]
Kermadec Isds.[25]
S. Pacific Isds.[15]
Tonga.[14, 26], Samoa.[33]
Brazil, Obidos.
French Guiana,
W. Indies:[24, 33, 26]
St. Thomas.[18]
Grenada.[18]
W. Africa:[15]
Gambia.[16]
Congo.[11, 12]
S. Africa:[16, 24]
Caffraria.[1, 2]
Natal.[13]
C. Africa.[25]
E. Africa.[29]
Mauritius.
Palestine:
Haifa.[32]

6. Oxyptilus, Zeller 1841.

Oxyptilus, Zell., Isis XXXV 765 (1841); Herr.-Schäff., Schmett. Eur. V 370 (1855) [redescr.], t. 8 ff. 12-14 [*hieracii*] (1849); Wlgn., K. Svensk. Vet. Akad. III, No. 7, p. 14 (1862) [? 1859] [redescr.]; Wocke, Hein., Kleinschm. Deutschl. II. ii. 789 (1876) [redescr.]; Snell., Vlind. Ned., Microlep., pp. 1025-1026, ff. 452, 459-462 (1882) [redescr.]; Meyr., T. E. S. 1886. 8 (1886) [redescr.]; Meyr., T. E. S. 1890. 485 (1890) [redescr.]; Meyr., Handb., p. 431 (1895) [redescr.]; Tutt, Pteroph. Brit., pp. 60-62 (1895) [redescr.]; Hofmann, Ber. Nat. Ver. Regensb. V 119-124 (1896) [redescr.]; Fernald, Pteroph. N. Amer., pp. 16-17, t. 2 ff. 3, 4 [neur.] (1898) [redescr.]; Rebel, Cat. Pal. Lep. II. 71 (1901);

Type: *piloselle, Z.* [Europe.]

Fernald, U. S. Nat. Mus. Bull. 52, pp. 441-442 (1903); Barrett, Lep. Brit. Isds. IX 362-363 (1904) [redescr.]; Tutt, Ent. Rec. XVII 35, 37 (1905); Tutt, Brit. Lep V 95, 413-419 (1906) [redescr.]; Spuler, Schmett. Eur. II 323, f. 98 [neur.] (1910) [redescr.]; Meyr., Wyts. Gen. Ins., fasc. 100, pp. 3, 6-7 (1910) [redescr.]; Meyr., Cat. Pteroph., pp. 5-8 (1913); Wlsm., Biol. Centr. Am., Het. IV 435 (1915); Forbes, Lep. N. York, pp. 644-645 (1924) [redescr.]; Meyr., Rev. Handb., p. 449 (1928) [redescr.].

Capperia, Tutt, Brit. Lep. V 470-471 (1906); Tutt, Ent. Rec. XVII 37 (1905) [non-descr.] — Type: *heterodactyla,* Vill. [Europe.]

Geina, Tutt, Brit. Lep.V 411 (1906) [non-descr.] — Type: *"didactylus,"* Linn. [Europe.]

Crombrugghia, Tutt, Brit. Lep V 499-451 (1906). — Type: *distans, Z.* [Europe.]

Pterophorus [nec Geoffr.], Barnes and Lindsey, Contr. N. H. Lep. N. Am. IV 297-299 (1921). — Type: *"Alucita didactyla,* Linn." [Europe.]

causodes, Meyrick 1905.

Oxyptilus causodes, Meyr., B. J. XVI 582 (1905) ♂ ♀[1]; Fletcher, Spol. Zeylan. VI 24-25, t. A. f. 4, t. E. f. 9 (1909)[2]; Meyr., Wyts. Gen. Ins., fasc. 100, p. 7 (1910)[3]; Meyr., Cat. Pteroph., p. 8 (1913)[4]; Fletcher, Ind. Agric. Ent. Mem. VI 15 (1921)[5]. Larva descr.[2, 5], in fruits of *Dillenia retusa.*[2, 5], *D. indica.*

Ceylon:[3, 4]
Peradeniya.[1, 2, 5]
India:
South India.[4]
Karwar., Pusa.[5]
Dehra Dun.

chordites, Meyrick 1913.

Oxyptilus chordites, Meyr., Exot. Micr. I 106 (1913) ♂ ♀[1]; Meyr., Cat. Pteroph., p. 5 (1913)[2]; Fletcher, Ind. Agric. Ent. Mem. VI 14 (1921)[3].

Ceylon:[2]
Colombo.[1]
India: N. Kanara.[2]
Karwar.[1]

clarisignis, Meyrick 1924.

Oxyptilus clarisignis, Meyr., Exot. Micr. III 93 (1924) ♂ ♀[1].

Muktesar.[1]
(Kumaon; 7,000 ft).

epidectes, Meyrick 1908.

Oxyptilus epidectes, Meyr., T. E. S. 1907 476-477 (1908) ♂ ♀[1]; Fletcher, Spol, Zeylan. VI 26, t. A. f. 5

Ceylon:[3]
Kandy.[2, 4]

(1909)[2]; Meyr., Cat. Pteroph., p. 5 (1913)[3]; Fletcher, Ind. Agric. Ent. Mem. VI 14 (1921)[4].
Larva on *Biophytum sensitivum*[2, 4].

Maskeliya.[1, 2]
Haldummulla.[4]
Madulsima[2, 4]
India:[3]
Coorg.[1], Pollibetta.[4]
Nilgiris (3,500 ft.)[1]
Burma:
Mone:[1]

Mauritius.[1, 3]

lactucae, Fletcher 1921.

Oxyptilus lactucac, Fletcher, Ind. Agric. Ent. Mem. VI 13-14 (1921)[1].©
Larva on Lettuce[1].

Dehra Dun.[1]

pelecyntes, Meyrick 1908.

Oxyptilus pelecyntes, Meyr., T. E. S. 1907. 477 (1908) ♂ ♀[1]; Meyr., Cat. Pteroph., p. 6 (1913)[2]; Fletcher, Ind. Agric. Ent. Mem. VI 14 (1921)[3].
Larva on *Scutellaria discolor*[3].

Ceylon:[2]
Haldummulla.[3]
India: Assam.[2]
Khasis.[1]
Shillong.[3]

regalis, Fletcher 1909.

Oxyptilus regalis, Fletcher, Spolia Zeylan. VI 25-26 (1909) ♂[1]; Meyr., Cat. Pteroph., p. 8 (1913).[2]

Ceylon:[2], Colombo.[1]
India: Bombay.
Pusa.

regulus, Meyrik 1906.

Oxyptilus regulus, Meyr., B. J. XVII 135 (1906) ♂[1]; Flethcer, Spol. Zeylan. VI 25 (1909)[2]; Meyr., Wyts Gen. Ins., fasc. 100, p. 7, tab. f. 6 (1910)[3]; Meyr., Cat. Pteroph., p. 8 (1913)[4]; Meyr., Ark. Zool. XIV, No. 15, p. 1 (1922)[5].

Ceylon:[3, 4]
Maskeliya.[1]
India:[3, 4]
Palni Hills.
N. Kanara.
Bombay.
Khasis.

Tonkin.
N. Australia.[4]
Queensland:
Cedar Creek.[5]

zanclistes, Meyrick 1905.

Oxyptilus zanclistes, Meyr., B. J. XVI 581-582 (1905) ♂ ♀[1]; Meyr., Cat. Pteroph., p. 5 (1913)[2]; Corbett

Ceylon:[2], Kandy.
Maskeliya.
India:[2], Coorg.

and Gates, Dept. Agric. F. M. S. Bull. 38, p. 11 (1926)[3].

Larva on flowers of cowpea[3].

Khasis., Burma:[2]
Fort Stedman.[1]
Malay States.
N. Australia.[2]
Port Darwin.

7. Xyroptila, Meyrick 1908.

Xyroptila, Meyr., T. E. S. 1907. 479-480 (1908); Meyr., Wyts. Gen. Ins., fasc. 100, p. 8 (1910) [redescr.].

Type: *ænophanes,* M.

caminites, Meyrick 1908.

Oxyptilus caminites, Meyr., T. E. S. 1907. 478-479 (1908) ♂[1].

Xyroptila caminites, Meyr., Wyts. Gen. Ins., fasc. 100, p. 8 (1910)[2]; Meyr., Cat. Pteroph., p. 9 (1913)[3].

Assam:[2, 3]
Khasis.[1]

œnophanes, Meyrick 1908.

Xyroptila œnophanes, Meyr., T. E. S. 1907. 480-481 (1908) ♂ ♀[1]; Meyr., Cat. Pteroph., p. 9 (1913)[2].

Bombay:[2]
Bandora.[1]

predator, Meyrick 1910.

Oxyptilus predator, Meyr., Rec. Ind. Mus. V 217 (1910) ♂[1]; Meyr., Cat. Pteroph, p. 9 (1913)[2].

E. Himalaya:[2]
Sukna (500 ft.)[1]

tectonica, Meyrick 1914.

Xyroptila tectonica, Meyr., Ent. Mitteil., Suppl. III, p. 46 (1914) ♂[1]; Meyr., Exot. MICF. II 420 (1921)[2]. Larva on *Bridelia [tomentosa]*[2].

India:
Ranchi., Pusa.
Burma: Minbu.

Java.[2], Tonkin.
Formosa., Kankau.[1]
Tailhorin.[1]
Rhodesia.
Nyasaland.
Mt. Mlanje.

vaughani, Fletcher 1909.

Oxyptilus vaughani, Fletcher, Spol. Zeylan. VI 23-24 (1909) ♂[1]; Meyr., Cat. Pteroph., p. 8 (1913)[2].

Xyroptila raughani, Fletcher, Ind. Agric. Ent. Mem. VI 15-16 (1921)[3].
Larva apparently on *Dimorphocalyx glabellus.*[1, 3]

Ceylon:[2]
Haldummulla.[1]
Madulsima.[1]
Aiáwa.[1]
Alutnuwara.[1]
Trincomali.[1]
India: Shillong.

8. Deuterocopus, Zeller 1852.

Deuterocopus, Zeller, Linn. Ent. VI 402 (1852); Zell., Bull. Soc. Nat. Mosc. XXVI, pt. ii. t. 4 ff. 17-20 (1853); Wlk., Cat. XXX 951 (1864); Meyr., T. E. S. 1886. 8 (1886) [redescr.]; Rebel, Denkschr. Math.-Nat. K. K. Akad. Wiss. Wien. LXXI ii. 85-86, fig. (1907); Fletcher, Spol. Zeylan. VI 16-19, t. B. f. C. (1909); Meyr., Wyts. Gen. Ins., fasc. 100, pp. 7-8 (1910) [redescr.]; Fletcher, T. E. S. 1910. 109-112 (1910) [redescr.].

Type: *tengstrœmi*, Z. [Java; Queensland.]

Deuteroscopus [orror orthogr.], Hofmann, Iris XI 329 (1898); Pagenstecher, Abh. Ges. Zoologica XXIX 241 (1900); Spuler, Schmett. Eur. II 318, f. 92 [neur.] (1910) [redescr.].

alopeeodes, Meyrick 1911.

Deuterocopus alopecodes, Meyr., B. J, XXI 105-106 (1911) ♂ ♀[1]; Meyr., Cat. Pteroph., p. 9 (1913)[2]; Fletcher, Ind. Agric. Ent. Mem. VI 16 (1921)[3]; Meyr., Ann. Transv. Mus. VIII 49 (1921)[4].
Larva on *Vitis* sp.[1]

N. Kanara.[2]
Karwar.[1, 3]

Rhodesia:
Umvuma.[4]

atrapex, Fletcher 1909.

Deuterocopus atrapex, Fletcher, Spol. Zeylan. VI 19 (1909) [♂][1]; Fletcher, T. E. S. 1910. 118-120, t. 44 f. 5 (1910) [redescr.].

Deuterocopus tengstrœmi [nec Zell.], Meyr., T. E. S. 1907. 414 (1908)[3].

? *Deuteroscopus tengstrœmi* [nec Zell.], Pagenstecher, Zoologica XXIX 141. (1900)[4].

Deuterocopus atrapex, Meyr., Cat. Pteroph., p. 9 (1913)[5]; Meyr., Ark Zool. XIV, No. 15, p. 1 (1922)[6].

Ceylon:[5]
Kandy.[1, 2], Galle.[2]
India:[5]
Khasis.[2], Burma:
Mergui.[2]

Selangor.[2]

S.E. Borneo:[5]
Pulo Laut.[2]
Moluccas.[5], Ternate.[2]
Amboyna.[2], Batian.[2]
S. Celebes.[2, 5]
Macassar.[2], Sanguir.[2]
Ombay (Allor).[2]
Halmahera.[2]
Neu Pomern.[1]
Kinigunang.[2]

atrapex, Fletcher 1909.—

New Guinea: Dorey.[2]
Kei Islands.[2]
Fergusson Isd.[2]
N. Australia:[5] Port Darwin.
Queensland: Bellenden Ker.[6]

dorites, Meyrick 1913.

Deuterocopus dorites, Meyr., Exot. Micr. I 108-109 (1913) ♂[1]; Meyr., Cat. Pteroph., p. 9 (1913).[2]

S. India.[2], Nadgani.[1] (Nilgiris; 2500 ft.)

fervens, Meyrick 1913.

Deuterocopus fervens, Meyr., Exot. Micr. I 108 (1913) ♂[1]; Meyr., Cat. Pteroph., p. 9 (1913).[2]

S. India.[2], Nilgiris.[1] (West Slopes; 2000ft.)

lophopteryx, Fletcher 1910.

Deuterocopus lophopteryx, Fletcher, T. E. S. 1910. 121-122, t. 44 f. 6 (1910) ♀[1].

Deuterocopus atrapex [part], Fletcher, Spol. Zeylan. VI 18-19 (1909)[2].

Deuterocopus lophopteryx, Meyr., Cat. Pteroph., p. 9 (1913).[3]

Ceylon.[3]
Kandy.[1, 2]
Yala.[2],

Moluccas.[3]
Amboyna.[1]
Sanguir.[1]
S. Celebes.[1, 3]
Halmahera.[1]

planeta, Meyrick 1908.

Deuterocopus planeta, Meyr., T. E. S. 1907. 473-474 (1908) ♂[1].

Deuterocopus rubrodaciylus [nec Pag.], Fletcher, Spol. Zeylan. VI 20, t. E. f. 7 [neur.] (1909) [biol.][2]

Deuterocopus planeta, Fletcher, T. E. S. 1910 131-134, f. 5 [pupa], t. 44 f. 10 (1910) [redescr.; biol.][3]; Meyr., Wyts. Gen. Ins., fasc. 100, p. 8, tab. f. 5 (1910)[4]; Meyr. Cat. Pteroph., p. 9 (1913)[5]; Fletcher, Ind., Agric. Ent. Mem. VI 17-18 (1921)[6].
Biology[2, 3]. Egg[2]. Larva descr.[2, 3], on flowers of *Leela sambucina*[2, 3, 6]. Pupa descr.[2, 3].

Ceylon.[5]
Galle.[2, 3]
India., Coorg.[1]
Pollibetta.[6]
Bababudins.[6]
Khasis.[1, 3], Burma.[5]
Bernardmyo.[3]

Portuguese Timor.[3]
Tenimber.[3]
New Guinea.[5]
Little Kei Isd.[3]

ritsemae, Walsingham 1884.

Deuterocopus ritsemae, Wlsm., Notes Leyden Mus. VI 243 (1884)[1].
Deuteroscopus rubrodactylus, Pagenstecher, Zoologica XXIX 241 (1900)[2].
? *Deuterocopus rubrodactylus,* Meyr. T. E. S. 1907. 473 (1908)[3].
Deuterocopus rubondactylus, Fletcher, Spol. Zeylan. VI, t. B f. C [nec p. 20] (1909)[4].
Deuterocopus ritsemae, Fletcher, Spel. Zeylan. VI 21 (1909)[5]; Fletcher, T. E. S. 1910. 134-138, t. 44 ff. 11, 12 (1910) [redescr., biol, text fig. 6 (pupa)][6]; Meyr., Cat. Pteroph., p. 9 (1913)[7]; Fletcher, Ind. Agric. Ent. Mem. VI 18 (1921)[8].
Biology[6]. Pupa figured[6], on leaf of *Leea sambucina.*[6]

Ceylon.[7] Peradeniya.[6] Kandy.[8] Galle.[6] India.[7] Coorg. Pollibetta.[8], Assam. Margherita.[6] Burma., Mergui.[6]

Java.[1] S.E. Borneo.[7] Pulo Laut.[6] Riam Kiwa.[6] Tambora.[6], Batian.[6] Bali.[6], Ternate.[6] Moluccas.[7] Halmahera.[6] N. New Guinea.[7] Humboldt Bay.[1] Neu Pommern. Kinigunang.[2, 6]

socotranus, Rebel 1907.

Deuterocopus socotranus, Rebel, Denkschr. Math. Nat. Klasse K. Akad. Wiss. Wien. LXXI. ii. 85-87, fig. (1907) ♂ ♀; Fletcher, T. E. S. 1910. 124-130, ff. 3 [larva], 4 [pupa], t. 44 f. 8, t. 45 f. 1 [♂ genit.] (1910) [redescr. biol.][2].
Deuterocopus tengstroemi [nec Zell.], Meyr., T. E. S. 1886. 8 (1886) [part][3]; Meyr., B. J. XVII 134 (1906)[4]; Fletcher, Spol. Zeylan. VI 16-20 (1909)[5]; Lefroy, Ind. Ins. Life, p. 528 (1909)[6].
Deuterocopus viticola, Meyr., B. J. XXI 104-105 (1911) ♂ ♀[7]; Meyr., Cat. Pteroph., p. 9 (1913)[8].
Deuterocopus triannulatus, Meyr., Exot. Micr. I. 107-108 (1913) ♂ ♀[9]; Meyr., Cat. Pteroph., p. 9 (1913)[10].
Deuterocopus socotranus, Meyr., Cat. Pteroph. p. 9 (1913)[11]; Fletcher, Ind. Agric. Ent. Mem. VI 16-17 (1921) [=vificola].[12]
Biology[2, 12] Larva descr.[2, 12], on flowers of *Vitis quadrangularis*[2, 7, 12] *V. trifolia*[12]. Pupa descr.[2, 12]

Ceylon:[8, 12] Puttalam.[2, 7] Trincomali.[2] Hambantota.[2, 7, 18] Yala.[2], Madulsima.[2] India.[12] Coimbatore.[12] Ganjam.[2], Pusa.[12] Surat.[2, 12], Burma. Moulmein.[12] Minbu.[12]

Sumba.[2, 12] Tambora.[2, 12] Amboyna.[2, 12] New Guinea. Port Moresby.[2, 9] Humboldt Bay.[2] Queensland.[10, 12] Townsville.[2, 9] Sokotra.[1, 2, 11, 12] E. Africa.[12]

socotranus, Rebel 1907.—

Ibea.[2], Netal.[12]
Delagoa Bay.[2]
W. Africa.[12]
Gambia.[2]

9. Koremaguia, Hampson 1891.

Koremaguia, Hmp., Ill. Het. VIII 142 (1891); Meyr., Wyts. Gen. Ins., fasc. 100, p. 8 (1910) [redescr.].

Type: [*alticola,* Feld.=] *aurantidactyla,* Hmp.

alticola, Felder 1875.

Cnaemidophorus alticola, Felder, Reise Novara, Lep. Het., t. 140 f. 59 (1875)♀[1].

Koremaguia avrantidactyla, Hmp., Ill, Het. VIII 142, t. 156 f. 20 (1891)[2].

Koremaguia alticola Meyr., T. E. S. 1907. 481 (1908)[3]; Meyr., Wyts. Gen. Ins. fasc. 100, p. 8, tab. f. 9 (1910)[4]; Meyr., Cat. Pteroph., p. 10 (1913)[5].

India.[5]
Himalaya.[1]
Nilgiris.[2, 5]
Andamans.[3, 5]

10. Heptaloba, Walsingham 1885.

Heptaloba, Wlsm., E. M. M. XXI 175 (1885); Fletcher, Spol. Zeylan. VI 15, t. B f. B [neur.] (1909); Meyr., Wyts. Gen. Ins., fasc. 100, p. 7 (1910) [redescr.].

Type: *argyriodactyla,* Wlk.

argyriodactyla, Walker 1864.

Platyptilus argyriodactylus, Wlk., Cat. XXX 929-930 (1864) ♂[1]; Fletcher, T. E. S. 1925. 600-601 (1926) [type][2].
Heptaloba argyriodactyla, Wlsm., E. M. M. XXI 175, fig. (1885)[3]; Moore, Lep. Ceylon III 527, t. 209 f. 13 (1887)[1]; Fletcher, Spol. Zeylan. VI 15, t. B f. B [neur.] (1909)[5]; Meyr., Wyts. Gen. Ins., fasc. 100, p. 7, tab. f. 4 (1910)[6]; Meyr., Cat. Pteroph., p. 8 (1913)[7].

Ceylon.[1, 2, 6, 7]
Kandy.[5]
Haldummulla.
Haputale.
Colombo.

11. Platyptilia, Hübner 1826.

Platyptilia, Hb., Verz. p. 429 (1826); Wocke, Hein., Kleinschm. Deuts. II ii. 783 (1876) [redescr.]; Snell., Vlind. Ned., Microlep., pp. 1013-1014, f. 451 *a, b* (1882) [redescr.]; Meyr., T. E. S. 1886. 9 (1886) [redescr.]; Meyr., Tr. N. Z. Inst. XVII 127 (1885) [redescr.]; Meyr., T. E. S. 1890. 485 (1890) [redescr.]; Meyr., Handb., p. 432 (1895) [redescr.]; Tutt, Pteroph. Brit., p. 22 (1895) [redescr.]; Hofmann,

Type: *gonodactyla,* Schiff.

Ber. Nat. Ver. Regensb. V pp. 48, 60 (1896) [redescr.]; Fernald, Pteroph. N. Am., pp. 22-23; t. 2 ff. 1, 2 [neur.] (1898) [redescr.]; Rebel, Cat. Pal. Lep. II 72 (1901); Fernald, U. S. Nat. Mus. Bull. 52, p. 442 (1903); Tutt, Ent. Rec. XVII 35, 37 (1905); Tutt, Brit. Lep V 78-83, 86-90, 92-95, 105, 182-219 (1906) [redescr.]; Fletcher, Spol. Zeylan. VI 11, t. B. f. A [neur. *pusillidactyla*] (1909); Meyr., Wyts. Gen. Ins., fasc. 100, p. 9 (1910) [redescr.] [part]; Spuler, Schmett. Eur. II 319, f. 94 [neur. *tesseradactyla*]. (1910) [redescr.]; Meyr., Cat. Pteroph., p. 10 (1913); Wlsm., Biol. Centr. Am., Het. IV 437 (1915); Barnes and Lindsey, Contrib. N. H. Lep. N. Am. IV 308-313 (1921) [redescr.]; Forbes. Lep. N. York, p. 642, f. 402 [neur *cosmodactyla*] (1924) [redescr.]; Meyr. Rev. Handb. p. 450 (1928) [redescr.]; Hudson, Lep. New Zealand, p. 207. t. E ff. 7-9 [*falcatalis*] (1928) [redescr.].

Platyptilus, Zeller, Isis XXXV 764-765, 770, 772-788, 866 (1841) [*emend.*]; Her.-Schäff., Schmett. Eur. V 366 (1855) [redescr.], t. 8 ff. 8-11 (1849); Wlgn., K. Svensk. Vet. Akad. III, No. 7, p. 11 (1862) [? 1859] [redescr.]; Barrett, Lep. Brit. Isds. IX 346 (1904) [redescr.].

Amblyptilia, Hb., verz., p. 430 (1826); Hofmann, Ber. Nat. Ver. Regensb. V pp. 48, 82 (1896) [redescr.]; Tutt, Ent. Rec. XVII 35 (1905); Tutt, Brit. Lep. V 268-273 (1906) [redescr.]; Barrett, Lep. Brit. Isds. IX 358 (1904) [*Amblyptilus;* redescr.]. — Type: *acanthodactyla,* Hb. [Europe].

Cnaemidophorus [praeocc.], Wlgn., K. Svensk. Vet. Akad. III, No. 7, p. 10 (1862) [? 1859]; Tutt, Pteroph. Brit., p. 18 (1896); Tutt, Ent. Rec. XVII 35 (1905). — Type: *rhododactyla,* Schiff.

Cnemidophorus, Zeller, Stett. Ent. Ztg. XXVII 332 (1865) *[emend.];* Barrett, Lep. Brit. Isds. IX 343 (1904) [redescr.].

Eucnaemidophorus, Wlgn., Ent. Tidskr. II 96 (1881); Hofmann, Ber. Nat. Ver. Regensb. V pp. 48, 57 (1896) [redescr.]; Tutt. Ent. Rec. XVII 36 (1905); Tutt, Brit. Lep. V 254-256 (1906) [redescr.]. — Type: *rhododactyla,* Schiff.

Mariana, Tutt, Brit. Lep. V 160 (1906) — Type: *metzneri, Z.* [Europe.]

Fredericina, Tutt, Brit. Lep. V 160-161 (1906); Tutt, Ent. Rec. XVII 37 (1905) [non-descr.]. — Type: *calodactyla,* Schiff. [Europe].

Gillmeria, Tutt, Brit. Lep. V 219-221 (1906); Ent. Rec. XVII 37 (1905) [non-descr.]. — Type: *ochrodactyla,* Schiff. [Europe].

? *Plctyptiliodes*, Strand, Arch. f. Naturg. 1912, A. 12, p. 65 (1912).

Type: *albisignatula*, Strand [Spanish Guinea].

cacaliae, Fletcher 1921.

Platyptilia cacaliœ, Fletcher, Ind. Agric. Ent. Mem. VI 23, t. 6 [fig. of pupa] (1921) ♂ ♀[1].
Larva in flower-heads of *Cacalia coccinea*[1]. Pupa figd.

Combatore.[1]

citropleura, Meyrick 1908.

Platyptilia citropleura, Meyr., T. E. S. 1907 482-483 (1908) ♂ ♀[1]; Fletcher, Spol. Zeylan. VI 15 (1909)[2]; Meyr., Cat. Pteroph., p. 10 (1913)[3]; Meyr., Ent. Mitt. Suppl. III. 46 (1914)[4]' Fletcher, Ind. Agric. Ent. Mem. VI 18-19, t. 6 [fig., pupa] (1921)[5].
Larva descr.[5], in see-pods of *Begonia* spp.[5] Pupa descr.[5]

Ceylon.[3]
Maskeliya.[1, 2, 5]
India.[3], Assam.
Khasis.[1]

Formosa.[4]

direptalis, Walker 1864.

Oxyptilus direptalis, Wlk., Cat. XXX 934 (1864)[1]; Fletcher, T. E. S. 1925. 609-611, f. 5 (1926) [type redescr.][2].

Pterphorus acanthodactylus [nec Hb.], Kollar, Hügel's Kaschmir IV. ii. 495 (1844)[3].

Amblyptilus cosmodactylus [nec Hb.], Wlsm., T. E. S. 1881. 277-278 (1881)[4].

Platyptilia direptalis, Meyr., T. E. S. 1907. 485-486 (1908) [♂ ♀ redescr.][5]; Fletcher, Spol. Zeylan. VI 12 (1909)[6]; Meyr., Wyts. Gen. Ins., fasc. 100, p. 11, tab. f. 10 (1910)[7]; Meyr., Cat. Pteroph., p. 12 (1913)[8]; Fletcher, Ind. Agric. Ent. Mem. VI 21-22, t. 5 ff. 1, 2, t. 6 [fig. pupa] (1921) [biol.][9].
Larva figured[9], on *Teucrium quadrifarium*[9], *Scutellaria discolor*[9]. Pupa figured[9].

Ceylon.[8]
Pattipola.[5], India.
Palni Hills.[5]
Kodaikanal.
Nilgiris[5] (6000 ft.)
Pusa.[9]
Shillong.[2, 7]
Masuri.[3]
Muktesar (Kumaon: 7000 ft.).
Simla.[6], Dungagali[9] (Hazara: 8000 ft.).

S. Africa.[8]
Cape.[1, 2, 5], Natal.[2]
Durban.[4]
Spring Vale.[4]
W. Africa.
Congo.[5]*

euctimena, Turner 1913.

Platyptilia euclimena, Turner, Proc. R. Soc. Queensl. XXIV 160 (1913) ♂ ♀[1]; Meyr., Cat. Pteroph., p. 16 (1913)[2].

Shillong.

China.

* If this record refers to Walker's specimen *b*, it is *Sphenarches caffer* and not *P. direplalis*.

euctimena, Turner 1913.—

Queensland.[2]
Kuranda.[1]
Brisbane.[1]
Toowoomba.[1]

exaltata, Zeller 1867.

Platyptilus exaltatus, Zell., Stett. Ent. Ztg. XXVIII. 411-412 (1867) ♂ ♀[1].
Platyptilia exaltata, Cotes and Swinhoe, Cat. Moths India, p. 668, No. 4541 (1889)[2]; Wlsm., Swinheo's Cat. Het. Mus. Oxon. II 541 (1900)[3]; Meyr., Cat. Pteroph., p. 12 (1913)[4].

Assam.[3], Shillong.
Cherrapunji.
Sikkim.[4], Darjiling.[1]
Muktesar (Kumaon: 7000 ft.).
Dharmsala.[3]
Simla.[2]
Kukti (N.W. Him.).[3]

forcipata, Zeller 1867.

Pterophorus (Oedematophorus?) forcipatus, Zeller, Stett, Ent. Ztg. XXVIII 412-414, t. 2 f. 7 (1867)[1].
Aedematophorus forcipatus, Cotes and Swinhoe, Cat. Moths India, p. 669, No. 4551 (1889)[2].
Platyptilia forcipata, Meyr., Cat. Pteroph., p. 12 (1913)[3].

Sikkim.[2, 3]
Darjiling.[1]

gonodactyla, Schiffermüller 1775.

Alucita gonodactyla, Schiff., Verz. Schmett. Wien, p. 320 (1775)[1]; Goeze, Ent. Beytr. III, iv, 177 (1783)[2]; Illiger, Verz. Wien. ii. 130-131 (1801)[3]; Zeller, Isis XXXV 882 (1841)[4].
Alucita diptera, Sulzer, Gesch. Ins., p. 163, t. 23 f. 19 (1776)[5].
Alucita megadactyla, Hb., Samml. Eur. Schmett. Aluc., t. 2 f. 6 (1804)[6].
Platyptilia megadactyla, Hb., Verz., p. 429 (1826)[7].
Alucita trigonodactyla, Hw., Lep. Brit., p. 478 (1811)[8].
Alucita tesseradactyla [nec Linn.], Treits., Schmett. Eur. IX ii 230-232 (1833)[9].
Pterophorus tesseradactylus [nec Linn.], Dup., Lep. France XI 647, t. 313 f. 5 (1838)[10]; Dup., Lep. France, Suppl. IV 500, t. 88 f. 8 (1842)[11].
Alucita tesseradactyla var. *c,* Zett., Ins. Lapp., p. 1012 (1840)[12]; Eversm. Faun. Lep. Volg. Ural., p. 605 (1844)[13].
Pterophorus zetterstedti [part.] Zell., Isis XXXV 777 (1841)[14]; Zeller, Isis p. 300 (1846)[15]; Dup., Cat. Meth. Lep. Eur., p. 381 (1846)[16].

India., Himalaya.[69]
Darjiling.[67]
Rawalpindi.[67]
[?Murree.]

C. Asia.
Ak Chehir.[21]
Altai Mts.[21]
N. Asia., Uralsk.[71]
Kasakewitsch.[71]
Altai. Mts.[38]
Asia Minor.[69]
Brussa.[66]
Bithynia.[64]
Europe.[64, 69]
N. Europe.[66]
C. Europe.[66]
Brit. Isds.[66]
Holland.[53]
Belgium.[66]
France.[64, 66], Italy.[64, 66]

Pterophorus ochrodactylus [nec Schiff.], Sand., Cat. Lep. Auv., p. 203 (1879)[17].

Platyptilia farfara, Gregson, Entom. XVIII 151 (1885)[18]; Tutt, Entom. 169-171, 195-196 (1885)[19].

Platyptilia farfarella [nec Zeller 1841], South, Entom. XVIII 172 (1885)[20].

Platyptilia gonodactyla var. *albidior*, Caradja, Iris XXXIV 80 (1920) ♂ ♀[21].

Pterophorus trigonodactylus, Samouelle, Ent. Comp,, p. 409 (1819)[22]; Curtis, Brit. Entom., fo. 161 (1827)[23]; Rennie, Conspectus, pp. 230-231 (1832)[24]; Stephens, Ill. Brit. Ent., Haust. IV 375 (1835)[25]; Wood, Index Ent. p-237, t. 51 f. 1644 (1838)[26]; Westw., Brit. Moths II 261, t. 124 f. 4 (1845)[27]; Stainton, Syst. Cat. Brit. Tin. Pter., p. 31 (1849)[28]; Stainton, List. Brit. Anim. B. M. XVI 174 (1854)[29]; Westw., Wood's Index Ent., p. 237, t. 51 f. 1644 (1854)[30]; Stainton, Zool. p. 3064 (1858)[31]; Stt., Manual II 440, fig. (1859)[32]; Peers, Entom. II 38 (1864) [larva][33]; Machin, Entom. XIII 283 (1880)[33a].

Platyptilus trigonodactylus, Jordan, E. M. M. VIII 137-138 (1871)[34].

Platyptilia trigonodactyla, South, Entom. XV 31-33, 145, t. 3 ff. 1[a-c] (1882)[35].

Pterophorus gonodactylus, Herr.-Schäff., Schmett. Eur. V, Pteroph. t. 2 f. 9 (1850)[36]; Zeller, Linn. Ent. VI 330-333 (1852) [redescr.][37]; Led., Verh. z-b. Ges. Wien V 120 (1855)[38]; Frey, Tin. Pteroph. Schweiz. pp. 402-403 (1856) [redescr.][39]; Morris, Brit. Moths IV 290, t. 131 f. 21 (1872)[40].

Platyptilus gonodactylus, Herr.-Schäff., Schmett. Eur. V 368-369 (1855) [re-descr.][41]; Wlgn., K. Svensk. Vet. Akad. Handl. III, No. 7, p. 12 (1862) [? 1859][42]; Jordan, E. M. M. VI 121 (1869)[43]; Nolcken, Lep. Faun. Estl. p. 800 (1871)[44]; Gregson, Entom. VI 426-427 (1873) [larva][45]; Barrett. Lep. Brit. Isds. IX 352-354, t. 414 ff. 1, 1[a] (1904) [redescr.; biol.][46].

Platyptilia gonodactyla, Wocke, Hein., Kleinschm. Deuts. II. ii. 784 (1876) [redescr.][47]; Frey, Lep. Schweiz., p. 428 (1880)[48]; Barrett. E. M. M. XVIII 177 (1882)[49]; Tutt, Entom. XVIII 169-171, 195-196 (1885)[50]; Sorh., Kleinschm. Brand., p. 2 (1886)[51];

Switzerland.[39, 48, 62, 66, 70, 72]
Austria.[1, 62, 66]
Germany.[62, 66]
Denmark.[66]
Norway.[42, 66], Latvia.[66]
Russia.[64, 66]

Leech, Brit. Pyr., p. 53, t. 16 f. 6 (1886)[52]; Snell., Vlind. Ned., Microlep., pp. 1021-1023 (1882)[53]; Tutt, Entom. XX 211-212 (1887) [larva][55]; Tutt, E. M. M. 104-105, 105-107 (1888)[56]; Tutt, Young Nat. X 163 (1889)[57]; Tutt. Young Nat. XI 123 (1890)[58]; Meyr., T. E. S. 1890. 486 (1890)[59]; Meyr., Handb., p. 434 (1895)[60]; Tutt, Pteroph. Brit., pp. 40-47 (1895)[61]; Hofmann, Ber. Nat. Ver. Regensb. V 66-68 (1896)[62]; Hofm., Ill, Zeits. Ent. III 135, 152 (1898)[63]; Rebel, Cat. Pal. Lep. II 72, No. 1332 (1901)[64]; Porritt, Buckler's Larvae IX 347 (1901)[65]; Tutt, Brit. Lep. V 201-219 (1906) [redescr.; biol.].; Meyr., Rec. Ind. Mus. V 217 (1910)[67]; Spuler, Schmett. Eur. II 320, t. 82 f. 56 (1910)[68]; Meyr., Cat. Pteroph., p. 14 (1913)[69]; Müller-Rutz, Schmett. Schweiz. II 442-443, No. 2021 (1914)[70]; Caradja, Iris XXXIV 80 (1920)[71]; Fletcher, Ent. Rec. XXXIX 35 (1927)[72]; Meyr., Rev. Handb., pp. 452-453 (1928) [redescr.][73].

Biology[53, 56, 62, 66]. Larva descr.[35, 42, 45, 46, 56, 52, 66] on *Tussilago farfara*[35, 42, 46-48, 53, 62, 66, 68], *T. nivea*[66], *T. alba*[66], *Petasites*[62, 68, 70], *Petasites farfara*[45]. Pupa descr.[33, 35, 46, 56, 62, 66].

ignifera, Meyrick 1908.

Platyptilia ignifera, Meyr., T. E. S. 1907. 481-482 (1908) ♂[1]; Meyr., Wyts. Gen. Ins., fasc. 100, p. 10 (1910)[2]; Meyr., Cat. Pteroph., p. 10 (1913)[3].	Assam:[3] Khasis.[1]

isocrates, Meyrick 1924.

Platyptilia isocrates, Meyr., Exot. Micr. III 94 (1924) ♂ ♀[1].	Killanmarg[1] (Kashmir; 10,000ft.).

molopias, Meyrick 1906.*

Platyptilia molopias, Meyr., B.J. XVII 135-136 (1906)♂ ♀[1]; Fletcher, Spol. Zeylan. VI 12-13, t. A f. 1, t, E f. 4 [larva] (1909)[2]; Meyr., Wyts. Gen. Ins. fase. 100, p. 11 (1910)[3]; Meyr., Ann. Transv. Mus. II 218 (1911)[4]; Meyr., Cat. Pteroph., p. 13 (1913)[5]; Meyr.,	Ceylon.[3, 5, 8] Kandy.[2] Maskeliya.[1, 2, 7] Nuwara Eliya.[2] Pattipola.[2, 7] Haldummulla.[7]

* In view of Meyrick's remark[4] that the African specimens ascribed to *molopias* show constant differences from Indian and Ceylonese examples, it seems doubtful whether the African form is really *molopias.*

Voyage Alluaud Afr. Orient., Lep. p. 38 (1920)[6]; Flecher, Ind. Agric. Ent. Mem. VI 22, t. 6 [fig. pupa] (1921)[7].

Platyptilia benitensis, Strand, Arch. Naturg. LXXVIII, A. 12, p. 64 (1912)[8].

Biology[2, 7].Larva Desor.[2, 7], on *Teucrium tomentosum*[2, 7] Pupa descr.[2, 7]

Diyatalawa.[2]
Madulsima.[2]
India., S. India.[5]
Dibidi (Coorg).

Madagascar.
E. Africa.[5, 6]
Mauritius.
S. Afirca.[5]
Natal., Transvaal.[4]
W. Africa.
Spanish Guinea.[8]

pusilidactyla. Walker 1864.

Oxyptilus pusillidactulus, Wlk., Cat. XXX 933 (1864) ♀[1]; Fletcher, T.E.S. 1925. 607=608 (1926) [type redescr.][2].

Platyptilia brevipennis, Zeller, Verh. z-b. Ges. Wien. XXIV 442-443, t. 12 f. 12 (1874)♂ ♀[3]; Meyr., Wytsm. Gen. Ins., fasc. 100, p. 10 (1910)[4]; Meyr. Cat. Pteroph., p. 11 (1913)[5]; Wlsm., Biol. Centr. Am., Het. IV 437 (1915)[6].

Platyptilia tecnidion, Zeller, H.S. E. R. XIII 468-469, t. 6 f. 162 (1877) ♂?[7].

Platyptilia hemimetra, Meyr., T. E. S. 1886. 18-19 (1886) ♂[3]; Meyr., B. J. XVII 135 (1906)[9].

Platyptilia lantana, Busek, Insec. Inscit. Menstr. II 103-104 (1914)[10]; Wlsm., Biol. Centr. Am., Het. IV 437 (1915)[11].

Platyptilia pusillidactyla, Wlsm., . P. Z. S. 1891. 495 (1892)[12]; Wlsm., P. Z. S. 1897. 57 (1897)[13]; Meyr., T. E. S. 1907. 483 (1908) [=*hemimetra*][14]; Fletcher, Spol. Zeylan. VI 13-14, t. A F. 2, t. E ff. 5, 6 -1909) [biol.[[15]; Fletcher, Tr. Linn. Soc. (Zool.ii) XIII 313 (1910)[16]; Fletcher, *t. c.* pp. 399-400 (1910)[17]; Meyr., Cat. Pteroph., p. 13 (1913) [=*tecnidion* = *hemimetra*][18]; Fletcher, S. Ind. Ins., p. 444, f. 321 (1914)[19]; Fletcher, Entl. Note 73 (1916)[20]; Fletcher, Proc. Second Entl. Meeting, p. 39 (1917)[21]; Ramachandra Rao, Ind. Agric. Ent. Mem. V 280-281, t. 24, 34 f. 1 (1920) [biol.][22]; Fletcher, Ind. Agrtic. Ent. Mem. VI 19-21, t. 4, t. 6 [fig. pupa] (1921) [biol.][23].

Ceylon.[14, 18]
Wherever *Lantana* grows.[15]
Puttalam.[9]
Maturatta.[9] Kandy.[9]
Maskeliya.[9]
India[14, 15, 18]
Quilon.
Coimbatore.[23]
Ootacamund[23]
Coorg.[23]
Bababudins.[23], Pusa.[23]
Shillong.[23]
Hazaribagh.
Kodaikanal.
Burma., Maymyo.[23]

Java.[6], Hongkong.
Fiji.
Hawaiian Isads.[18]
Honolulu,[10, 11, 15]
Peru.[4, 6], Lima.
Payta.[3], Colombia.[6]
Brazil., Santarem.
Ceara.[6], Guinea.[18] C. America.[10, 11]
Mexico.[6], Florida.[6]
W. Indies.[14, 15, 18]
Jamaica.[1, 2, 12, 13, 14]
St. Tomas.[7, 12, 13]

Biology[15, 22, 23]. Larva descr. [15, 22, 23] in flowers of *Lantana camara*[10, 15, 22, 23], *L. induca*[22, 23], *Lippia geminata*[22, 23], *Uticularia*. Pupa descr.[15, 22, 23].

St. Croix.[13]
Grenada.[13]
Bermuda Isds.[1]
W. Africa., Gambia.[6]
S. Africa., Transvaal Pretoria.
Réunion.[8, 9, 14, 15, 18]
Syechelles.[18]
Mahé.[15]

rhododactyla, Schiffermüller 1775.

Alucita rhododactyla, Schiff., Wien Verz., p. 146 (1775)[1]; Goeze, Ent. Beytr. III. iv. 177 (1783)[2]; Gmelin, Linn. Syst. Nat. (ed. xiii) I. v. 2615, No. 1299 (1788) [redescr.][3]; Villers, Linn. Ent. Faun. Suec. IV 547 (1789)[redescr.][4]; Illiger, Verz. Schmett. Wien II 130 (1801)[5]; Hübner, Samml. Eur. Schmett., Alue., t. 2 f. 8 (1804)[6]; Hb., Rauper IX, Aluc. i, t. 13 ff. 1[a-b] (circ. 1800)[7]; Haworth, Lep. Brit., p. 478 (1811)[8]; Treits., Schmett. Eur. IX ii. 228 (1833) [redescr.][9].

Pterophorus rhododactylus, Fabr., Mantisa Ins. II 258-259, No. 6 (1787) [redescr.][10]; Fabr., Ent. Syst. III. ii. 347, No. 7 (1794) [redescr.][11]; Latr., Hist. Nat. Crust. Ins. XIV 257-258 (1805) [redescr.][12]; Samouelle, Ent. Comp., p. 409 (1819)[13]; Curtis, Brit. Ent., folio 161 (1827)[14]; Stephens, Ill. Brit. Ent., Haust. IV 375-376, t. 41 f. 3 (1834) [redescr.][15]; Dup., Lep. France XI 644, t. 313 f. 4 (1838) [redescr.][16]; Wood, Index Ent., p. 237, t. 51 f. 1645 (1838)[17]; Zeller, Isis XXXV 772 (1841)[18]; Westw., Brit. Moths II 261, t. 124 ff. 5, 6 [larva], 7 [pupa] (1845) [redescr.][10]; Dup., Cat. Meth. Lep. Eur., p. 381 (1846)[20]; Tengstr., Finl. Fjär., p. 154 (1847)[21]; Richter, Stett. Ent. Ztg. XI 28 (1850) [larva][22]; Zell, Linn. Ent. VI 326 (1852)[23]; Stainton, List Brit. Anim. B. M. XVI 173 (1854)[24]; Westw., Wood's Index Ent., p. 237, t. 51, f. 1645 (1854)[25]; Frey, Tin. Pter. Schweiz., pp. 401-402 (1856) [redescr.][26]; Stainton, Manual II 440 (1859) [redescr.][27]; Lederer, Wien. Ent. Mon. VII 46 (1863)[28]; Morris, Brit. Moths IV 288, t. 131 f. 17 (1872)[29]; Porritt, Entom. VIII 183-184 (1875) [larv. habits][30]; Porritt, E. M. M. XII 88-89 (1875) [larva][31].

Kashmir.[66], Srinagar.

Europe.[61, 64, 66]
C. Europe.[57]
S. Europe.[57]
Sweden.[32, 50]
Norway.[50]
Germany.[10, 32, 50]
Belgium.[50]
S. E. England.[8, 15, 17, 19,25, 29, 45, 50, 66]
France.[12, 32, 50]
Spain.[50, 63], Italy.[50]
Sicily.[50]
Austria.[1, 48, 50]
Hungary.[32, 50]
Switzerland.[26, 48, 50, 62]
Bosnia.[50]
Herzegowina.[50]
Bulgaria.[28, 50]
Rumelia.[50]
Rumania.[50]
Finland.[57], Livonia.[57]
N. Africa.
Asia Minor.
Bithynia.[57]
Brussa.[37, 50]
Amasia.[50, 63]
S. Caucasus.[37, 50]
Armenia.[57], C. Asia.
Ak Chehir.[63]
N. Asia., E. Siberia.[66]
Kasakewitsch.[63]
Radde.[63]

Platyptilus rhododactylus, Herr-Schäff., Schmett. Eur. V 366 (1855) [redescr.][32]; Dyar, Jl. N. Y. Ent. Soc. III 21 (1895) [larva][33].

Cnaemidophorus rhododactylus, Wlgn., K. Svensk. Vet. Akad. Handl. III, No. 7, pp. 10-11 (1862) (?1859) [redescr.][24]; Nolcken, Lep. Faun. Estl., p. 799 (1871)[35]; Wocke, Hein. Kleinschm. Deutschl. II. ii. 782-783 (1876) [redescr.][36]; Staudinger, H. S. E. R. XV 423 (1880)[37]; Frey, Lep. Schweiz., pp. 427-428 (1880)[38]; South, Entom. XVIII 275-277, t. 1 ff. 3, 3[a-d] (1885) [redescr.; biol][39]; Sorhagen, Kleinschm. Brand., p. 1 (1886)[40]; Leech, Brit. Pyral., p. 51 (1886)[41]; Tutt, Young Nat. X, pp. 163, 201 (1889)[42]; Tutt, Pter. Brit., pp. 19-22 (1896) [redescr.; biol.][43].

Cnemidophorus rhododactylus, Zeller, Stett. Ent. Stg. XXVIII 332 (1867)[44]; Barrett, Lep. Brit. Isds. IX 343-345, t. 413 ff. 2, 2[a-b] (1904) [redescr.; biol.][45].

Cnoemidophorus rhododactylus, Jordan, E. M. M. VI 120 (1869)[46].

Eucnemidophorus rhododactylus, Wlgn., Ent. Tidskr. II 96 (1881)[47]; Hofmann, Ber. Nat. Ver. Regensb. V 58-60 (1896) [redescr.][48]; Hofm. Ill. Zeits. Ent. III 129-131, tab. f. 4 [larva] (1898)[49]; Tutt, Brit. Lep. V. 256-257 (1906) [redescr.; biol][50]; Spuler, Schmett. Eur. II 319, t. 82 f. 55, Nachtr. t. 10 f. 7 [larva[(1910) [redescr.][51]

Platyptilia rhododactyla, Hb., Verz. p. 429 (1826)[52]; Steph., Ill. Brit. Ent. Haust. IV, App. p. 424 (1835)[53]; Meyr., T. E. S. 1890. 486 (1890)[54]; Meyr., Handb., p. 435 (1895) [redescr.][55]; Dyar, Ent. Rec. XI 39-40, t. 1 f. 4 (1899) [larva][56]; Rebel, Cat. Lep. Pal. II 72, No. 1325 (1901)[57]; Porritt, Buckler's Larvae IX 337, t. 163 f. 2 (1901)[58]; Murtfeldt, Canad. Entom. XXXVI 334-335 (1904) [biol.][59]; Meyr., Wyts. Gen. Ins., fasc. 100, p. 10 (1910)[60]; Meyr., Cat. Pteroh., p. 11 (1913)[61]; Müller-Rutz, Schmett. Schweiz. II 442, No. 2020 (1914)[62]; Caradja, Iris XXXIV 79 (1920)[63]; Barnes and Lindsey, Contrib. N. H. Lep. N. Am. IV 313-314, t. 43 f. 3 (1921) [redescr.][64]; Forbes, Lep. N. York, p. 644 (1924) [redescr.][65]; Meyr., Rev. Handb., p. 451 (1928) [redescr.][66].

N. America.[66]
N. Atlantic States.[61]
Missouri.[65]
St. Louis.[50, 59, 67]
Kirkwood.[64]

Biology[31, 43, 45, 48, 50, 59]. Larva[7, 9, 22, 31, 33, 43, 45, 48, 49, 50, 51, 56, 59, 64, 65, 66], *Rosa canina*[1, 9, 31, 43, 45, 50], *R. centifolia*[9, 50], Garden Roses[22], in flower-buds, leaf-buds, on young shoots. Pupa[7, 9, 31, 43, 45, 48, 50, 51, 59, 64].

sythoffi, Snellen 1903.

Platyptilia (Amblyptilia) sythoffi, Snellen, Tijds. Ent. XLVI 54-56, t. 5 ff. 15, 16 (1903)[1].

Platyptilia taprobanes [nec Felder], Meyr., T. E. S. 1907. 482 (1908)[2]; Fletcher, Spol. Zeylan. VI 14 (1909)[3]; Meyr., Cat. Pteroph. p. 10 (1913)[4]; Fletcher, Ind. Agric. Ent. Mem. VI 19, t. 3 f. 1 [larva]. t. 6 [fig. pupa] (1921)[5].
Biology[5]. Larva[5], on *Scutellaria discolor*[5]. Pupa[5].

Ceylon.[4]
Maskeliya.[2, 3]
Pattipola.[2, 3]
Nuwara Eliya.[3]
Madulsima.[2, 3]
Peradeniya.[2, 3]
India.[4], Palnis[2]
Shevaroys.[5]
Coimbatore.[5]
Pusa.[5], Khasis.[2]
Shillong.[5]
Cherrapunji.

Java.[4], Preanger[1]
(W. Java:5000 ft.).

taprobanes, Felder 1875.

Amblyptilia taprobanes, Felder, Reise Novara, Lep. Het., t. 140 f. 54 (1875) ♂[1]; Moore, Lep. Ceylon III 527 (1887)[2]; Cotes and Swinhoe, Cat. Moths India, p. 668, No. 4540 (1889)[3].

Platyptilia brachymorpha, Meyr., T. E. S. 1888. 240-241 (1888) ♂?[4].

Amblyptilia seeboldi, Hofmann, Iris XI 31-34 (1898)[5]; Rebel, Cat. Pal Lep. II 73, No. 1340 (1901) [*Platyptilia*][6].

Platyptilia brachymorpha, Meyr. B. J. XVII 135 (1906)[7]; Wlsm., Faun. Hawaii. I 474 (1907) [redescr.][8]; Meyr., T. E. S. 1907. 483 (1908) [=*seeboldi*][9]; Fletcher, Spol. Zeylan. VI 12, t. A f. 3 (1909)[10]; Lefroy, Ind. Ins. Life, p. 528 (1909)[11]; Meyr., Wyts. Gen. Ins., fasc. 100, p. 10 (1910)[12]; Fletcher, Tr. Linn. Soc. (Zool ii) XIII 401 (1910)[13]; Meyr., Cat. Pteroph., p. 11 (1913) [=*seeboldi*][14]; Fletcher, Ind. Agric. Ent. Mem. VI 21, t. 3 f. 2, t. 6 [fig. pupa] (1921)[15]; Meyr., Entom. LIV 275 (1921) [=*crenulata*][16]; Meyr.,

Ceylon.[1-3, 14, 15]
Low-country.[10]
Puttalam.[10]
Galle.[10], Trincomali.[10]
Dambulla.[10]
Madulsima.[10]
Arawa.[10]
Alutnuwara.[10]
Mankulam.[10]
Anuradhapura.[10]
Kegalle.[10, 15]
Haldummulla.[10]
India.[9, 10, 14]
Kodaikanal.
Sidapur (Coorg).[15]
Pollibetta.[15]
Chakradharpur.[15]
Pusa.[15], Burma.
Pyinmana.[15]

Queensland.
Bellenden Ker.[18]

Entl. News XXXIII 213 (1922) [=*crenulata*][17]; Meyr., Ark. Zool. XIV, No. 15, p. 1 (1922)[18].

Platyptilia crenulata, Barnes and Mc-Dunnough, Contrib. N.H. Lep. N. Am. II 185, t. 3 f. 8 (1913)♂[19]; Grossbeck, Bull. Amer. Mus. N. H. XXXVII 135 (1917)[20]; Barnes and Lindsey, Contrib. N. H. Lep. N. Amer. IV 316, t. 41 f. 15, t. 50 f. 5 (1921) [redescr.][21].

Platyptilia acanthodactyla var. *phoenicodactyla,* Chrétien, Ann. S. E. Fr. LXXXIV 295 (1915) [22].

Platyptilia terlizzii, Turati, Ann. Soc. Ital. Sci. Nat. Milano LXV 67-68, f. 28 (1926)⊙[?♂][23].

Biology[15]. Larva descr.[15], on flowers of *Celsia coromandeliana* ([11, 15]), boring fruits of *Limnophila heterophylla,* in seeds of *Pentstemon,* in unripe fruits of *Veronica Anagallis.* Pupa descr[15].

Hawaiian Isds.[4, 9, 10, 14, 15], Oahu.[8] Hawaii.[8] Florida.[19, 21] Arizona.[21] S. California.[21] Brazil., Algeria. Gafsa.[22] Cyrenaica. Derna.[23], Cyprus. Syria.[6, 10, 14, 15] Akbes.[5] S. Africa.[9, 10, 14, 15] Mauritius.[14] Aldabra.[13, 15]

12. Stenoptilia, Hübner 1826.

Stenoptilia, Hb., Verz. p. 430 (1826); Meyr. T. E. S. 1890. 487, t. 15 ff. 4, 12 [neur. *pterodactyla*] (1890) [predescr.]; Meyr., Handb., p. 440, fig. [neur. *pterodactyla*] (1895) [redescr.]; Hofmann, Ber. Nat. Ver. Regensb. V 48, 92-97 (1896) [redescr.]; Fernald, Pteroph. N. Amer., p. 57 (1898) [part]; Rebel, Cat. Pal. Lep. II 76 (1901); Fernald, U. S. Nat. Mus. Bull. 52, p. 447 (1903); Tutt, Ent. Rec. XVII 35 (1905); Tutt, Brit. Lep. V 360-361 (1906); Fletcher, Spol. Zeylan. VI 10 (1909); Meyr., Wytsm. Gen. Ins., fasc. 100, p. 18 (1910) [redescr.]; Spuler, Schemtt. Eur. II 321, f. 95 [neur, *coprodactyla*] (1910) [redescr.]; Meyr., Cat. Pteroph., p. 28 (1913); Barnes and Lindsey, Contrib. N. H. Lep. N. Amer. IV 349-350 (1921)[redescr.]; Forbes, Lep. N. York, p. 646, f. 403 (1924) [redescr.]; Lindsey, Denison Univ. Bull. XX 192, t. 20 f. 9 [neur. *pterodactyla*] (1924); Meyr., Rev. Handb., p. 459, fig. [neur. *pterodactyla*] (1928) [redescr.]; Hudson, New Zeal. Lep. p. 210 (1928) [redescr.].

Type : *pterodactyla,* Linn. [Europe; Asia Minor; C. Asia; N. America].

Mimaeseoptilus, Wallengren, K. Svensk. Vet. Akad. III, No. 7, p. 17 (1862) [?1859]; Jordan, E. M. M. VI 123 (1869) [redescr.]; Wocke, Hein. Kleinschm.

Type : *pterodactyla,* Linn.

Deutschl. II. ii. 793 (1876) [redescr.]; Tutt, Pteroph. Brit., p. 84 (1896).

Mimeseoptilus, Zeller, Stett. Ent. Stg. XXVIII 332, 337 (1867) [emend.]; Wlsm., Pteroph. Calif. Oregon, p. 32 (1880); Meyr., T. E. S. 1886. 11 (1886) [redescr.].

Mimaesioptilus, Barrett, Lep. Brit Isds. IX 373 (1904) [redescr.].

Doxosteres, Meyr., T. E. S. 1886. 10-11 (1886).

Type : [*zophodactyla,* Dup.=] canalis, Wlk.

Adkinia, Tutt, Brit. Lep V. 318-319 (1906); Tutt, Ent. Rec. XVII 37 (1905) [non-descr.].

Type : *bipunctidactyla,* Scop.
[Europe; Asia Minor; Persia; Siberia].

petraea, Meyr. 1908.

Stenoptilia petraed, Meyr. T. E. S. 1907. 504 (1908)♂[1]; Fletcher, Spol. Zeylan. VI. 11 (1909)[2]; Meyr., Cat. Pteroph., p 29 (1913)[3]; Meyr., Voyage Alluaud Afr. Orient., Lep. pp. 43-44 (1920)[4].

Ceylon.[3, 4]
Diyatalawa.[2]
S. India.[3, 4]
Palnis (6000 ft.)[1]
Kodaikanal.
Nilgiris (6000 ft.)[1]

Brit. E. Africa (6400 ft.).[4]
Nairobi.[4]

zophodactyla, Duponchel 1838.

Pterophorus zophodactylus, Dup., Lep. France XI 668, t. 314 f. 4 (1838)[1]; Dup., Cat. Meth. Lep. Eur., p. 382 (1846)[2]; Zeller, Linn. Ent. VI 365 (1852) [perhaps=*loewii*][3]; Gregson, Entom. IV 350 (1869) [larva][4]; Morris, Brit. Moths IV 294-295, t. 132 f. 8 (1872)[5]; Porritt, E. M. M. XX 228 (1884) [larva][6].

Mimaeseoptilus zophodactylus, Wocke, Cat, Lap. Pal., p. 343, No. 3153 (1871) [=*loewii*][7]; Wocke, Hein, Kleinschm. Deutschl. II ii. 795 (1876) [redescr.][8]; Frey, Lep. Schweiz., p. 430 (1880)[9]; Barrett, E. M. M. XVIII 180 (1882) [=*loewii*][10]; South, Entom. XXII 35 (1889) [redescr.; larva; = *loewii*][11]; Tutt, Entom. XXII 104 (1889) [biol.][12]; Tutt, Young Nat. X. 165 (1889)[13]; Tutt, Brit. Nat. II pp. 111, 157

Ceylon.[47, 53, 59]
Hakgala.[48]
Diyatalawa.[48]
Madulsima.[48]
Lunugala.
India.[46, 47, 52, 53, 59]
Ootacamund.[55]
Bababudin Hills.[55]
Pusa.[55]
Khasi Hills.
Shillong.
Cherrapunji.[55]
Peshawar.[55]
Hangu.[55]

(1892)[14]; Tutt, Pteroph. Brit., pp. 93-100 (1896) [biol.;=*loewii*= *hodgkinsoni*][15].

Mimeseoptilus zophodactylus, Leech, Brit. Pyral., p. 60, t. 17 f. 6 (1886) [larva][16].

Mimaesioptilus zophodactylus, Barrett, Lep. Brit. Isds. IX 377-380, t, 416 f. 1 (1904) [biol.][17].

Pterophorus loewii, Zeller, Isis XL 38-39 (1847)♂[18]; Zell., Isis XL 904 (1847) [redescr.][19]; Zell., Linn. Ent. VI 364-365 (1852)♂ ♀[20]; Herr.-Schäff., Schmett. Eur. V (1855) [redescr.][21]; Stainton. Ent. Annual 1858, p. 98[22]; Stt., Manual II 442 (1859) [redescr.][23]; Stt., Ent. Annual 1860, p. 143[24]; Schmid, Berl. Ent. Zeits. VIII 65 (1864)[25]; de Graaf, Tijds. Ent. XI 79-81, t. 2 f. 5 (1868)[26]; Snell., Vlind. Ned., Microlep., pp. 1043-1044 (1882) [redescr.][27].

Mimaeseoptilus loewii, Jordan, E. M. M. VI pp. 123, 125 (1869)[28]; Porritt, Buckler's Larvae IX 358 (1901)[29].

Pterophorus nyctidactylus, Kollar Ms. [non-descr.; quoted by Herrich-Schäffer (Schmett. Eur. V 375: 1855) under *loewii* as "*nyctidactylus*, Kollar," but is apparently a MS. name, as it does not occur in Kollar's *Verz. der Oestreich. Schmett*][30].

Pterophorus canalis, Wlk., Cat. XXX 944 (1864)♂ ♀[31]; Fletcher, T. E. S. 1925. 623 (1926) [type][32].

Doxosteres canalis, Meyr., T. E. S. 1885. 425 (1885) ♂ ♀[33]; Meyr., T. E. S. 1886. 10-11 (1886)[34].

Mimeseoptilus semicostatus, Zeller, Verz. z.-b. Ges. Wien XXIII 323-324 (1873)♂[35].

Stenoptilia semicostala, Fernald, Smith's List Lep. Bor. Amer., p. 87, No. 4565 (1891)[36]; Fern., Pteroph. N. Amer., p. 60 (1898)[37]; Fern., U. S. Nat. Mus. Bull. 52, p. 447, No. 4995 (1903)[38]; Meyr., Cat. Pteroph., p. 30 (1913)[39].

Adkinia zophodactylus, Tutt, Brit. Lep. V 319-334 (1906) [biol.][40].

Stenoptilia zophodactyla, Meyr., T. E. S. 1890, 487 (1890)[41]; Meyr., Handb., p. 440 (1895) [redescr.][42]; Hofmann, Ber. Nat. Ver. Regensb. V 105-106 (1896) [redescr.; biol.][43]; Rebel, Cat. Pal. Lep. II 76, No. 1405 (1901)[44]; Lambillion, Rev. Mens. Soc. Ent. Namur 1904, pp. 50 *et seg.* (1904)[45]; Meyr., E. M. M. XLIII 146 (1907) [=*canalis*][46]; Meyr., T.

Parachinar.[55]
Abhottabad.[55]
Kashmir.[55], Srinagar.

Europe.[46, 53]
France.[1, 40]
Brit. Isds.[40, 59]
Holland.[26, 27, 40]
Germany.[8, 40, 43]
Switzerland.[8, 9, 40, 43, 54, 58]
Italy.[20]
Sicily.[19, 40]
Spain.[40]
Asia Minor.[40, 44, 53]
Rhodes.[18, 20, 40]
Syria.[40], Armenia.[44]
S. Africa.[52, 53, 59]
Transvaal.[49]
Cape Colony.[50]
Tonkin.
Australia.[31, 46, 52, 53, 59]
Queensland.[33, 57]
N. S. Wales.[33]
Sydney.[31]
S. Australia.[33]
New Zealand.[53, 59]
Wellington.[52, 60]
Christchurch.[60]
N. America.[59]
U. S. America.
Texas.[35, 37-39]
California.[56]
Florida.[56]
S. America.[52, 53, 59]
Argentina.
Parana.[47]

E. S. 1907. 504-505 (1908)[47]; Fletcher, Spol. Zeylan. VI 10-11, t. E. f. 3 [larva] (1909) [biol.][48]; Meyr., Ann. Transv. Mus. II 4 (1910)[49]; Meyr., Ann. S. Afr. Mus. V 411 (1910)[50]; Spuler, Schmett. Eur II 322 (1910) [redescr.; biol.][51]; Meyr., Trans. N. Z. Inst. XLIV 120 (1912)[52]; Meyr., Cat. Pteroph., p. 29 (1913) [=*loewii*=*canalis*][53]; Müller-Rutz., Schmett. Schweiz. II 445, No. 2031 (1914)[54]; Fletcher, Ind. Agric. Ent. Mem. VI 24-25 (1921)[55]; Barnes and Lindsey, Contrib. N. H. Lep. N. Amer. IV 354-356, t. 42 f. 16, t. 49 f. 15 (1921) [redescr.; biol.;=*loewii*=*canalis*= *semicostatus*][56]; Meyr., Ark. Zool. XIV, No. 15, p. 1 (1922)[57]; Müller-Tutz, Mitt. Schweiz. Ent. Ges. XIII 234 (1922)[58]; Meyr., Rev. Handb., pp. 459-460 (1928) [redescr.; biol.][59] Hudson, New Zeal. Lep., p. 211, t. 23, f. 12 (1928)[60].

Biology[6, 12, 15, 17, 40, 43, 48, 51, 56, 58]. Larva[4, 6, 11, 13, 15, 16, 17, 26, 29, 40, 42, 43, 48, 51, 56, 59] on flowers of *Erythrae centaurium*[4, 6, 8, 9, 11, 12, 15, 17, 26, 27, 40, 43], *E. pulchella*[17, 41], *E. littoralis*[26, 27, 40], *E. venusta*[56] *Gentiana germanica*[40, 43], *Chlora perfoliata*[11, 17, 40], *Sopubia trifida*[47, 48], Pupa descr.[4, 6, 15, 17, 40, 43, 48]

13. Exelastis, Meyrick 1907.

Exelastis, Meyr., B. J. XVII 730 (1907); Fletcher, Spol. Zeylan. VI 32-33, t. C. f. F. [neur. *phlyctaenias*] (1909); Barnes and Lindsey, Contrib. N. H. Lep. N. Amer. IV 347-348, t. 48 f. 8[a] [neur. *cervinicolor*] (1921).

Type : *atomosa*, Wlsm.

Marasmarcha [part], Meyr., Wyts. Gen. Ins., fasc. 100, p. 17 (1910); Barnes and Lindsey, Contrib. N. H. Lep. N. Am. IV 348, t. 48 f. 8 [neur. *pumilio*] (1921); Forbes, Lep. N. York, p. 646 (1924).

atomosa, Walsingham 1885.

Aciptilia atomosa, Wlsm., P. Z. S. 1885. 885 (1885)[1].

? *Pterophorus congrualis* [nec Wlk.], Swinhoe, P. Z. S. 1884, 526 (1884)[2].

? *Pterophorus ischnodactylus* [noc Tr.], Cotes and Swinhoe, Cat. Moths India, p. 669, No. 4546 (1889)[3].

Ceylon.[?].[12, 13]
Peradeniya.
[Meyr. Coll.;? ex. Pusa.]

India.[12, 13]
Throughout Plains.
Bombay.[1]

Pterophorus atomosus, Cotes and Swinhoe, Cat. Moths India, p. 668, No. 4543 (1889)[4].

Exelastis parasita, Lefroy, Ind. Ins. Pests, pp. 140-142 ff. 157-160 (1906) [biol.][5].

Alucita ebalensis, Rebel, Denkschr. Mathnat, Klasso K. Akad. Wiss. Wien LXXI. ii. 84 (1907)♂[6].

Exelastis atomosa; Lefroy, Ind. Agric. Ent. Mem. I 219, ff. 67, 68 (1907)[7]; Meyr., B. J. XVII 730 (1907)[8]; Lcfroy, Ind. Ins. Life, p. 527, t. 53 (1909)[9].

Marasmarcha atomosa, Meyr., Wyts. Gen. Ins., fasc. 100, p. 18 (1910)[10]; Meyr., Ann. Transv. Mus. II 220 (1911)[11]; Meyr., Cat. Pteroph., p. 27 (1913)[12]; Meyr., Voyage Alluaud Afr. Orient., Lep. p. 42 (1920)[13].

Exelastis atomosa, Woodhouse, Crop Pest Handb. Behar & Orissa, p. [39], t. 17 (1914)[14]; Fletcher, S. Ind. Ins., pp. 444-445, t. 38 (1914)[15]; Fletcher, Proc. Second Entl. Meeting, pp. 44, 56 (1917)[16]; Flethcer, Proc. Third Entl. Meeting, p. 137 (1920)[17]; Fletcher, Ind. Agric. Ent. Mem. VI 26-28, t. 7 (1921) [biol.][18].

Biology[5, 7, 9, 14, 15, 18]. *Larva descr.*[5, 7, 9, 14, 15, 18] on *Cajanus indicus*[5, 7, 9, 16, 17, 18], *Dolichos lablab*[5, 7, 9, 16, 17, 18]. Pupa descr.[5, 7, 9, 14, 15, 18].

Poona., Karachi.[2, 3]
Bhopal.
Hoshangabad.
Bilaspur, C.P.
Baroda.
Bellary distr.
Coimbatore., Pusa.
Abbottabad.
Peshawar.
Khasis., Nilgiris.
N. Kanara.

Aden.[6], E. Africa.
Kilimandjaro[13]
(2500 ft.)
S. Africa.[13] Natal.
Transvaal.
Moluccas.[12]
New Guinea.[12, 13]
Little Kei Isd.

crepuscularis, Meyrick 1910.

Pterophorus crepuscularis, Meyr., Anr. Transv. mus. II 4, t. 2 f. 2 (1910)♂[1].

Marasmarcha crepuscularis, Meyr., Ann. Transv. Mus. II 219 (1911)[2]; Meyr., Cat. Pteroph., p. 27 (1913)[3].

Exelastis pumilio [nec Zeller], Fletcher, Sci. Rept. Agric. Res. Inst. Pusa 1920-1921, p. 46, t. 5 f. 1 *A*-C [larva, pupa, imago] (1921)[4].

Shillong.[4]

Transvaal.[3]
Warmberg.[1], Pretoria.
Barberton.
N. Australia.

phlystaenias, Meyrick 1911.

Marasmarcha phlyctaenias, Meyr., B. J. XXI 106 (1911) ♂ ♀[1]; Meyr., Cat. Pteroph., p. 27 (1913)[2]; Meyr., Voyage Alluaud Afr. Orient., Lep. p. 43 (1920)[3].

Exelastis phlyctaenias, Fletcher, Ind. Agric. Ent. Mem. VI 26 (1921)[4].

Ceylon.[2], Puttalam.[1]
Trincomali.[1, 5]
Colombo.[4]
Kegalle.[4]
Haldummulla.[4]
Anuradhapura.[5]
Galle.[5], Weligama.[5]

Exelastis atomosa [nec Wlsm.], Fletcher, Spol. Zeyl. VI 33, t. A f. 11, t. C f. F [neur.] (1909)[3].

Larva perhaps on flowers of *Anacardium occidentale*[5].

Undugoda.[5]
Madulsima.[5]
S. India.[2]
N. Coorg.[1]

E. Africa.
Kilimandjaro.[3]
(3000 feet).

pumilio, Zeller 1873.

Mimeseoptilus pumilio, Zeller, Verh. z.-b. Ges. Wien XXIII 324 (1873)♂[1].

Marasmarcha liophanes, Meyr., T. E. S., 1886. 19 (1886) ♀[2]; Meyr., B. J. XVII 136 (1906)[3]; Meyr., B. J. XVII 730 (1907)[4]; Meyr., Wyts. Gen. Ins., fasc. 100, p. 18, tab. f. 15 (1910)[5]; Meyr. Cat. Pteroph., p. 27 (1913) [=*griseodactylus*][6]; Meyr., Ent. Mitteil., Suppl. III 47 (1914)[7].

Stenoptilia? pumilio, Wlsm., P. Z. S. 1891. 495 (1892)[8]; Wlsm., P. Z. S. 1897. 58 (1897)[9].

Mimoeseoptilus gilvidorsis [nec Zeller], Hedemann, Stett. Ent. Ztg. LVII 8-9 (1896) [part][10].

Stenoptilia pumilio, Fernald, Smith's List Lep. Bor. Amer., p. 87, No. 4566 (1891)[11]; Fern., Pteroph. n. Amer., p. 58 (1898) [redescr.][12]; Fern., U. S. Nat. Mus. Bull. 52, p. 447, No. 4991 (1903)[13]; Meyr., Cat. Pteroph., p. 28 (1913)[14].

Leioptilus? griseodactylus, Pagenstecher, Zoologica XXIX 240 (1900)[15].

Exelastis liophanes, Fletcher, Spol. Zeylan. VI 33-34, t. A f. 12 (1909)[16]; Fletcher, Trans. Linn. Soc. (Zool. ii) XIII 403 (1910)[17]; Fletcher, Ind. Agric. Ent. Mem. VI 26 (1921) [part][18].

Marasmarcha pumilio, Barnes and Lindsey, Contrib. N. H. Lep. N. Amer. IV 348-349, t. 42, f. 7, t. 49 f. 13, t. 48 f. 8 [neur.] (1921) [redescr.;-*liophanes*][19]; Forbes, Lep. N. York, p. 646 (1924) [redescr.][20]; Meyr., Ins. Samoa III 69 (1927)[21]; Meyr., T. E. S. LXXVI 491 (1929)[22].

Ceylon.[6]
Throughout to 2000 feet.[16]
India.[6, 19]
Coimbatore.[18]
Coorg.[18], Cuttack.[18]
Hoshangabad.[18]
Palamau.[18], Pusa.[18]
Lumding (Assam).[18]
Andaman Isds.
Burma.
Lashio.[18], Tatkon.[18]
Myitkyina.[18]

China.[6, 18], Formosa.[7]
Borneo.[5, 6], Sarawak.
Bismarck Isds.[6, 15]
New Hebrides.[21]
Fiji.[21]
Samoa.[21]
Austral Isds.[22]
Society Isds.[22]
Marquesas.[22]
N. America.
Texas.[1, 12, 13, 14, 19]
Missouri.[19, 20]
Florida.[19]
S. Carolina.[19]
New Jersey.[19, 20]
W. Indies.
St. Vincent.[8]
Jamaica.[9], St. Croix.[10]

Larva on *Oxalis*[18],? *Ambrosia artemisiaefolia*[19], *Desmodium* sp.*, *Alysicarpus vaginalis**.

Grenada.[9]
Barbados.[18], Africa.[19]
S. Africa.[21], Natal.[18]
Madagascar.[21]
Reunion.[2, 6]
Seychelles.[6, 17]
Mahé.[16]

ALUCITINAE.

24. Oidaematophorus, Wallengren 1862.

Oidaematophorus, Wlgn., K. Svensk. Vet. Akad. Handl. III, No. 7, p. 19 (1862) [? 1859]; Jordan, E. M. M. VI 125 (1869) [redescr.]; Wocke, Hein. Kleinschm. Deutschl. II. ii. 799 (1876) [redescr.]; Tutt, Pteroph. Brit., p. 103 (1896); Tutt, Ent. Rec. XVII 35 (1905); Barnes and Lindsey, Contrib. N. H. Lep. N. Amer. IV 374-380 (1921) [redescr.]; Forbes, Lep. N. York, p. 647, f. 400 (1924) [redescr.]

Type : *lithoda tylrs.,* Tr. [Europe].

Ocdematophorus, Zeller, Stett. Ztg. XXVIII pp. 331, 337 (1867) [emend.]; Wlsm., Pteroph. Calif. Oregon, p. 34 (1880); Meyr., T. E. S. 1886. 10 (1886); Hofmann, Ber. Nat. Ver. Refensb. V pp. 49, 148-151 (1896) [redescr.]; Barrett, Lep. Brit. Isds. IX 382 (1904) [redescr.].

Pterophorus [nec Geoffroy], Wlgn., K. Svensk. Vet. Akad. Handl. III, No. 7 p. 20 (1862) [?1859]; Jordan, E. M. M. VI 149 (1869) [redescr.]; Wlsm., Pteroph. Calif. Oregon, p. 39 (1880); Meyr., T. E. S. 1886. 10 (1886) [redescr.]; Hofmann, Ber. Nat. Ver. Regensb. V pp. 49, 155-157 (1896) [redescr.]; Tutt, Pteroph. Brit., p. 110 (1896); Wlsm. and Durrant, E. M. M. XXXIII 42 (1897); Fernald, Pteroph. N. Amer., p. 39 (1898); Rebel, Cat. Pal. Lep. II 75 (1901); Fernald, U. S. Nat. Mus. Bull. 52, p. 444 (1903); Barrett, Lep. Brit. Isds. IX 384 (1904) [redescr.]; Meyr., Wyts. Gen. Ins., fasc. 100, p. 15 (1910) [redescr.]; Spuler, Schmett. Eur. II 325, f. 101 [neur. *lithodactylus*] (1910) [redescr.]; Meyr., Cat. Pteroph., p. 21 (1913); Wlsm., Biol.

Type : *monodactylus,* L.

---

* Specimens bred from larvæ on *Desmodium* and *Alysicarpus vaginalis* at Pusa are perhaps not *pumilio.*

Centr. Amer., Het. IV 439-440 (1915) [bibliogr.]; Meyr., Rev. Handb., p. 456 (1928) [redescr.].

Leioptilus, Wlgn., K. Svensk. Vet. Akad. Handl. III, No. 7, p. 21 (1862) [?1859]; Zeller, Stett. Ent. Ztg. XXVIII pp. 331, 338 (1867) [redescr.]; Jordan E. M. M. VI 149-150 (1869) [redescr.]; Wooke, Hein. Kleinsohm. Deutschl. II ii. 802 (1876) [redescr.]; Hafmann, Ber. Nat. Ver. Regensb. V pp. 49, 160-164 (1896) [redescr.]; Tutt., Pteroph. Brit., p. 114 (1896); Barrett, Lep. Brit Isds. IX 387 (1904) [redescr.]; Tutt, Brit. Lep. V 96 (1906) II 326, f. 102 [neur. *carphodactylus*] (1910) [redescr.].

Type : *tephradactylus,* H. [Europe].

Lioptilus, Zeller, Stett. Ent. Ztg. XXVIII 331 (1867) [emend.]; Wlsm., Pteroph. Calif. Oregon, p. 41 (1880).

Alucita [nec. Linn.], Meyr., T. E. S. 1890. 487-488 (1890); Meyr., Handb., p. 438 (1895) [redescr.].

Type : *monodactyla,* L.

Ovendenia, Tutt, Ent. Rec. XVII 37 (1905) [non-descr.].

Hellinsia, Tutt, Ent. Rec. XVII 37 (1905) [non-descr.].

Emmelina, Tutt, Ent. Rec. XVII 37 (1905) [non-descr.].

harpactes, Meyrick 1908.

Pterophorus harpactes, Meyr., T. E. S. 1907. 495 (1908)♂[1]; Meyr., Wyts. Gen. Ins., fasc. 100, p. 15 (1910)[2]; Meyr., Cat., Pteroph., p. 22 (1913)[3].

Assam.[3]
Khasis.[1]

lienigianus, Zeller 1852.

Pterophorus lienigianus, Zeller, Linn. Ent. VI 380-381 (1852)♂ ♀[1]; Herr.-Schäff., Schmett. Eur. V 379-380 (1855) [redescr.][2]; de Graaf, Tijds. Ent. II 50 (1859)[3]; Stainton, Manual II 443 (1859) [redescr.][4]; Schleich, Stett. Ent. Ztg. XXIX 392 (1868) [larva][5]; Morris, Brit. Moths IV 298-299, t. 132 f. 16 (1872)[6]; Snellen, Vlind. Ned., Microlep., pp. 1045-1046, f. 470 [palpus] (1882) [redescr.][7]; Leech, Brit. Pyral. Pter., p. 63, t. 17 f. 10 (1886)[8]; Rebel, Cat. Pal. Lep. II 75, No. 1389 (1901)[9]; Meyr., T. E. S. 1907. 497 (1908) [=*serindibanus*][10]; Fletcher, Spol. Zeylan. VI 34-35 (1909)[11]; Meyr., Wyts Gen. Ins., fasc. 100, p. 15 (1910)[12]; Meyr., Ann. Transv. Mus. II

Ceylon.[10, 14, 16, 19, 24]
Anuradhapura.[11]
Galle.[11]
Trincomali.[11]
Maskeliya.[11]
Pundaluoya.[11]
Pattipola.[11]
Nuwara Eliya.[11]
Madulsima.[11]
India.[10, 14, 16, 19, 24]
Coimbatore.[20]
Gooty., N. Kanara.
Godavari Dist.[20]
Pusa., Peshawar.[20]

221 (1911)[13]; Meyr., Cat. Pteroph., p. 22 (1913) [=*victorianus*][14]; Fletcher, S. Ind. Ins., p. 445, f. 322 (1914)[15]; Wlsm., Biol, Centr. Amer., Het. IV 443 (1915)[16]; Fletcher, Proc. Second Entl. Meeting, p. 288 (1917)[17]; Fletcher, Proc. Third Entl. Meeting I 137 (1920)[18]; Meyr., Voyage Alluaud Afr. Orient., Lep. pp. 41-42 (1920)[19]; Fletcher, Ind. Agric. Ent. Mem. VI 29 (1921)[20]; Meyr., Entom. LIV 276 (1921) [=linus][21]; Meyr., Ark. Zool. XIV, No. 15, p. 1 (1922)[22]; Meyr., Entl. News XXXIII 213 (1922) [=*linus*][23]; Meyr., Rev. Handb., p. 457 (1928) [redescr.][24].

Leioptilus lienigianus, Wocke, Hein. Kleinschm. Deutschl. II ii. 802-803 (1876) [redescr.][25]; South, Entom. XV 105, t. 2 f. 3 (1882) [larva; pupa][26]; Sheldon, Entom. XXI 113 (1888)[27]; Tutt, Pteroph. Brit., pp. 117-121 (1896) [redescr.; biol.][28]; Hofmann, Ber. Nat. Ver. Regensb. V 171-172 (1896) [redescr.][29]; Barrett, Lep. Brit. Isds. IX 388-390, t. 416 ff. 5, 5[a]; (1904) [redescr., biol][30] Spuler, Schmett. Eur. II 327, Nachtr. t. 10 f. 13 [larva] (1910) [redescr.][31].

Pterophorus (Leioptilus) lienigianus, Barrett, E. M. M. VIII 156-158 (1871) [larva; biol.][32].

Alucita lienigiana, Meyr., T. E. S. 1890. 488 (1890)[33]; Meyr., Handb., p. 439 (1895) [redescr.][34].

Pterophorus melinodactylus, Herrich-Schäffer, Schmett. Eur. V, Pteroph. t. 6 f. 33 (1853)[35].

Pterophorus scarodactyla [nec Hb.], Becker, Ann. S. E. Belg. V. 56-57, t. 2 ff. 3, 3[a-b] (1861) [biol.][32].

Leioptilus serindibanus, Wlsm., Moore's Lep. Ceylon III 527, t. 209 f. 14 (1887)[37]; Cotes and Swinhoe, Cat. Moths India, p. 669, No. 4552 (1889)[38].

Pterophorus sericeodactylus, Pagenstecher, Zoologica XXIX 240 (1900)[39].

Ovendenia septodactyla [nec Treits.], Tutt. Ent. Rec. XVII. 37 (1905)[40]; Tutt, Brit. Lep. V, pp. 77-82, 84-85, 87-90, 93-94, 97, 99, 101, 111, 118, 153, 253, 264, 285, 386, 395-396, 398, 409, 484-485, 487-488 (1906) [redescr.; biol.][41].

Shillong.
Burma : Maymyo.[20]

China.[24], Japan.
Nagasaki.[16], Korea.
Gensan.[16], Siberia.
Kasakewitsch.[43]
Europe.[16, 19]
C. Europe.[9, 14, 24, 34]
S. England.[4, 6, 27, 28, 30, 31, 34]
Holland.[3, 7]
Livonia.[1, 2, 9]
Silesia.[1, 2]
Germany.[25, 29]
Austria.[2, 29], Styria.[1]
Italy.[1, 2, 9], Africa.[19, 14]
W. Africa.[14]
Transvaal.[13]
Brit. E. Africa.
Rurunga.[19] (4700 feet)
Mauritius.
N. America.[19, 24]
Arizona.[16]
Ohio.[44]
Pennsylvania.[44]
New Hampshire.[44]
Maryland.[44]
Massachusetts.[44]
Mexico.[16]
C. America.[19]
Guatemala.[16]
S. America.[23, 24]
Bermuda.[24]
Queensland.
Cedar Creek.[22]
Bismark Isds.[24, 30]

Pterophorus victorianus, Strand, Arch. Naturg. LXXVIII. A 12., p. 130 (1913)[42].

Pterophorus lienigianus var. *catharodactylus,* Caradja, Iris XXXIV 86 (1920)[43].

Oidamatophorus linus, Barnes and Lindsey, Contrib. N. H. Lep. N. Amer. IV 409-410, t. 47 f. 2, t. 52 f. 9 (1921)♂ ♀[44].

Biology.[5, 28, 29, 30, 32, 36]. Larva descr.[5, 7, 24, 26, 28, 29, 30, 31, 32, 36] on *Artemisia*[24], *A. vulgaris*[5, 6, 7, 25, 26, 28, 29, 30, 31, 32], *A. campestris*[36], *A. maritima*[6], *Chrysanthemum*[24], *Pluchea*[24], *Solanum*[24], *S. melongena,*[15, 17, 18, 20], "*Khagra*" (*Xanthium strumariam*)[20]. Pupa descr.[26, 28, 29, 30].

monodactylus, Linnaeus 1758.

Phalaena Aucita monodactyla, Linn., Syst., Nat. (ed. X) I. 542 (1758)[1]; Linn., Faun. Suec., p. 370, No. 1452 (1761)[2]; Linn., Syst. Nat. (ed. XII) I. ii. 899, No. 453 (1767) [redescr.][3]; Vill., Linn. Ent. Faun. Suec. II 530-531, No. 1083 (1789) [redescr.; larva, pupa][4]; Gmelin, Linn. Syst. Nat. (ed. XIII) I. v. 2615, No. 453 (1790) [redescr.][5].

Pterophorus ["Le pterophore brun"], Geoffroy, Hist. abrégée des Insectes II 92 (1762)[6].

Pterophorus monodactylus, Fabr., Syst. Ent., p. 671, No. 1 (1775) [redescr.][7]; Fb., Spec. Ins. II 310, No. 1 (1781) [redescr.][8]; Fb., Mantissa Ins. II 258, No. 1 (1787) [redescr.][9]; Fb., Ent. Syst. III. ii. 345, No. 1 (1794) [redescr.][10]; Wlgn., K. Svensk. Vet. Akad. Handl. III, No. 7, pp. 20-21 (1862) [redescr.][11]; Zell., Stett. Ent. Ztg. XXVIII 338 (1867)[12]; Zell., Verh. z.-b. Ges. Wien. XXIII (326) (1873)[13]; Zell., Verh. z.-b. Ges. Wien XXV 355-356 (1875)[14]; Wocke, Hein. Kleinschm. Deutschl. II. ii. 801-802 (1876) [redescr.][15]; Wlsm., Pter. Calif. Oregon, pp. 39-41, t. 2 f. 16, t. 3 f. 1 (1880)[16]; Snell., Vlind. Ned., Microlep. pp. 1048-1050 (1882) [redescr.][17]; South, Entom. XVIII 277-279, t. 1 ff. 4, 4[a-c] (1885) [redescr.; biol.][18]; Leech, Brit. Pyr. Pter., t. 17 f. 10 (1886)[19]; Alph., Rom. Mem. Lep. V. 231 (1889)[20]; Holt-White, Butts. Moths

Kashmir.[39, 44]
Srinagar., Hazara.
Parachinar.[42]
Masuri.[57]

Japan.[39], Siberia.[41]
W. Thian Shan.[37]
C. Asia.[36]
N. Persia.[27]
Turkestasn.[27]
Armenia.[27]
Asia Minor.[27, 39]
Europe.[27, 32, 36, 39, 44]
Russia.[59]
Sweden.[2, 11, 59]
Germany.[15, 25]
Austria.[25]
Switzerland.[25, 63]
France.[59], Holland.[17]
Brit. Isds.[26, 44, 59]
Sicily.[59]
Madeira Isds.[27, 73]
Canary Isds.[21-24, 27, 31-32]
N. Africa[27, 32, 36, 39, 44]
Algeria.[39], Tunis.[39]
Cyrenaica.[43]

Tenerife, p. 95 (1890)[21]; Rebel, Ann. Hofmus. Wien. VII. 263, 282 No. 38 (1892)[22]; Rebel, Ann. Hofmus. Wien. IX. 16, 81 No. 140 (1894)[23]; Rebel, Ann. Hofmus. Wien XI pp. 115, 146 No. 153 (1896)[24]; Hofmann, Ber. Nat. Ver. Regensb. V 157-160 (1896) [redescr.; biol.][25]; Tutt, Pteroph. Brit., pp. 111-115 (1896) [redescr.; biol.][26]; Rebel, Cat. Pal. II 75, No. 1387 (1901)[27]; Fernald, Pteroph. N. Amer., pp. 51-52, t. 1 ff. 1-13, t. 2 ff. 3, 4, t. 7 ff. 1-5 (1898) [redescr.][28]; Fernald, U. S. Nat. Mus. Bull. 52, p. 446, No. 4981 (1903)[29]; Barrett, Lep. Brit. Isds. IX 385-387, t. 416 ff. 4, 4[a] (1904) [redescr.; biol.][30]; Rebel, Ann. Hofmus Wien XXI 43, No. 179 (1906)[31]; Wlsm., P. Z. S. 1907. 919-920 (1908)[32]; Meyr., Wyts. Gen. Ins. fasc. 100, p. 17 (1910)[33]; Spuler, Schmett. Eur. II 326, t. 82 f. 63, Nachtr. t. 10 f. 15 [larva] (1910) [redescr.][34]; Rebel, Iris XXVI 84 (1912)[35]; Meyr., Cat. Pteroph., p. 26 (1913)[36]; Rebel, Iris XXVIII 272 (1914)[37]; Müller Rutz, Schmett. Schweiz. II 489, No. 2050 (1914)[38]; Wlsm., Biol Centr. Amer., Het. IV 442 (1915)[39]; Fletcher, Ind. Agric. Dept. Bull. 89, p. 81, Note 184 (1919).[40]; Meyr., Voyage Alluaud Afr. Orient., Lep. p. 42 (1920)[41]; Fletcher, Ind. Agric. Ent. Mem. VI 29 (1921)[42]; Turati, Atti. Soc. Sci. Nat. Milano LXVI 336 (1927)[43]; Meyr., Rev. Handb., p. 458 (1928) [redescr.][44].

Phalaena Alucita galactodactyla [nec Schiff. 1775] Hübner, Beitr. Gesch. Schmett., Part 1, pp. 28-29, t. 4 f. U (1786)[45]; Hb., *t.c.*, Nachtrag p. 109 (1790)[46].

Pterophorus albodactylus, Fabr., Ent. Syst. III. ii. 348 (1974)[47]; Coquebert, Illustr. Icon. Insect. I 31, t. 7 ff. 14[a-C] (1798) [redescr.; fig. of type][48]; Latreille, Hist. Nat. Crust. Ins. XIV 258 (1802) [redescr.][49]; Zeller, Isis, p. 856 (1841)[50].

Alucita pterodactyla [nec Linn.], Hb., Samml. Eur. Schmett., Aluc. t. 1 f. 4 (1804)[51]; Treitschke, Schmett. Eur. IX. ii. 242-244 (1833) [part.][52].

Pterophorus pterodactylus [nec Linn.], Stephens, Ill. Brit. Ent., Haust. IV 372-373 (1835) [redescr.][53]; Wood, Index Ent., p. 235, t. 51 f. 1631 (1838)[54];

Egypt., Alexandria.[74]
Cairo.[35], E. Africa.[44]
Kenya41 (7000 ft.)
N. America.[27, 32, 36, 44]
Atlantic States.[82]
Massachusetts.[14]
Maine.[28], Ohio.[13]
Texas.[13], Arizona.[77]
Oregon.[16, 39]
California.[16, 28, 39, 77, 80, 81]
Pacific States.[82]
S. Canada.[82]
Mexico.[82]
Amula.[39]
(Guerrero; 6000 ft.)

Duponchel, Lep. France XI 663, t. 314 f. 2 (1838); Zeller, Isis, pp. 846-848 [part], t. 4 f. 7 (1841)[56]; Kollar, Hügel's Kaschmir IV. ii. 496 (1844)[57]; Westw., Brit. Moths II 259, t. 123 ff. 7-9 (1845) [? f. 8. larva on thistle] [redescr.][58]; Zeller, Linn. Ent. VI 377-378 (1852)[59]; Stainton, List Brit. Anim. B. M. XVI 177-178 (1854) [part][60]; Westw., Wood's Index Ent., p. 235, t. 51 f. 1631 (1854)[61]; Herrich-Schäffer, Schmett. Eur. V 379, Pter. t. 5 f. 27 (1855) [redescr.][62]; Frey, Tin. Pter. Schweiz., p. 415 (1856) [redescr., larva][63]; Stainton, Manual II 443 (1859) [redescr.][64]; Jordan, E. M. M. VI 149 (1869) [*monodactylus*=[65]; Morris, Brit. Moths IV 298, t. 132 f. 15 (1872)[66]; Buckler, Larvae Brit. Butt. Moths IX 365 (1901) [larva][67].

Pterophorus cineridactylus, Fitch, Tran. N. Y. Agric. Soc. XIV 848 (1855)[68].

Pterophorus naevosidactylus, Fitch, Trans. N. Y. Agric. Soc. XIV 849 (1855).[69]

Pterophorus pergracilidactylus, Panckard, Ann. Lyc. N. H. New York X 265-266 (1873)[70].

Alucita monodactyla, Meyr., T. E. S. 1890, 488 (1890)[71]; Fernald, Smith's List Lep. Bor. Amer., p. 87 No. 4577 (1881)[72]; Wlsm., T. E. S. 1894. 539 (1894)[73]; Bethune-Baker, T. E. S. 1894. 50 (1894)[74]; Meyr., Handb., 439 (1895) [redescr.][75].

Pterophorus barberi, Dyar Proc. E. S. Wash. V. 228 (1903)[76]; Meyr., Cat. Pteroph., p. 24 (1913)[77].

Emmelina monodactyla, Tutt, Ent. Rec. XVII 37 (1905)[78]; Tutt. Brit. Lep. V pp. 70-75, 85-90, 92-93, 97, 101, 105-106, 110, 115, 118, 120-121, 338, 360, 379, 386 (906)[79].

Pterophorus pictipennis, Grinnell Chand. Ent. XL 320 (1908)[80]; Meyr., Cat. Pteroph, p. 25 (1913)[81].

Oidaemataphorus monodactylus, Barnes and Lindsey, Contrib. N. H. Lep. N. Amer. IV 443-445, t. 46 ff. 13-15, t. 54 f. 13 (1921) [redescr.; biol.][82]; Forbes, Lep. N. York, p. 651 (1924) [redescr.][83].

Biology[18, 25, 26, 30, 82]. Larva descr.[11, 18, 25, 26, 28, 30, 44, 63, 67, 82] on Convolulus.[6, 44], *C. floridus*.[23], *C. major*[26, 82] *C. sepium*[26, 28, 30, 82] *Ipomoesa purpurea*[82], *Chenopodium album*,[28, 82] *Atriplex patula*,[28, 82], *Eupatorium purpureum*[82], *Calluna vulgaris*[82].

Pupa descr.[18, 25, 26, 30]

procontias, Meyrick 1908.

Pterophorus procontias, Meyr., T. E. S. 1907. 496 (1908)♂ ♀[1]; Meyr., Wyts. Gen. Ins., fasc. 100, p. 15, tab. f. 14 (1910)[2]; Meyr., Cat. Pteroph., p. 22 (1913)[3].

Assam.[3] Khasis.[1] Shillong.

sematias, Meyrick 1908.

Pteroporus sematias, Meyr., T. E. S. 1907. 496-497 (1908)♀[1]; Fletcher, Spol. Zeylan. VI 35 (1909)[2]; Meyr., Wyts. Gen. Ins., fasc. 100, p. 15 (1910)[3]; Meyr., Cat. Pteroph., p. 22 (1913)[4].

Ceylon.[4] Maskeliya.[1] S. India.[4] Nilgiris.

triacias, Meyrick 1908.

Pterophorus triadias, Meyr., T. E. S. 1907. 494-495 (1908)♂ ♀[1]; Meyr., Wyts. Gen. Ins., fasc. 100 p. 15 (1910)[2], Meyr., Cat. Pteroph., p. 22 (1913)[3].

Assam.[3], Khasis.[1]

Nyasaland: Mt. Mlanje. [Meyr. Coll.; *?? eadem species*].

15. Pselnophorus, Wallengren 1881.

Pselnophorus, Wlgn., Ent. Tidskr. XI 96 (1881); Meyr., Handb., p. 437 (1895) [redescr.]; Hofmann, Ber. Nat. Ver. Regensb. V. pp. 50, 178-179 (1896) [redescr.]; Rebel, Cat. Pal. Lep. II 74 (1901); Meyr., T. E. S. 1907. 491 (1908); Meyr., Wyts. Gen. Ins., fasc. 100, p. 14 (1910) [redescr.]; Spuler, Schmett. Eur. II 327, t. 103 [neur*brachydactylus*] (1910) [redescr.]; Meyr., Cat. Pteroph., p. 20 (1913); Barnes and Lindsey, Contrib. N. H. Lep. N. Amer., pp. 361-362, t. 48 f. 4 [neur. *belfragei*] (1921) [redescr.]; Meyr., Rev. Handb., p. 455 (1928) [redescr.].

Type : *brachydactylus*, Tr. [Europe].

Gypsochares, Meyr., T. E. S. 1890. 488 (1890); Rebel, Cat. Pal. Lep. II 75 (1901); Spuler, Schmett. Eur. II 323 (1910) [redescr.]. — Type : *baptodactylus* Zell. [Europe].

Crasimetis, Meyr., T. E. S. 1890. 489 (1890). — Type : *brachydactylus,* Tr.

albitarsellus, Walsingham 1900.

Alucita albitarsella, Wlsm., Swinhoe's Cat. Het. Mus. Oxon. II 542 (1900)♂[1].

Pselnophorus albitarsellus, Meyr., T. E. S. 1907. 492 (1908)[2]; Meyr., Wyts. Gen. Ins., fasc. 100, p. 14 (1910)[3]; Meyr., Cat. Pteroph., p. 20 (1913)[4]; Fletcher, Ind. Agric. Ent. Mem. VI 30 (1921)[5].

Larva descr.[1], boring in shoots of "a common jungle plant." Pupa suspended by tail from underside of a leaf[1].

S. India.[4]
Ootacamund.[1]
Kodaikanal.

catharotes, Meyrick 1908.

Pselnophorus catharotes, Meyr., T. E. S. 1907. 491-492, (1908)♂ ♀[1]; Meyr., Wyts. Gen. Ins., fasc. 100, p. 14, tab. f. 8 (1910)[2]; Meyr. Cat. Pteroph., p. 20 (1913)[3].

Pterophorus baptodactylus [nec Zeller 1850], Zeller, Stett. Ent. Ztg. XXVIII 414 (1867)[4]; Cotes and Swinhoe, Cat. Moths India, p. 668, No. 4544 (1889)[3]. [nec *Ps. catharotes,* Meyr., Ann. Transv. Mus. III 267 (1913) [Transvaal]; Meyr., Voyage Alluaud Afr. Orient., Lep. p. 42 (1920) [Brit. E. Africa]. Both these records refer to *Ps. pachyceros,* Meyr. 1921. See Meyr., Ann. Transv. Mus. VIII 50 (1921)].

Assam.[3],
Khasis.[1]
Shillong.
Laitlyngkot.
Mawphlang.
Cherrapunji.
Sikkim.[4, 5]
Kurseong.
Kumaon.
Bhim Tal.
Muktesar.

16. Alucita, Linnaeus 1758.

Phalaena Alucita, Linn., Syst. Nat. (ed. X) I. pp. 496, 542 (1758); Poda, Ins. Mus. Graec. p. 94, t. 2 f. 14 (1761); Linn., Faun. Suec., p. 370 (1761); Linn., Syst. Nat. (ed. XII) I. ii. pp. 809 [redescr.], 899 (1767); Gmelin, Linn. Syst. Nat. (ed. xiii) I. v. p. 2615 (1790) [redescr.]. — Type : *pentadactyla,* Linn. [Europe].

Pterophorus, Geoffroy, Hist. abrégée des Insectes II. pp. 24-25, 90-93 (1762); Schluga, Primae Lineae cognit. Ins., p. 37 (1767) [redescr.]; Fabr., Syst. — Type : *pentadactyla,* Linn.

Ent., pp. [XXV], 671 (1775); Fb., Genera Ins., p. 168 (1777) [redescr.]; Fb., Spec. Ins. II 310 (1781); Fb., Mantissa Ins. I p. xviii (1787), II. 258 (1787) [redescr.]; Fb., Ent. Syst. III. i. [p. vi] (1793) [redescr.], III. ii. 345 (1794) [redescr.]; Lamarck, Syst. Anim. sans Vert., p. 288 (1801) [redescr.; *pentadactyla* cited as type]; Latreille, H. N. Crust. Ins. III 418 (1802) [redescr.]; Latr. H. N. Crust. Ins. V 325 (1803) [redescr.]; Latr., H. N. Crust. Ins. XIV 257 (1805) [redescr.]; Latr., Consid. gén Ins., pp. 570 [charact.], 442 ["*didactylus*, F.", cited as type] (1810); Leach, Brewster's Edinb. Encycl. IX 135 (1815) [redescr.]; Curtis, Brit. Entom. IV 161 (1827) [redescr.; type *pentadactyla*]; Stephens, Ill. Brit. Entom., Haust. IV 370-371 (1835) [redescr.]; Westwood, Synopsis Brit. Ins., p. 115 (1840) [redescr.; type *pentadactyla*]; Meyr., T. E. S. 1890. 489, t. 15 ff. 6, 13 [neur. *spilodactyla*] (1890) [redescr.]; Meyr., Handb. p. 435 (1895) [redescr.; fig. neur. *spilodactyla*].

Alucita, Scopoli, Introd. Hist. Nat., p. 428 (1777) [redescr.]; Schrank, Fauna Boica II. i. 157 (1801) [redescr.]; Haworth, Lep. Brit., p. 475 (1811); Oken, Lehrbuch III. i. 679 (1815) [charact.; *Pterophorus* quoted as syn.]; Wlsm. and Durrant, E. M. M. XXXIII 41-42 (1897); Rebel, Cat. Pal. Lep. II 73 (1901); Tutt, Ent. Rec. XVII pp. 34, 37 (1905); Tutt, Brit. Lep. V 70-76, 92-95, 115, 118, 409, 412 (1906); Meyr., Wyts. Gen. Ins., fasc. 100, pp. 3, 12-14 (1910) [redescr.]; Spuler, Schmett. Eur. II 328, t 104 [neur. *baliodactyla*] (1910) [redescr.]; Meyr., Cat. Pteroph, p. 16 (1913); Wlsm., Biol. Centr. Amer., Het. IV 439 (1915); Forbes, Lep. New York p. 640, f. 401 [neur. *pentudactyla*] (1924); Meyr., Rev. Handb., pp. 453-454, fig. [neur. *tetradactyla*] (1928) [redescr.]; Hudson, New Zeal. Lep., p. 209, t. E. ff. 4-6 [*monospilalis*] (1928) [redescr.].

Type : *pentadactyla*, Linn.

Aciptilia, Hübner, Verz. p. 430 (1826); Wocke, Hein. Kleinschm. Deutschl. II. ii. 806 (1876) [part]; Snellen, Vlind. Ned., Microlep., pp. 1052-1053, ff. 455, 480-483 (1882) [redescr.; part.]; Meyr., T.

E. S. 1886. 9 (1886) [redescr.]; Tutt, Pteroph. Brit., p. 137 (1896); Hofmann, Ber. Nat. Ver. Regensb. V pp. 50, 183-186 (1896) [redescr.]; Barnes and Lindsey, Contrib, N. H. Lep. N. Amer. IV 359-360 (1921) [redescr.].

Aciptilus, Zeller, Isis XXXV 768 (1841) [emend.]; Herrich-Schäffer, Schmett. Eur. V 381-382 (1855) [redescr.; part], t. 8 f. 18 [*pentadactyla*] (1849); Wlgn., K. Svensk. Vet. Akad. Handl. III, No. 7, p. 23 (1862) [redescr.]; Zeller Stett. ent. Ztg. XXVIII. 332, 338 (1867) [redescr.]; Jordan, E. M. M. VI 150 (1869) [redescr.]; Barrett, Lep. Brit. Isds. IX 398 (1904) [redescr.].

Acyptilus, de Graaf, Tijds. Ent. II 37 (1859) [redescr.; emend.].

Merrifieldia, Tutt, Ent. Rec. XVII 37 (1905) [non-descr.]. Type : *tridactyla,* Linn. [Europe].

Porrittia, Tutt, Ent. Rec. XVII 37 (1905) [non-descr.]. Type : *galactodactyla,* Schiff. [Europe].

Wheeleria, Tutt, Ent. Rec. XVII 37 (1905) [non-descr.]. Type : *spilodactyla,* Curtis. [Europe].

candidalis, Walker 1864.

Aciptilus candidalis, Wlk., Cat. XXX 948 (1864)♀[1]; Fletcher, T. E. S. 1925. 626 (1926) [type redescr.][2].

Aciptilus leucadactylus, Wlk., Cat. XXX 949 (1864)♂ ♀[3]; Moore, Lep. Ceylon III 528, t. 209 f. 15 (1887) [redescr.][4]; Fletcher, T. E. S. 1925. 628 (1926) [type redescr.][5].

Pterophorus leucadactylus, Cotes and Swinhoe, Cat. Moths India, p. 669, No. 4548 (1889)[6].

Alucita candidalis, Wlsm, T E. S. 1881. 282 (1881)[7]; Wlsm., T. E. S. 1897. 36 (1897)[8]; Meyr., T. E. S. (1907), 490 (1908) [=*leucadactylus*][9]; Fletcher, Spol. Zeylan. VI 36 (1909)[10]; Meyr., Wyts. Gen. Ins., fasc. 100, p. 13 (1910)[11]; Meyr., Cat. Pteroph., p. 17 (1913)[12]; Meyr., Ent. Mitt., Suppl. III, p. 47 (1914)[13]; Meyr., Voyage Alluaud Afr. Orient., Lep. p. 40 (1920)[14]; Meyr., Ins. Samoa III 69 (1927)[15].

Ceylon.[3-6, 12, 14]
Low-country.[10]
Kegalle.[10]
Labugama.[10]
Weligama.[10]
Colombo.[5], Kandy.[10]
Maskeliya.[10]
Haldummulla.[10]
Lunugala.[10]
Puttalam.
India.[9, 12, 14]
Coimbatore.
Santikoppa (Coorg).
Khasis.

Malaya.[14]
Philippines.[9, 12]
Manila.
Formosa.[13]

(?) *Pterophous millierei,* Montrouzier, Ann. Soc. Linn. Lyon XI 253 (1864)[16].

Larva on *Argyreia.*

New Guinea.[9, 12]
New Caledonia.[16]
Queensland.[9, 12, 15]
Samoa.[15]
Tonga.[15]
New Hebrides.[15]
Brit. E. Africa.[14]
C. Africa.[12]
W. Africa.
Sierra Leone.[1, 2]
French Congo.[8]
S. Africa.[9, 12]
Natal.
Spring Vale.[7]
Transvaal.
Pretoria.

elæopa, Meyrick 1908.

Alucita elæopa, Meyr., T. E. S. 1907. 490 (1908)♂ ♀[1]; Meyr., Wyts. Gen. Ins., fasc. 100, p. 13, tab. f. 11 (1910)[2]; Meyr., Cat. Pteroph., p. 17 (1913)[3].

Assam.[2, 3]
Khasis[1]

lacteipennis, Walker 1864.

Aciptilus lacteipennis, Wlk., Cat. XXX 949 (1864)♀[1]; Fletcher, T. E. S. 1925. 627, f. 12 (1926) [type redescr.][2].

Pterophorus lacteipennis, Cotes and Swinhoe, Cat. Moths India, p. 669, No. 4547 (1889) ["India-*error loc.*][3].

Alucita lacteipennis, Meyr., T. E. S. 1907. 488-489 (1908)♂ ♀[4]; Meyr., Wyts. Gen. Ins., fasc. 100, p. 13 (1910)[5]; Meyr., Cat. Pteroph., p. 16 (1913)[6]; Durrant, Report B. O. U. Exped., Lep. p. 149 (1915)[7].

Burma[4, 5, 6]
Moulmein[1, 2]

Borneo[4, 5, 6]
Formosa
New Guinea[5, 6, 7]
Woodlark Isd.[4]
Solomon Isds.[4, 6]
Queensland[5, 6]
Cairns[4]

melanopoda, Fletcher 1907.

Alucita melanopod, Fletcher, Entom. XL 284 (1907)♂[1]; Meyr., T. E. S. 1907. 489 (1908)[2]; Fletcher, Spol. Zeylan. VI 36, t. A. f. 13 (1909)[3] Meyr., Wyts. Gen. Ins., fasc. 100, p. 13 (1910)[4]; Meyr., Cat. Pteroph., p. 16 (1913)[5].

Ceylon[2, 5]
Madulsima[1, 2]
Haragama[1, 3]
Kandy[1, 3]
Coorg
Mercara
Assam[2, 3, 5]
Khasis[1]

niveodactyla, Pagenstecher 1900.

Alucita niveodactyla, Pag., Zoologica, XXIX 240 (1900)[1]; Meyr., T. E. S. 1907. 490 (1908) [=*nivea*][2]; Fletcher, Spol. Zeylan. VI 36, t. F. f. 9 [larva] (1909)[3]; Poulton, Proc. E. S. London (1909), pp. 39-40 (1909)[4]; Meyr., Wyts. Gen. Ins., fasc. 100, p. 13 (1910)[5]; Meyr., Cat. Pteroph., p. 17 (1913)[6]; Fletcher, Ind. Agric. Ent. Mem. VI 30 (1921)[7], Corbett and Gates, Dept. Agric. F. M. S. Bull. 38, p. 11 (1926)[7a].

Aciptilia nivea, Snellen, Tijds. Ent. XLVI 56, t. 5 f. 17 (1903)♂[8].

Aciptilus leucadactylus, Wlk. [part.], Fletcher, T. E. S. 1925. 628 (1926)[9].

Biology[3]. Larva descr.[3], on *Ipomœ*[3, 7], *Ipomœa batatas*[7a] Pupa descr.[3].

Ceylon[2, 6, 9]
Maskeliya[3]
Haputale[3]
Haldummulla
Diyatalawa[3]
Madulsima[3]
Lunugala[3]
Kandy
India[6]
Travancore
Khasis
Shillong[7]
Cherrapunji[7]
Assam
Lumding
Tinsukia

Malay States[7a]
Borneo
Sarawak
Philippines[2, 3, 6]
Cochin China[4]
Java[3, 6]
Batavia[8]
Tegal[8]
Pekalongan[8]
Sumatra[6]
Solomon Isds.[2]
BismarckArchip.[1, 3, 6]
Queensland[6]

rhyparias, Meyrick 1908.

Alucita rhyparias, Meyr., T. E. S. 1907. 489-490 (1908)♂[1]; Meyr., Cat. Pteroph., p. 17 (1913)[2]; Mer., Vogage Alluaud Afr. Orient., Lep. p. 41 (1990)[3].

"India"
[Meyr. M.S.]

Transvaal[2, 3]
Pretoria[1]
Orange Free State[2, 3]
Zululand[2, 3]
Brit. E. Africa
Rurunga (4,700 ft.)[3]

sesamitis, Meyrick 1905.

Alucita sesamitis, Meyr., B. J. XVI 582-583 (1905)♂[1]; Meyr., Wyts. Gen. Ins., fasc. 100, p. 14 (1910)[2]; Meyr., Cat. Pteroph., p. 20 (1913)[3].

Burma[2, 3]
Koni[1]
Assam
Khasis

AGDISTINÆ.

17. Steganodactyla, Walsingham 1891.

Steganodactyla, Wlsm., E. M. M. XXVII 241 (1891); Wlsm., Novit. Lep. I, p. 3 [sep.], t. 12 ff. 3, 3[a-b] (1892) [redescr.].

Ochyrotica [part.], Meyr., Wyts. Gen. Ins., fasc. 100, p. 20 (1910); Meyr., Cat. Pteroph., p. 31 (1913).

Type : *concursa*, Wlsm.

concursa, Walsingham 1891.

Steganodactyla concursa, Wlsm., E. M. M. XXVII 241-242 (1891)♂[1]; Wlsm., Novit. Lep. I, p. 3 [sep.], t. 12 ff. 3, 3[a] [head], 3[b] [larva] (1892) [redescr.][2]; Hering, Stett. Ent. Ztg. LXIV 96 (1903)[a]; Fletcher, Spol. Zeylan. VI 9-10, t. E ff. 1, 2 [larva, pupa] (1909) [biol][4]; Fletcher, Ind. Agric. Dept. Bull. 89, p. 81, Note 183 (1919)[5]; Fletcher Ind. Agric. Ent. Mem. VI 30-31 (1921)[6].

Ochyrotica concursa, Meyr., Wyts. Gen. Ins., fasc. 100, p. 20 (1910)[7]; Meyr., Cat. Pteroph., p. 31 (1913)[8]

Biology[4]. Larva descr.[1, 2, 4], on *Argyreia*[1, 2, 4] *Ipomœa populifolia*[4] *Lantana*.[6] Pupa descr.[1, 2, 4].

Ceylon[7, 8]
Galle[4]
Weligama[4]
Colombo[4]
Kandy[4]
Matalé[6]
Maskeliya
Pundaluoya[1, 2, 4]
Haputale[4]
Diyatalawa[4]
Madulsima[4]
S. India[8]
Coorg[5, 6]
Belgaum[6]
N. Kanara

Sumatra[3]
Philippines[8]
Mindoro
Tonkin
Comoro Isds.
Mayotte
Mauritius

connexiva, Walsingham 1891.

Steganodactyla connexiva, Wlsm., E. M. M. XXVII 242-243 (1891)♀[1]; Wlsm., Novit. Lep. I, p. 4 [sep.], t.

India[4, 5]
Khasis

12 f. 4 (1892) [redescr.]; Durrant, Rept. B. O. U. Expedn., Lep. p. 149 (1915)[3].

Ochyrotica connexiva, Meyr., Wyts. Gen. Ins., fasc. 100, p, 20, tab. f. 17 (1910)[4]; Meyr., Cat. Pteroph., p. 31 (1913)[5].

Burma[4, 5]
Thandsung[1, 2] (Karen Hills : 4,000 ft.)
Philippines[5]
New Guinea[3]

18. Agdistis, Hübner 1826.

Agdistis, Hb., Verz., p. 429 (1826); Zeller, Linn. Ent. VI 321 (1852) [redescr.; =*Adactyla*]; Stainton, List Brit. Anim. B. M. XVI 172 (1854); Herrich-Schäffer, Schmett. Eur. V 364 (1855) [redescr.], t. 8 ff. 4-7 [*huebneri*]; Wocke, Hein. Kleinschm. Deutschl. II ii. 780-781 (1876) [redescr.]; Snellen, Vlind. Ned., Microlep, p. 1011, t. 13 f. 450 (1882); Meyr., T. E. S. 1890. 486 (1890) [redescr.]; Meyr., Handb., p. 441, fig. [neur., *bennetii*] (1895) [redescr.]; Tutt, Pteroph. Brit., p. 14 (1896) [redescr.]; Hofmann, Ber. Nat. Ver. Regensb. V, pp. 47, 50-53 (1896) [redescr.]; Rebel, Cat. Pal. Lep. II 77 (1901); Tutt, Ent. Rec. XVII 35 (1905); Tutt, Brit. Lep. V 127-135 (1906) Redescr.]; Wlsm., Ent. Rec. XIX 53-55 (1907) [remarks on spp.]; Spuler, Schmett. Eur. II 318, f. 93 [neur. *adactyla*] (1910) [redescr.]; Meyr., Wyts. Gen. Ins., fasc. 100, pp. 20-21 (1910) [redescr.]; Meyr., Cat. Pteroph, p. 31 (1913); Barnes and Lindsey, Contrib. N. H. Lep. N. Amer. IV 447, t. 42 t. 20, t. 52 f. 16 [*americana*] (1921) [redescr.]; Meyr., Rev. Handb., p. 460, fig. 9 [neur. *bennetii*] (1928) [redescr.].

Type : *adactyla,* Hb. [Europe]

Agdistes, Stephens, Ill. Brit. Ent., Haust. IV 370 (1835) [emend.]; Westwood, Synopsis Brit. Ins., p. 115 (1840) [redescr.]; Westw., Brit. Moths II 258 (1845) [redescr.]; Barrett, Lep. Brit. Isds. IX 341 (1904) [redescr.].

Adactylus, Curtis, Brit. Entom. X, Expl. t. 471 (1833); Curtis, Genera Brit. Lep., t. 15 f. 193 (1858); Tutt, Brit. Lep. V 135-136 (1906).

Type : *bennetii,* Curtis [W. Europe]

Adactyla, Zeller, Isis XXXV 763-764, t. 4 ff. 1, 2, 14, 15 [*huebneri*] (1841); Duponchel, Cat. Meth. Lep. Eur., p. 381 (1845) [redescr.]; Zeller, Isis XL 898 (1847)

Type : [*adactyla,* Hb.=] *huebneri,* Curtis [Europe]

[remarks]; Stainton, Manual II 440 (1859) [redescr.].

Ernestia, Tutt, Brit. Lep. V 128 (1906). — Type : *lerinsis,* Mill. [S. W. Europe]

Herbertia, Tutt, Brit. Lep. V 129 (1906). — Type : *tamaricis,* Zell.

nanodes, Meyrick 1906.

Agdistis nanodes, Meyr., B. J. XVII 136 (1906) ♂♀[1]; Fletcher, Spol. Zeylan. VI 8 (1909)[2]; Meyr., Wyts. Gen. Ins., fasc. 100, p. 21, tab, f. 16 (1910)[3]; Meyr., Cat. Pteroph., p. 31 (1913)[4].

Ceylon[3,4]
Puttalam[1,2]
Elephants Pass

tamaricis, Zeller 1847.

Adactyla tamaricis, Zeller, Iris XL 899 (1847) [♂][1].

Agdistis tamaricis, Zeller, Linn. Ent. VI 325-326 (1852) [redescr.][2]; Herrich-Schäffer, Schmett. Eur. V 365 (1855) [redescr.], Pteroph. t. 7 ff. 41, 42 ["*paralia*"], ? 3 ["*tamaricis*"] (1854)[3]; Milliere, Icones III 237, t. 126 ff. 5-7 (1869)[4]; Hofmann, Stett. Ent. Ztg. XXXII 125 (1871)[5]; Hofmann, Mitt. Schweiz. ent-ges. III 291 (1871)[6]; Milliére, Cat. Lep. Alpes-Marit., p. 378 (1875)[6a]; Wocke, Hein, Kleinschm. Deutschl. II. ii. 781-782 (1876) [redescr.][7]; Stdgr., H. S. E. R. XV 422 (1880)[8]; Frey, Lep. Schweiz, p. 427 (1890)[9]; Meyr., E. M. M. XXII 106 (1885)[10]; Bethune-Baker, T. E. S. 1894. 50 (1894)[11]; Hofmann, Ber. Nat. Ver. Regensb. V 56-57 (1896) [redescr.; biol.][12]; Rebel, Ann. Hofmus. Wien XI 115 (1896)[13]; Rebel, Ann. Hofmus. Wien XIII 376 (1898)[14]; Rebel, Cat. Pal. Lep. II 78, No. 1428 (1901)[15]; Rebel, Ann. Hofmus. Wien XXI 182 (1906)[16]; Wlsm., Ent. Rec. XIX pp. 54, 55 (1907)[17]; Wlsm., P. Z. S. 1907. 925 (1908)[18]; Meyr., Wyts. Gen. Ins., fasc. 100, p. 21 (1910)[19]; Spuler, Schmett. Eur. II 319 (1910) [redescr.][20]; Rebel, Iris XXVI 85 (1912)[21]; Meyr., Cat. Pteroph., p. 32 (1913)[22]; Müller-Rutz, Schmett. Schweiz II 442, No. 2019 (1914)[23]; Fletcher, Ind. Agric. Ent. Mem. VI 31 (1921)[24]; Barraud, Entom. LVI 59 (1923)[25]; Turati, Atti Soc. Sci. Nat. Milano LXIII 149 (1924)[26].

India[22]
Karachi[17]
Peshawar[24]

W. C. Asia[10]
Persia[8, 20, 22]
Asia Minor[20, 22]
Brussa[8]
Syria[20, 22]
Haifa[25]
C. Europe[19, 22]
Germany[2, 12, 15, 20]
Switzerland[6, 7, 9, 12, 20, 23]
France[4, 15, 20, 6a]
Spain[15, 20]
S. Europe[18, 19, 22]
Dalmatia[20]
Canary Isds.
Tenerife[14, 16, 18, 20]
Cd. de Verde Isds.
St. Vincent[10]
Africa[22], Algeria[17, 18]
N. Africa[19]
Cyrenaica[26]
Egypt[11, 21], W. Africa[17]
Cape Colony[17]
Arabia[17, 22]

Herbertia tamaricis, Chapman and Tutt, Brit. Lep. V 129-130, 132 [larva], 135, 253 (1906) [biol.][27].

Biology[4, 12, 27]. Larva descr.[4, 8, 20, 27] on *Tamarix gallica,*[1, 4, 12, 18], *T. germanica**[6a, 9], *Myricaria germanica**[4, 7, 12], Myrica germanica*[20] Pupa descr.[1, 12, 20].

---

78 Species.

Species omitted.

(1) *Platyptilia acanthodactyla,* Hb.; see under *P. direptalis,* Wlk.

(2) *Stenoptilia pterodactyla,* Linn.; see under *Oidœmatophorus monodactylus,* Linn.

(3) *Pselnophorus baptodactylus,* Zeller; see under *Ps. catharotes,* Meyr.

(4) *Alucita ischnodactyla,* Tr.; see under *Exelastis atomosa,* Wlsm.

(5) *Agdistis sinhala,* Fletcher.

Agdistis sinhala, Fletcher, Spol. Zeylan. VI 8 (1909) ♂ [Kanday].

Agdistis fortissimus, Wlsm. MS. [♂ Kelani Valley, Ceylon].

Agdistopis petrochroa, Hmp., Novit. Zool. XXIV 44 (1917) ♂ [Kandy (type); Singapore; Formosa].

This species, which must be known as **Agdistopis sinhala, Fletcher 1909,** is not an Alucitid. According to Sir George Hampson, it belongs to the Galleriadae.

---

* NOTE - Milliére([6a]) refers to this plant as *Tamarix (Myricaria) germanica.*

PART XXI - LYCIDAE

BESTIMMUNGSTABELLE DER TRIBUS

1. Auf dom Prothorax fehlen die Areolen, es ist eine vortiefte Mittellinie oder ein erhöhter Mittelkiel vorhanden, nur selten sind rudimentare Anzeichen von Seitenareolen vorhanden, Elytren mit oder ohne Gitterung. — 2

   Auf dem Prothorax finden sich immer Areolen die zwischen 3-7 wechseln konnen, Elytren immer mit Gitterung. — 7

2. Prothorax mit vertiefter Mittelfurche. — *Lygistopterinae.*

   Prothorax mit mehr oder weniger deutlichen Mittelkiel. — 3

3. Fühler wenigstens so lang wie der ganze Körper, meist länger. — *Atelinae.*

   Fühler nicht auffallend verlängert, immer kürzer als der Körper. — 4

4. Klauen des ♂ sehwach, des ♀ stark gezähnt. — *Macrolycinae.*

   Klauen einfach, ungezähnt. — 5

5. Mittelgrosse (15 mm.) bis sehr grosse (30 mm.) Arten, Elytren nach hinten erweitert. — *Lycinae.*

   Kleine Arten (5-10 mm.), Elytren parallel. — 6

6. Auf den Elytren sind Primär-und Sekundärrippen immer entwickelt, wenn auch zuweilen durch Behaarung verdeckt. — *Platerodinae.*

Es sind nur die Primärrippen stark, wenn auch zuweilen verkürzt, Sekundärrippen fehlen ganz oder sind rudimentär. *Dilophotinae.*

7. Auf den Elytren its die erste neben der Sutura liegende Rippe verkürzt und findet sich nurim basalen Drittel. *Trichalinae.*

Alle Rippen sind voll entwickelt. 8

8. Männliche Fühlerglieder ohne Lamellen, Primär- und Sekundärrippen immer voll entwickelt, Prothorax immer mit 7 Areolen. *Metriorrhynchinae.*

Männliche Fühlerglieder mit Lamellen, oder es dehlen auf den Elytren die Sekundärrippen order der Prothorax mit rhomboidaler Areole. 9

9. Prothorax mit rhomboidaler Areole, männliche Fühler nicht mit Lamellen. *Dictyopterinae.*

Prothorax nur mit discoidaler oder mit 5-7 Areolen, Fuhler meist beim ♂ pectinat, wenn nicht, so können die Sekundärrippen auf den Elytren fehlen. *Cladophorinae.*

BESTIMMUNGSTABELLE DER GATTUNGEN.

1. Macrolycinae.

Fühler des ♂ mit Lamellen, des ♀ gezahnt, Elytren mit 4, zuweilen undeutlichen, Rippen. *Macrolycus* Gorham.

Fühler in beiden Geschlechtern einfach, ohne Lamellen, Elytren mit 3 deutlichen Rippen. *Mesolycus* Gorham.

2. Lycinae.

Nur eine Gattung *Lycostomus* Motschoulsky.

3. Dictyopterinae.

1. 2. und 3. Fühlerglied etwa gleich lang, das 3. immer viel kürzer als das 4. 2

3. Fühlerglied länger als das 2., wenn auch nicht immer so lang wie das 4. 5

2. Fühler mit 10 Gliedern. 3

Fühler mit 11 Gliedern.	4
3. Fühler zart, lang behaart.	*Coloberos* Bourgeois.
Fühler kraftig, kurz behaart.	*Microcoloberos* Pic.
4. Elytren ohne Sekundärrippen.	*Pyropterus* Mulsant.
Elytren mit Sekundärrippen.	*Parataphes* Kleine.
5. Fühler gedrungen, die Fühler nur wenig länger als breit.	*Dictyoptera* Latreille.
Fühler schlank, lang, die Glieder vom 3 ab viel länger als breit.	*Potaphes* Kleine.

4. Cladophorinae.

1. Auf den Elytren sind nur die Primärrippen entwickelt, die Sekundärrippen sind auch nicht in Rudimenten angedeutet.	2
Auf den Elytren sind entweder die Sekundärrippen ganz vorhanden oder an der Basis und auf dem übrigen Teil durch die unterbrochene Gitterung erkennbar.	3
2. Prothorax mit rhomboidaler Areole die mit den Seitenrändern verbunden ist, das ganze Tier mehr oder weniger behaart.	*Xylobanoides* Kleine.
Prothorax mit 3-7 Areolen, woven die discoidale niemals rhomboidal sondern immer gegen den Vorderrand gestielt nach hinten offen ist, Behaarung fehlt.	*Xylobanus* C. O. Waterhouse.
3. Die discoidale Areole auf dem Prothorax ist niemals rhomboidal, sondern immer schlank und mit dem Vorderrand stielartig verbunden, order die Areole ist vorn und hinten offen.	4
Die discoidale Areole ist immer rhomboidal.	9
4. Prothorax mit 4 Areolen.	*Falsoconderis* Pic.
Prothorax nur mit discoidaler Areole oder mit 5-7 Areolen.	5
5. Prothorax nur mit discoidaler Areole.	*Bulenides* C. O. Waterhouse.
Prothorax mit 5-7 Areolen.	6

6. Fühler des ♂ einfach, ohne Lamellen. — *Prometanoeus* Kleine.

   Fühler des ♂ immer mit Lamellen. — 7

7. Die Oberseite des Körpers ist dicht behaart und verdeckt die Skulptur. — *Metanoeus* C. O. Waterhouse.

   Körper unbehaart. — 8

8. Prothorax mit 7 Areolen. — *Cladophorus* Guérin Méneville.

   Prothorax mit 5 Areolen. — *Cautires* C. O. Waterhouse.

9. Fühler kurz, gedrungen. — 10

   Fühler schlank, zuweilen pectinat. — 11

10. Discoidale Areole des Prothorax gegen den Vorderrand undeutlich, gegen den Hinterrand kräftig entwickelt. — *Lycoprogenthes* Pic.

    Discoidale Areole nach vorn und hinten kräftig entwickelt. — *Taphes* C. O. Waterhouse.

11. Alle Areolen unscharf, die discoidale zusammengedrückt, am Vorder und Hinterrand gestielt. — *Pseudoconderis* Pic.

    Areolen scharfkantig, entwickelt, discoidale nicht gestielt.

12. Fühler schlank, Glieder nicht gezahnt. — *Pseudosynchonnus* Pic.

    Fühley gedrungen, vom 3-10. Glied tief gezahnt, fast pectinat. — *Conderis* C. O. Waterhouse.

5. Atelinae.

Fühlerglieder breit, tief gezähnt. — *Atelius* C. O. Waterhouse.

Fühlerglieder filiform, ungezähnt. — *Scarelus* C. O. Waterhouse.

6. Trichalinae

Prothorax viel länger als breit. — *Leptotrichalus* Kleine.

Prothorax breiter als lang oder quadratisch. — *Trichalus* C. O. Waterhouse.

7. Metriorrhynchinae.

Nur eine Gattung. — *Metriorrhynchus* C. O. Waterhouse.

8. Platerodinae.

1. 2. und 3. Fühlerglied sehr kurz und von gleicher länge. — *Dihammatus* C. O. Waterhouse.

   3. Fühlerglied immer länger als das 2. — 2

2. 3-10. Fühlerglied pectinat. — 3

   Nicht pectinat, höchstens gezähnt. — 4

3. Grössere (10-18 mm.) Arten meist von schwarzer Farbe und roten Elytren, Gitterung nicht durch Behaarung verdeckt immer sehr deutlich entwickelt. — *Lyponia* C. O. Waterhouse.

   Kleinere bis ganz kleine (4-9 mm.) Arten von wechselnder Fabre, Gitterung und Rippenbildung auf den Elytren häufig unscharf und durch kurze Behaarung verdeckt. — *Ditoneces* C. O. Waterhouse.

4. Fühler robust, 3-10. Glied tief gezähnt. — *Melampyrus* C. O. Waterhouse.

   Fühler schlank, filiform oder nur ganz schwach gezähnt. — 5

5. Elytren mit 9 Rippen. — *Plateros* Bourgeois.

   Elytren mit 10 Rippen. — *Planeteros* Gorham.*

9. Lygistopterinae.

Kopf mehr oder weniger rüsselartig verlängert. — *Lygistopterus* Mulsant.

Kopf ohne Rüssel. — *Calochromus* Guérin Ménéville.

10. Dilophotinae.

1. Prothorax quer, breiter als lang. — 2

   Prothorax länger als breit. — 4

2. Das ganze Tier oberseits dicht pubescent, Fühler von gleicher Beschaffenheit, die Glieder dicht zusammenstehend, grossere Arten. — *Lyropaeus* C. O. Waterhouse.

---

* Diese beiden Gattungen sind nicht sicher auseinander zu halten.

Oberseite nakt oder nur schwach behaart, kleinere bis schr kleine Arten, Fühlerglieder immer deutlich getrennt. 3

3. Fühlerglieder dicht stehend. *Libnetis* C. O. Waterhouse.

Fühlerglieder locker stehend. *Libnetisia* Pic.

4. 3-10. Fühlerglied pectinat. *Flabellodilophotes* Pic.

Fühler nicht pectinat. *Dilophotes* C. O. Waterhouse.

TRANSLATION.

Key to Tribes.

1. On the prothorax areolae are wanting, an impressed median line or a raised median carina is present, rudimentary traces of lateral areolae are only rarely present, elytra with or without trellis-like reticulation. 2

On the prothorax areolae are always present and may vary between 3 to 7, elytra always with trellis-like reticulation. 7

2. Prothorax with an impressed median furrow. *Lygistopterinae.*

Prothorax with a more or less evident median carina. 3

3. Antenna at least as long as the whole body, usually longer. *Atelinae.*

Antenna not conspicuously prolonged, always shorter than the body. 4

4. Claws of the ♂ weakly, of the ♀ strongly toothed. *Macrolycinae.*

Claws simple, not toothed. 5

5. Moderately large (15 mm.) to very large (30 mm.) species, elytra widened posteriorly. *Lycinae.*

Small species (5-10 mm.), elytra parallel. 6

6. Primary and secondary ribs always developed on the elytra, even though occasionally concealed by pubescence. *Platerodinae.*

Only the primary ribs are strong, even though occasionally shortened, the secondary ribs are entirely wanting or rudimentary. *Dilophotinae.*

7. On the elytra the first rib lying near the suture is shortened and occurs only in the basal third. *Trichalinae.*

All ribs are fully developed. 8

8. Male antennal joints without lamellae, primary and secondary ribs always fully developed, prothorax always with 7 areolae. *Metriorrhynchinae.*

Male antennal joints with lamellae, or the secondary ribs are wanting on the elytra, or the prothorax with a rhomboidal areola. 9

9. Prothorax with a rhomboidal areola, male antenna without lamellae. *Dictyopterinae.*

Prothorax only with a discoidal or with 5-7 areolae, antenna usually pectinate in the ♂, if not, then the secondary ribs on the elytra may be absent. *Cladophorinae.*

KEY TO GENERA.

1. Macrolycinae.

Antenna of the ♂ with lamellae, of the ♀ toothed, elytra with 4, occasionally obscure, ribs. *Macrolycus* Gorham.

Antenna in both sexes simple, without lamellae, elytra with 3 evident ribs. *Mesolycus* Gorham.

2. Lycinae.

Only one genus. *Lycostomus* Motschoulsky.

3. Dictyopterinae.

1. 2nd and 3rd antennal joints of about the same length, the 3rd always much shorter than the 4th. 2

3rd antennal joint longer than the 2nd, even if not always as long as the 4th. 5

2. Antenna with 10 joints. 3

Antenna with 11 joints. 4

3. Antenna delicate, with long hairs.	*Coloberos* Bourgeois.
Antenna stout, with short hairs.	*Microcoloberos* Pic.
4. Elytra without secondary ribs.	*Pyropterus* Mulsant.
Elytra with secondary ribs.	*Parataphes* Kleine.
5. Antenna thickset, the antenna only a little longer than broad.	*Dictyoptera* Latreille.
Antenna slender, long, the joints from 3 onwards much longer than broad.	*Protaphes* Kleine.

4. Cladophorinae.

1. On the elytra only the primary ribs are developed, the secondary ribs are moreover not indicated by rudiments.	2
On the elytra either the secondary ribs are entirely present or are recognisable at the base and on the remaining part by means of the interrupted reticulation.	3
2. Prothorax with a rhomboidal areola which is connected with the lateral margins, the whole insect more or less hairy.	*Xylobanoides* Kleine.
Prothorax with 3-7 areolae, of which the discoidal is never rhomboidal but always stalked towards the anterior margin and is open posteriorly, pubescence absent.	*Xylobanus* C. O. Waterhouse.
3. The discoidal areola on the prothorax is never, rhomboidal but always slender and connected with the anterior margin by a stalk, or the areola is open anteriorly and posteriorly.	4
The discoidal areola is always rhomboidal.	9
4. Prothorax with 4 areolae.	*Falsoconderis* Pic.
Prothorax with only a discoidal areola or with 5-7 areolae.	5
5. Prothorax with only a discoidal areola.	*Bulenides* C. O. Waterhouse.
Prothorax with 5-7 areolae.	6

6. Male antenna simple, without lamellae.	*Prometanoeus* Kleine.
Male antenna always with lamellae.	7
7. The dorsal surface of the body is closely pubescent and [the pubescence] conceals the sculpture.	*Matanoeus* C. O. Waterhouse.
Body not pubescent.	8
8. Prothorax with 7 areolae.	*Cladophorus* Guérin Ménéville.
Prothorax with 5 areolae.	*Cautires* C. O. Waterhouse.
9. Antenna short, thickset.	10
Antenna slender, occasionally pectinate.	11
10. Discoidal areola of the prothorax obsolete opposite the anterior margin, strongly developed opposite the posterior margin.	*Lycoprogenthes* Pic.
Discoidal areola strongly developed anteriorly and posteriorly.	*Taphes* C. O. Waterhouse.
11. All areolae not sharp, the discoidal compressed, stalked on the anterior and posterior margins.	*Pseudoconderis* Pic.
Areolae sharp edged, developed, discoidal not stalked.	12
12. Antenna slender, joints not toothed.	*Pseudosynchonnus* Pic.
Antenna thickset, joints 3-10 deeply dentate, almost pectinate.	*Conderis* C. O. Waterhouse.

5. Atelinae.

Antennal joints broad, deeply dentate.	*Atelius* C.O. Waterhouse.
Antennal joints filiform, not toothed.	*Scarelus* C.O. Waterhouse.

6. Trichalinae.

Prothorax much longer than borad.	*Leptotrichalus* Kleine.
Prothorax broader than long or quadrate.	*Trichalus* C.O. Waterhouse.

7. Metriorrhynchinae.

Only one genus.	*Metriorrhynchus* C.O. Waterhouse.

8. Platerodinae.

1. 2nd and 3rd antennal joints very short and of equal length.	*Dihammatus* C. O. Waterhouse.
3rd antennal joint always longer than the 2nd.	2
2. 3rd-10th antennal joints pectinate.	3
Not pectinate, at most dentate.	4
3. Larger (10-18 mm.) species mostly black in colour with red elytra, trellis-like reticulation not concealed by pubescence, always very clearly developed.	*Lyponia* C. O. Waterhouse.
Smaller to very small (4-9 mm.) species of variable colouration, trellis-like reticulation and rib formation on the elytra frequently obscure and concealed by short pubescence.	*Ditoneces* C. O. Waterhouse.
4. Antenna robust, joints 3-10 deeply dentate.	*Melampyrus* C. O. Waterhouse.
Antenna slender, filiform or only quite weakly dentate.	5
5. Elytra with 9 ribs	*Plateros* Bourgeois.
Elytra with 10 ribs	*Planeteros* Gorham.*

9. Lygistopterinae.

Head more or less produced into a rostrum.	*Lygistopterus* Mulsant.
Head without rostrum.	*Calochromus* Guérin Méneville.

10. Dilophotinae.

1. Prothorax transverse, broader than long.	2
Prothorax longer than broad.	4
2. The whole insect closely pubescent on the upper surface, antenna of similar appearance, the joints closely approximate, larger species.	*Lyropaeus* C. O. Waterhouse.
Upper surface bare or only weakly pubescent, smaller to very small species, antennal joints always clearly separated.	3

* These two genera are not separable with certainty.

3. Antennal joints placed closely. — *Libnetis* C. O. Waterhouse.

Antennal joints placed loosely. — *Libnetisia* Pic.

4. Antennal joints 3-10 pectinate. — *Flabellodilophotes* Pic.

Antenna not pectinate. — *Dilophotes* C. O. Waterhouse.

C. F. C. BEESON,
Joint Editor.

LYCIDAE.

Bourgeois, Mon. Lye. Ane. Mond. L'Abeille XX, 1882, p. 53.

Lycusites Casteln. Hist. Nat. Ins. I, p. 261.

Lycites Blanch. Hist. Ins. II, 1845, p. 65; J. Duv. Gen. Col. d'Europ. III, p. 158.

Lycides Lacord. Gen. Col. IV. p. 289; Muls. Mollip., p. 33.

Lycoides Motsch. Schrenks Reisen II, p. 114.

Lycidœ v. Kiesenw. Naturg. Ins. Deutschl. IV, p. 433; J. L. Le Conte, Trans. Amer. Ent. Soc. IX, p. 16.

Lycina C.G. Thoms. Skand. Col. VI, p. 158.

Lycini Gemm. Har. Cat. Col. VI, 1869, p. 1626; Redtenb. Faun. Austr. 3, II, p. 1, Seidlitz, Faun. Balt., p. 331.

I. MACROLYCINAE Kleine.

I. F. R. XIII (V) 1929, P. 222.

I. Macrolycus C.O. Waterhouse.

Type: *coccineus* C.O. Wtrh.

C.O. Wtrh. Tr. E. S. L. 1878, p. 95; Ill. Typ. Sp., p. 1; Kln. D. E. Z., 1925, p. 325.

Cerceros Kraatz, D. E. Z., 1879, XXIII, p. 126.

coccineus C.O. Wtrh. Tr. E. S. L., 1878, p. 96; Kleine D. E. Z., 1925, p. 326, f. 1;[1] I. F. R. XIII (VI), 1929, p. 222[2].
Macrolycus bowringii C.O. Wtrh. Ill. Typ. Sp. p. I, t. 1, f. 1;[3] Gorh. Tr. E. S. L., 1890, p. 541[4]; Pic. F. I. Fr., 1923, No. 1, p. 6[5]; Bourg. A. S. E. Fr., LXX, 1901, p. 32[6].
Allahabad.[1, 3, 4] Darjeeling.[1] Himalaya.[1] Manipur.[2, 6] Upper Burma.[1] Ruby Mines.[2] Tonkin.[5]

dotatus, Kleine, D. E. Z., 1925, p. 328, f. 5, 6[1] — Tonkin.[1]

gracilis, Pic. F. I. Fr., 1923, No. 1, p. 10.[1] — Tonkin.[1]

testaceicollis, Pic. F. I. Fr., 1923, No. 1, p. 6[1]. — Laos.[1]

2. **Mesolycus** Gorham (1883).

Type: *atrorufus* Kiesw.

Gorh. Tr. E. S. L., 1883, p. 398.

Eros Kiesw. D. E. Z., 1879, p. 308 (pars).

vitalisi Pic. F. I. Fr., 1923, No. 1, p. 7[1]. — Tonkin.[1]

II. LYCINAE Kleine.

I. F. R. XIII (VI) 1929, p. 222

3. **Lycostomus** Motschoulsky (1861)

Type: *similis* Hope.

Motsch. B.M., 1861, p. 136; C.O. Wtrh. Ill.

Typ. Sp., 1879, p. 10; Gorh. Tr. E. S. L., 1890, p. 542; Pic. F. I. Fr., 1923, No. 1, p. 7; Kln. I. F. R. XIII, 1929, p. 222.

Lycus C. C. VI, 1869, p. 1626 (pars).

acutecostatus Fairm. C. R. B., 1891, p. LXXXIX.[1] — Kashmir.[1]

ambiguus C. O. Wtrh. Ill. Typ. Sp., 1879, p. 11, t. 3, f. 3;[1] Bourg. A. S. E. Fr., 1885, p. 74;[2] Gorh. Tr. E. S. L., 1890, p. 543; Kln. S. E. XLI, No. 1, p. 2; Kln. I. F. R. XIII, (VI), 1929, p. 222.
Cachar.[1, 2, 4] Nilgiri Hills.[4] Simla.[4], Punjab.[4] Dungangali.[4] Mussoorie.[4] Madura.[4] Trichinopoly.[4] Ranikhet.[4]

	Bodyar.[3] Chakrata.[2] India.[2]
analis Dalm. Schoenh. Syn. Ins. Ill. App. p. 30, t. 5, f. 10 (*Lycus*; C. O. Wat. Ill. Typ. Sp., 1879, p. 13[1]; Gorh. Tr. E. S. L., 1890, p. 543; Pic. F. I. Fr., 1923, No. 1, p. 8[2]; Bourg. A. S. E. Fr. LXX, 1901, p. 49[4]; Kln. I. F. R. XIII (VI), 1929, p. 223[3]; Kln. F. M. S. XIII (283), 1926, p. 181[5]. *Lycus divisus* F. Wlk. A. M. N. H. 1858, II, p. 282. *Lycus planicollis* F. Wlk., l. c., p.	N. Bengal.[1, 3, 4] Decca.[1, 3], Madras.[3] Punjab.[3] Ramandroog.[3] Tranquebar.[3] Coimbatore.[3] Anantapur.[3] Gooty.[3], Ceylon.[1] Cochin-China.[2] Cambodia.[2] Siam.[5], Mavaparai.
anceyi Pic. F. I. Fr. 1923, No. 1, p. 8.[1]	Cochin-China.[1]
atrimembris Pic. H. T. E. No. 425, 1926, p. 31[1].	Tonkin.[1]
binhanus Pic. H. T. E. No. 425, 1926, p. 31[1].	Tonkin.[1]
brevelineatus Pic. H. T. E. No. 425, 1926, p. 31[1].	Indo-China.[1]
cardinalis Kln. S. E. XLI, 1926 (1), p. 2;[1] Kln. I. F. R. XIII (VI), 1929, p. 224[2].	Dehra Dun.[1] Pondichery.[2]
curtithorax Pic. H. T. E. No. 425, 1926, p. 31[1].	Tonkin.[1]
elongatulus Beurg. A. S. E. Fr., 1885 (6), V, p. 73;[1] Kln. I. F. R. XIII (VI), 1929, p. 224.	Burma.[1, 2]
ferrugineus Febr. Ent. Syst. Suppl. p. 125; Kln. F. M. S. XIII (2 & 3), 1926, p. 125[1].	Pahang.[1], Sumatra. Borneo., Java. Timor.
hedini Kln. I. F. R. XIII (VI), 1929, p. 225, Pl. I. F. 3, p. 4[1].	South-east Tibet.[1]
honestus Bourg. A. S. E. Fr. 1885 (6), V, p. 73;[1] Kln. I. F. R. XIII (VI), 1929, p.224[2]. *modestus* C. O. Wat. Ill, Typ. Sp. 1879, p. 11, t. 2, f. 10;[3] Gorh. Tr. E. S. L., 1890, p. 543.	Rangirum. 6,000 ft. Bootan.[1, 2, 3] Chamba Chamba.[1], Manipur.[2] W. Almora.[2] Mussoorie.[2] Madura.[2] Trichinopoly.[2] Assam.[2] Amarkantak. 3,500 ft.
indicus Kln. I. F. R. XIII (VI), 1929, p. 224, P. I. F. I.[1].	W. Almora.[1]

internexus F. Wlk. A. M. N. H., 1858, II, p. 282;[1] C. O. Wat. Ill. Typ. Sp., 1879, p. 13, t. 3, f. 4;[2] Kln. I. F. R. XIII (VI), 1929, p. 224.
Ceylon.[1, 2]
Hambantota.[3]
Peradenyia.[3]

jeanvoinei Pic. Mel. IL, 1927, p. 4[1].
Indo-China.

lateritius Gorh. A. S. E. B. XXXIX, 1895, p. 299;[1] Kln. I. F. R. XIII (VI), 1929, p. 223[2].
Hillgrove.[1]
Nilgiris.[2]
Coimbatore.[2]
Simla.[2], Sadiya.[2]
Punjab.[2]
W. Almora Div.[2]
Lansdowne Div.[2]
Madras., Burma.[1, 2]

longicollis Bourg. A. S. E. Fr., 1890, p. 165[1]; Pic. F. I. Fr., 1923, No. 1, p. 8[2].
Cochin-China.[1, 2]

madurensis Pic. H. T. E. No. 425, 1926, p. 31[1].
Tonkin.[1]

marginatus Fabr. Syst. El. II, p. 110[1]; Kln. I. F. R. XIII (VI), 1929, p. 224.
Lycus melanurus Blanch. Voy. Pole Sud. IV, p. 74, t. 6, f. 2[3].
Lycus bivittatus Kirsch., M. Z. M. D. I, 1875, p. 34[4].
Siam.[2]
Singapore.[2, 3]
Selangor.[3]
Pahang.[2]
Perak.[2], Malacca.[2, 4]
Sumatra.[1], Borneo.

nigricollis Bourg. A. S. E. B. XL, 1905, p. 46[1].
Bengal.[1]

nigripes Fabr. Mant. Ins. I, 1787, p. 163 (*Pyrochroa*); Oliv Ent. II, 1790, No. 29, p. 11, t. 1, f. 12 (*Lycus*); C. C. VI, 1869, p. 1627;[1] C. O. Wat. Ill. Typ. Sp. 1879, p. 13, t. 3, f. 6;[2] Bourg. A. S. E. Fr. 1885 (6), V, p. 75;[3] 1890, p. 164[4] 1901, p. 49[5], Pic. F. I. Fr. 1923, No. 1, p. 8; Kln. I. F. R. XIII (VI), 1929, p. 222.
Darjeeling.[3]
Bombay.[7], Simla.[7]
Punjab.[7], Sikkim.[7]
Burma.[3, 4], Penang.[2, 4]
Siam.[1, 2, 3, 4, 6]
Cambodia.[5, 6]
Cochin-China.[3, 4, 5]

orientalis Kln. S. E. L. LXXXXI, 1930, p. 94.[1]
Mussoorie.[1]
Kashmir.[1]

patruelis Bourg. A. M. C. G. XVIII, 1883, p. 625.[1]
Debrepani. 6,000 ft.
Darjeeling.[1]
Nilgiris., Mussoorie.

perraudieri Bourg. A. S. E. Fr. 1890, p. 165[1]; Pic. F. I. Fr. 1923, No. 1, p. 8.[2]
Cambodia.[1, 2]
Laos.[2]

placidus C. O. Wat. Ill. Typ. Sp. 1879, p. 14, t. 3, f. 7, 7*a*[1]; Bourg. A. S. E. Fr. LXX, 1901, p. 48[2]; Kln. F. M. S. XIII (2 & 3), 1926, p. 181.[3]
Hong-Kong.[1]
Siam.[3], Pahang.[3]
Cochin-China.[2]
Tonkin.[2]

praeustus Fabr. Ent. Syst. 1792, p. 107 (*Lycus*); Gorh. Tr. E. S. L. 1890, p. 542; Bourg. A. S. E. Fr. LXX, 1901, p. 49[1]; Pic. F. I. Fr. 1923, No. 1, p. 8[2]; Kln. I. F. R. XIII (VI), 1929, p. 223.[3]

South India. Nagpur.[3], Kumaon.[3] Belgaum.[3], Madras.[3] Kashmir.[3], Ceylon.[3] Assam.[3], Burma.[3] Cambodia.[1, 2], Laos.[2]

purpureiventris Gorh. A. S. E. B. XXXIX, 1895, p. 299.[1]

Chamba.[1]

restituus Bourg. A. S. E. Fr. 1885 (6), V, p. 75[1]; Kln. I. F. R.

Lycostomus ambiguus C. O. Wat. Ill. Typ. Sp. 1879, p. 11 (♀).[3]

N. India.[1] Kumaon.[2] Mungphu.[2] Belgaum.[2], Cachar.[3]

rubrocinctus Fairm. A. S. E. Fr. 1886 (6), VI, p. 338[1] Kln. I. F. R. XIII (VI), 1929, p. 223.?

Sikkim.[2] 5-7,000 ft. Burma.[2], Yunnan.[1]

rufiventris C. O. Wat. Ill. Typ. Sp. 1879, p. 12[1]; Gorh. Tr. E. S. L. 1890, p. 543; Kln. I. F. R. XIII (VI), 1929, p. 223.

Sylhet.[2] Burma.[1, 2]

similis Hope, Gray's Zool. Miscell. 1831, p. 26 (♂); C. O. Wat. Ill. Typ. Sp. 1879, p. 10[1]; Gorh. Tr. E. S. L. 1890, p. 542; Bourg. A. S. E. Fr. LXX, 1901, p. 48[2]; Pic. F. I. Fr. 1923, No. 1, p. 8[3]; Kln. I. F. R. XIII (VI), 1929, p. 223.[4]

Lycus triangularis Hope, 1 c. p. 26 ♀.

Lycus suturalis Redtb. Hügels Kaschmir Ins. 1848, IV, p. 508, t. 23, f. 7; Bourg. A. S. E. Fr. 1885 (6), V, p. 73.[5]

Lycus geminus F. Wlk. A. M. N. H. 1858, II, p. 281.

Lycus cinnabarinus Candèze, Mém. Liége 1861, p. 355, t. 3, f. 1 c.

Lycostomus coccineus Motsch. B.M. 1861, I, p. 136, t. 9, f. 13.

Nepal.[1, 4] Allahabad.[1] Bootan.[1, 2] Ceylon.[1, 2, 4] Naducatura.[4] Sitapahar. Coimbatore. Dehra Dun. Kumaon.[4] Coimbatore.[4] Dehra Dun.[4] Sylhet.[4], Kashmir.[4, 5] Assam.[4], Darjeeling.[5] Siam.[1, 4], Tonkin.[2, 3, 4] Laos.[3], China.[3]

singularis C. O. Wat. Ill. Typ. Sp. 1879, p. 12, t. 2, f. 9[1]; Gorh. Tr. E. S. L. 1890, p. 543; Kln. I. F. R. XIII (VI), 1929, p. 223.[2]

South India.[1] W. Almora.[2] Sylhet.[2] Ranikhet.[2] Assam.[2]

stevensi Kln. I. F. R. XIII (VI), 1929, p. 225, Pl. I. f. 2.[1]

Upper Assam.[1]

striatus C. O. Wat. Ill. Typ. Sp. 1879, p. 15, t. 3, f. 11[1]; Gorh. Tr. E. S. L. 1890, p. 543.	India.[1]
thoracicus C.O. Wat. Ill, Typ. Sp. p. 16, t. 3, f. 1[1]; Bourg. A. S. E. Fr. 1885 (6), V, p. 76;[2] Gorh. Tr. E.S.L. 1890, p. 542; Pic. F.I. Fr. 1923, No. 1, p. 7.[3]	Bhutan.[1, 2] Darjeeling.[1, 2] Sikkim.[1, 2], Tonkin.[3]

III. DICTYOPTERINAE Kleine.

I. F. R. XIII (VI) 1929, p. 222.

4. PYROPTERUS MULSANT (1838).

Type: *nigroruber* Deg (*affinis* Payk.).

Muls. Ann. Soc. Agr. Lyon. 1838, I, p. 81, t. 5, f. 6.

gilvus Kln. I. F. R. XIII (VI), p. 226, Pl. I F. 5 u. 7.[1]	Himalaya.[1]
malaccanus Pic. H. T. E. No. 407, 1922, p. 13.[1]	Malacca.[1]
4-costatus Pic. M. IL. 1927, p. 6.[1]	Tonkin.[1]
sculpturatus C. O. Wat. Tr. E. S. L. 1878, p. 12; Ill. Typ. Sp. 1879, p. 63, t. 15, f. 8[1]; Kln. I. F. R. XIII (VI), 1929, p. 226.[2]	Ceylon.[2], Manipur.[2] Johore.[2], Singapore.[2] Borneo.[1], Tonkin.[1]
tricostatus Pic. M. IL. 1927, p. 6.[1]	Tonkin.[1]

5. Parataphes Kleine (1925).

Type: *dohertyi* Kln.

Kln. S. E. Z. LXXXV, 1925, p. 133.

dohertyi Kln. l. c. p. 134, f. 3.[1]	Burma.[1]
sanguinolentus Kln. I. F. R. XIII (VI), 1929, p. 237, Pl. V. F. 141-142.	Travancore.[1] Kodaikanal. Palneys.[2]

6. Protaphes Kleine (1926).

Type: *confertus* Kln.

Kln. S. M. J. III (3) No. 10, 1926, p. 363.

confirmatus Kln. S. M. J. III (3) No. 10, 1926, p. 365, fig. 4-6.[1]	Kedah. Sumatra.[1]
miniatus Kln. S. E. Z. LXXXXI, 1930, p. 95[1].	Perak.[1]

7. Microcoloberos Pic. (1913).

Type: *malaccanus* Pic.

Pic. M. VI, 1913, p. 2.

malaccanus Pic. l. c. p. 2.[1] — Malacca.[1]

8. Coloberos Bourgeois (1885).

Type: *tenellus* Bourg.

Bourg. A. S. E. Fr. 1885, p. 82.

longipennis Pic. B. S. E. Fr. 1911, p. 331.[1] — Ceylon.[1]

robusticornis Pic. M. 41, 1928, p. 17.[1] — Cochin-China.[1]

tenellus Bourg. l. c. p. 83.[1] — Darjeeling.[1]

9. Dictyoptera Latreille (1829).

Type: *aurora* Herbst.

Latr. Règne anim. ed. II, IV, 1829, p. 464.

braeti Bourg. A. S. E. B. IL, 1905, p. 46.[1] — Kurseong.[1]

IV. CLADOPHORINAE Kleine.

I. F. R. XIII (VI), 1929, p. 222.

10. Cladophorus Guérin Méneville (1830).

Type: *formosus* Guér. Ménév.

Guéri Ménév. Voy. Coqu. Ent. 1830, p. 72; C. O. Wat. Tr. E. S. L. 1878, I, p. 103; Ill. Typ. Sp. 1879, p. 64 (*partim*).

Odontocerus Guér. Ménév. l. c.

Metriorrhynchus C. O. Wat. Tr. E. S. L. 1878, I, p. 101; Ill. Typ. Sp. 1879, p. 47 (*partim*).

atricolor Pic. F. I. Fr. No. 1, 1923, p. 16.[1] — Tonkin[1]

atronotatus Pic. F. I. Fr. No. 1, 1923, p. 17.[1] — Tonkin[1]

atrithorax Pic. H. T. E. No. 419, 1925, p. 8.[1] — East India.[1]

binhanus Pic. H. T. E. No. 425, 1926, p. 32.[1] — Tonkin.[1]

castetsi Pic. H. T. E. No. 419, 1925, p. 8.[1] — East India.[1]

diversithorax Pic. M. XLII, 1924, p. 32.[1] — Tonkin.[1]

hoanus Pic. H. T. E. No. 425, 1926, p. 32.[1] — Tonkin.[1]

inapicalis Pic. H. T. E. No. 425, 1926, p. 32.[1] — Tonkin.[1]

laboisieri Pic. H. T. E. No. 450, 1925, p. 9.[1] — Tonkin.[1]

laoensis Pic. H. T. E. No. 426, 1926, p. 34.[1] — Laos.[1]

laticollis Pic. H. T. E. No. 425, 1926, p. 32.[1] — Laos.[1]

notatithorax Pic. F. I. Fr. No. 1, 1923, p. 16.[1] — Tonkin.[1]

orientalis C. O. Wat. Ill. Typ. Sp. 1879, p. 58, t. XIV, f. 8[1]; Pic. F. I. Fr. No. 1, 1923, p. 16.[2] — Siam.[1, 2]

Metriorrhynchus orientalis C. O. Wat. 1. c.

parallelus Pic. H. T. E. No. 426, 1926, p. 34.[1] — Tonkin.[1]

particularis Pic. H. T. E. No. 425, 1926, p. 32.[1] — Tonkin.[1]

reverandi Pic. H. T. E. No. 420, 1925, p. 9.[1] — Penang.[1]

rudeplicatus Pic. H. T. E. No. 419, 1925, p. 8.[1] — Sikkim.[1]

satrapa Bourg. A. S. E. B. IL, 1905, p. 47.[1] — Kurseong.[1]

vitalisi Pic. F. I. Fr. No. 1, 1923, p. 17.[1] — Tonkin.[1]

11. **Bulenides** C. O. Waterhouse (1879).

Type: *obsoletus* C. O, Wat.

C. O. Wat. Ill. Typ. Sp. 1879, p. 34.

aridus Kln. E. M. V., 1926, No. 2, p. 152, f. 10, 21, 36, 48.[1] — Penang.[1]

coccineus Kln. Treubia XI, 1929, p. 1. — Langharvi Isl.

imitator Kln. S. E. Z. LXXXXI, 1930, p. 95.[1] — Perak.[1]

indus Kirsch, M. Z. M. D. I., 1875, p. 36;[1] Kln. F. M. S. IXII (2 & 3) 1926, p. 186;[2] E. M. XV, No. 2, 1926, p. 143, f. 2, 4, 19, 26, 48;[3] I. F. R. XIII (VI) 1929, p. 226. *Caenia inda Kirsch,* 1. c. — Johore.[4], Malacca.[1] Penang.[1, 4] Pahang. Selangor.[2, 3]

Bulenides brevelineatus Pic. H. T. E. No. 419, 1925, p. 7. — Kuala Lumpur.[2, 3] Sumatra.[3]

Bulenides vicinus Pic. H. T. E. No. 420, 1925, p. 10. — Borneo.[3], Java.[3]

malayensis Kln. Treubia XI, 1930, p. 349.[1] — Langkarvi Isl.[1]

nebulosus Kln. l. c. p. 350.[1] — Penang.[1]

pauper C. O. Wat. Ill. Typ. Sp. 1879, p. 35, t. IX, f. 2,[1] Kln. E. M. XV, No. 2, 1926, p. 147, f. 2, 12, 22, 27, 42, 43;[2] F. M. S. XIII (2 & 3) 1926, p. 185;[3] I. F. R. XIII (VI) 1929, p. 227.[4]	Negri Sembilan.[4] Selangor.[2, 3] Perak.[2, 3], Pahang.[3] Kuala Lumpur.[2] Sumatra.[1, 2]
pauperulus Bourg. A. M. C. G. XVIII, 1883, p. 638. Kln. E. M. XV, No. 2, 1926, p. 144, f. 3, 11, 25, 37;[1] F. M. S. XIII (2 & 3) 1926, p. 186;[2] I. F. R. XIII (VI), 1929, p. 227.	Malacca.[1, 4] Perak.[1, 2] Pahang.[2], Selangor.[3] Sumatra.[1, 4, 6]
atrivollis Pic. H. T. E. No. 405, 1921, p. 8.[4]	Borneo.[1]
notaticollis Pic. 1. c.[5]	Java.[5]
medanensis Pic. 1. c. No. 419, 1925, p. 7.[6]	
purpureus Pic. A. S. L. L. LXIX, 1922, p. 74;[1] Kln. E. M. XV, 1926, No. 2, p. 153, f. 55, 56;[2] I. F. R. XIII (VI) 1929, p. 227.[3]	India.[3], Assam.[3] Tenasserim.[3] Formosa.[1, 2]
regalis Kln. I. F. R. XIII (VI) 1929, p. 227, Pl. 1 F. 9, u. 10.[1]	Sylhet.[1]
sijthoffi Kln. E. M. XV, 1926, No. 2, p. 149, f. 3, 9, 23, 28, 38;[1] I. F. R. XIII (VI) 1929, p. 227.[2]	Andamans.[2] Java.[1]
testaceus Pic. H. T. E. No. 405, 1921, p. 8;[1] F. J. Er. 1923, No. 1, p. 12;[2] Kln. E. M. XV, 1926, No. 2, p. 153, f. 3, 17, 50;[3] I. F. R. XIII (VI), 1929, p. 227.[4]	India.[1], Assam.[4] Burma.[4] Tenasserim.[4] Penang.[3, 4], Laos.[2]

12. **Cautires** C. O. Waterhouse (1879).

Type: *excellens* C. O. Wat.

C. O. Wat. Ill. Typ. Sp. 1879, p. 36.

astutus F. Wlk. A. M. N. H. (3) II, 1858, p. 281;[1] C. O. Wat. Ill. Typ. Sp. 1879, p. 55, t. XIII, f. 8; Kln. I. F. R. XIII (VI), 1929, p. 228.[2]	Ceylon.[1, 2]
Lycus astutus F. Walk. 1. c.	
congener C. O. Wat. Ill. Type Sp. 1879, p. 36, t. IX, f. 7;[1] Kln. F. M. S. XIII (2 & 3) 1926, p. 186;[2] I. F. R. XIII (VI) 1929, p. 228.[3]	Johore.[3], Selangor.[2] Kuala Lumpur.[2] Pahang.[2], Perak.[2] Sumatra., Java.[1]
dolosus Klu. I. F. R. XIII (VI) p. 230. Pl. I, F. 19, 20.	Ceylon.[1]
discordans Kln. I. F. R. l. c. p. 231, Pl. I, F. 22.[1]	Sylhet.[1]

differens Kln. I. F. R. l. c. p. 228, Pl. I, F. 10, 11.[1]	Ceylon.[1]
griseus Kln. S. E. Z. LXXXXI, 1930, p. 96.[1]	Perak.[1]
incorruptus Kln. I. F. R. XIII (VI), 1929, Pl. I, F. 16.[1]	India.[1], Burma.[1] Assam.[1] Tenasserim.[1]
indus Kln. I. F. R. XIII (VI), 1929, p. 231, Pl. V, F. 137-139.[1]	Kodaikanal.[1]
jacobsoni Kln. E. M. XV, No. 1, 1926, p. 19, f. 11.[1]	Siam.[1], Perak. Selangor., Perak. Sumatra.[1]
latior Pic. H. T. E. No. 419, 1925, p. 8.[1]	Ceylon.[1]
legitimus Kln. F. M. S. XIII (2 & 3) 1926, p. 188, f. 20-23.[1]	Siam.[1]
lineatus Hope, Gray's Zool. Miscell. 1831, p. 26;[1] Kln. I. F. R. XIII (VI) 1929, p. 228.[2] *Omalysus lineatus* Hope 1. c. *Metriorrhynchus lineatus* Hope, C. O. Wat. Ill. Typ. Sp. 1879, p. 54, t. XIII, f. 9;[3] Gorh. Tr. E. S. L. 1890, p. 545.	India.[1], Malabar.[2] Madura.[2] Trichinopoly.[2] Anamalais.[2] Nepal.[3] Coimbatore.[3]
miniatipennis Bourg. B. S. E. Fr. 1888 (6) VIII, p. XXV;[1] Kln. I. F. R. XIII (VI), 1929, p. 228.[2]	Hindostan.[1] Manipur.[2] Kodaikanal 5,000'. Palneys 3-6,000'.
mirandus Kln. I. F. R. XIII (VI), 1929,[1] p. 231, Pl. I, F. 21.	Tenasserim.[1]
nervosus Kln. F. M. S. XIII (2 & 3) 1926, p. 187, f. 16, 17.[1]	Pahang.[1], Burma. Perak.
nigrostriatus Kln. I. F. R. XIII (VI), 1929, p. 228, Pl. I, F. 12, 13.	Sikkim.[1]
nodosus Kln. F.M.S. XIII (2 & 3) 1926, p. 189, f. 26-28.[1]	Siam.[1]
notabilis Kln. F.M.S. XIII (2 & 3) 1926, p. 189, f. 45-25.[1]	Siam.[1]
postmaculatus Kln. Arch. Nat. LXXXXII 1926 (1928) A XII, p. 141, Pl. III, F. 89-90.[1]	Dehra Dun.[1]
rarus Kln., I.F.R. XIII (VI), 1929, p. 229, Pl. I, F. 14, 15.[1]	Assam.[1]
ratus Kln. I. F. R. l. c. p. 230, Pl. I. F. 17, 18.[1]	Assam.[1] North Malabar.

selangorensis Kln. F. M. S., XVI, 1930, p. 170.[1] — Selangor. Kuala Lumpur.[1]

simillimus Kln. F. M. S. XIII, (2 & 3) 1926, p. 1926, p. 187, f. 18, 19.[1] — Pahang.[1]

tenebricus Kln. S. E. Z. LXXXXI, 1930, p. 96.[1] — Perak.[1]

13. Xylobanus C. O. Waterhouse (1879)

Type: *costifer* F. Wlk.

C. O. Wat. Ill. Typ. Sp. 1879, p. 38.

alveolus Bourg. A. S. E. Fr. 1885, (6) V, 9. 80;[1] Kln. I. F. R. XIII (VI) 1929, p. 232.[2] — Debrepani 6,000'. Darjeeling.[1], Burma.[2] Tenasserim.[2]

approximans Bourg. A. S. E. Fr. 1885, (6) V, p. 80,[1] Kln. I. F. R. XIII (VI) 1929, p. 232.[2] — Himalaya.[1] Manipur.[2], Assam.[2] Burma.[2]

aterrimus Kln. F.M.S. XIII (2 & 3) 1926, p. 190, f. 29, 31.[1] — Pahang.[1]

atricollis Gorh. A. S. E. B. XXXIX, 1895, p. 300.[1] — Kanara.[1]

beesoni Kln. I.F.R. XIII (VI), 1929, p. 235, Pl. V, F. 141.[1] — Burma.[1]

beneficus Kln. I. F. R. l. c., p. 235, Pl. II, F. 42-44.[1] — Tonkin.[1]

binhanus Pic. H. T. E. No. 420, 1925, p. 10.[1] — Tonkin.[1]

cardoni Bourg. A. S. E. B. IL, 1905, p. 48.[1] — Bengal.[1]

causarius Kln. I. F. R. XIII (VI), 1929, p. 232, Pl. I, F. 23 u. 24.[1] — Malacca.[1] Perak.

coomani Pic. H. T. E. No. 423, 1926, p. 22.[1] — Tonkin.[1]

costifer F. Walk. A. M. N. H. 1858, II, p. 282;[1] Wat. Ill. Typ. Sp. 1879, p. 38, t. IX, f. 10.[2] — Ceylon.[1,2]

Lycus costifer F. Wlk. l. c.

diversepunctatus Pic. F. I. Fr. 1923, No. 1, p. 19.[1] — Laos.[1]

fastidiosus C. O. Wat. Ill. Typ. Sp. 1879, p. 39, t. X, f. 3;[1] Pic. F. I. Fr. 1923, No. 1, p. 18[3] Bourg. A. S. E. Fr 1885, (6) V, p. 80.[2] — Himalaya.[2], Siam.[3] Laos.[3], Cambodia.[3] Java.[1,2]

fenestratus Pic. H. T. E. No. 407, 1922, p. 15.[1] — Perak., Sumatra.[1]

foveatus C. O. Wat. Ill. Typ. Sp. 1879, p. 40, t. X, f. 2.[1] — India (?)[1]

fractus Kln. T. IX (4) 1926, p. 310, f. 44.[1]	Selangor., Pahang. Perak., Borneo.[1]
fumigatus C. O. Wat. Ill. Typ. Sp. 1879, p. 42, t. XI, f. 1.[1]	Perak., Borneo.[1]
gratiosus C. O. Wat. Ill. Typ. Sp. 1879, p. 39, t. IX, f. 11;[1] Gorh. Tr. E. S. L. 1890, p. 545.[2]	Andamans.[1, 2]
hirtus Kln. I. F. R. XIII (VI), 1929, p. 234, Pl. I, F. 36-37, Pl. II, F. 38.[1]	Ceylon.[1]
humerifer F. Wlk. A. M. N. H. 1859, III, p. 258;[1] C. O. Wat. Ill. Typ. Sp. 1879, p. 39, t. IX, f. 8.[2]	Ceylon.[1, 2]
Lycus humerifer F. Wlk. l. c.	
humilis C. O. Wat. Ill. Typ. Sp. 1879, p. 40, t. X, f. 5, 5*a*;[1] Kln. F. M. S. XIII (2 & 3) 1926, p. 192, f. 37.[2]	Pahang.[2] Borneo.[1]
incertus Kln. I. F. R. XIII (VI), 1929, p. 233, Pl. I, F. 28-31.[1]	Assam.[1]
indicus Pic. Mel. XXXV, 1922, p. 6.[1]	India.[1]
irregularis Pic. F. I. Fr. 1923, No. 1, p. 19.[1]	Tonkin.[1]
laoensis Pic. H. T. E. No. 426, 1926, p. 34.[1]	Laos.[1]
latefoveolatus Pic. H. T. E. No. 419, 1925, p. 5.[1]	Perak., Sumatra.[1]
montanus Kln. I. F. R. XIII (VI), 1929, p. 234, Pl. I, Fig. 32-35.[1]	Burma.[1] Tenasserim.[1] Bengal.
multiareolatus Pic. M. IL, 1927, p. 5.[1]	Tonkin[1]
nigripennis Pic. M. L., 1927, p. 35.[1]	Tonkin[1]
nigrofenestratus Pic. H. T. E. No. 426, 1926, p. 34.[1]	Tonkin[1]
nigropunctatus Pic. F. J. Fr. 1923, No. 1, p. 19.[1]	Tonkin[1]
nothus Kln. F. M. S. XIII (2 & 3) 1926, p. 191, f. 32-34.[1]	Pahang.[1], Siam.
novellus Kln. F. M. S. XIII, (2 & 3) 1926, p. 191, F. 35-36.[1]	Perak.[1]
parallelus Kln. T. IX, (4) 1826, p. 306, f. 54.[1]	Perak., Borneo.[1]
phungi Pic. H. T. E. No. 426, 1926, p. 34.[1]	Tonkin.[1]
piceithorax Pic. H. T. E. No. 407, 1922, p. 15;[1] Kln. F. M. S. XIII (2 & 3), 1926, p. 192.[2] Kln. I. F. R. XIII (VI), 1923, p. 232.[3]	Pahang.[1], Perak.[2, 3] Selangor.[2, 3] Sumatra.[1], Borneo. Negri Sembilan.[4]

recens Kln. I. F. R. XIII (VI), 1929, p. 234, Pl. II, F. 39-41.[1] — Tenasserim.[1] Assam.

recurvus Kln. I.F.R. XIII (VI) 1929, p. 233, Pl. I.F. 25-27.[1] — Assam[1]

reticulatus Gorh. N. L. M. IV, 1882, p. 96.[1] — Perak., Sumatra.[1]

rubens C.O. Wat. Ill. Typ. Sp. 1879, p. 40, t. X, f. 4;[1] Pic. F. I. Fr. 1923, No. 1, p. 18.[2] — Siam.[1, 2] Bengal.

testaceus Pic. E. 1922, Nr. 407, p. 16.[1] — Assam.[1]

vitalisi Pic. F. I. Fr. 1923, No. 1, p. 19.[1] — Tonkin.[1]

14. **Xylobanellus** Kleine (1930).

Type: *atricolor* Kln.

Kln. F. M. S., XVI, 1930, p. 171.

atricolor Kln. F. M. S., XVI, 1930, p. 171.[1] — Selangor.[1]

15. **Pseudosynchonnus** Pic. (1922).

Type: *testaceus* Pic

Pic. H. T. E. No. 407, 1922, p. 13.

testaceus Pic. l. c. p. 13.[1] — Malacca.[1]

16. **Conderis** C. O. Waterhouse (1879)

Type: *signicollis* Kirsch.

C. O. Wat. Ill, Typ. Sp. 1879, p. 59.

becolor Kln. I.F.R. XIII (VI), 1929, p. 236.[1] Pl. II, F. 45-46. — Ceylon.[1]

brevelineatus Pic. F. I. Fr. 1923, No. 1, p. 14.[1] — Tonkin.[1]

lajoyei Pic. H. T. E. No. 405, 1921, p. 7[1]; F. I. Fr. 1923, No. 1, p. 14.[2] — Tonkin[1, 2]

laoensis Pic. H. T. E. No. 423, 1926, p. 23.[1] — Laos.[1]

major C. O. Wat. Ill. Typ. Sp. 1879, p. 60, t. XV, f. 6.[1] — Allahabad.[1] Darjeeling.[1]

minor Pic. H. T. E. No. 423, 1926, p. 23.[1] — India.[1]

minutus Pic. F. I. Fr. 1923, No. 1, p. 15.[1] — Laos.[1]

notatipennis Pic. H. T. E. No. 421, 1925, p. 13.[1] — Tonkin[1]

signicollis Kirsch, M. Z. M. D. I, 1875, p. 36;[1] C. O. Wat. Ill. Typ. Sp., 1879, p. 60;[2] Pic. F. I. Fr., 1923, No. 1, p. 14;[3] Kln. F. M. S. XIII (2 & 3) 1926, p. 193;[4] I. F. R. XIII, VI, 1929, p. 236.
Calopteron signicolle Kirsch. l. c.
Metriorrhynchus festivus C. O. Wat. Tr. E. S. L. 1878, p. 102 & 110.
Metriorrhynchus velutinus C. O. Wat. l. c. p. 110.
— Sylhet.[o], Manipur.[5] Darjeeling.[4], Siam.[4] Assam.[5], Burma.[5] Pahang.[4, 5], Perak.[4, 5] Selangor., Laos.[3] Cambodia.[3], Tonkin.[3] Malacca.[1, 3, 5] China.[3], Sumatra.[2, 3] Borneo.[4], Java.[3]

sordidus Pic. H. T. E. No. 405, 1921, p. 8;[1] F. I. Fr. 1923, No. 1, p. 14.[2] — Cochin-China.[1, 2] Tonkin.[2], Laos.[2]

tonkineus Pic. F. I. Fr. 1923, No. 1, p. 14.[1] — Tonkin[1]

trinotatus Pic. H. T. E. No. 405, 1921, p. 8.[1] — Burma.[1]

velutinus C. O. Wat. Tr. E. S. L. 1878, p. 110;[1] Bourg. A. S. E. Fr., 1885 (6) V, p. 81;[2] l. c., 1890, p. 166.[3] — Sibsagar.[2], Sikkim.[1] Darjeeling.[2], Burma.[2] Cambodia.[3], Tonkin.[3]

waterhousei Bourg. A. S. E. Fr., 1885, (6) V, p. 81.[1] — Allahabad.[1] Darjeeling.[1]

17. **Pseudoconderis** Pic. (1921).

Type: *gorhami* Pic.

Pic. H. T. E. No. 405, 1921, p. 8.

gorhami Pic. l. c., p. 8.[1] — India.[1]

18. **Lycoprogentes** Pic. (1915).

Type: *pouilloni* Pic.

Pic. Mel. XVI, 1915, p. 6.

pouilloni Pic. l. c., p. 6.[1] — Darjeeling.

19. **Falsoconderis** Pic. (1926).

Type: *pubescens* Pic

Pic. H. T. E. No. 423, 1926, p. 22.

pubescens Pic. H. T. E. No. 423, 1926, p. 23.[1] — Laos.[1]

20. **Xylobanoides** Kleine (1929).

Type: *unicolor* Gorh.

Kln. I. F. R. XIII (VI) 1929, p. 237.

unicolor Gorh. A. S. E. B. XLVII, 1903, p. 324. — Kashmir.[1]

Xylobanus unicolor Gorh. 1. c.

21. **Taphes** C.O. Waterhouse (1878).

Type: *brevicollis* C. O. Wat.

Tr. E. S. L. 1878, p. 102; Ill. Typ. Sp., 1879, p. 62

brevicollis C. O. Wat. Tr. E. S. L., 1878, p. 110;[1] Ill. Typ. Sp. 1879, p. 62, t. XV, f. 9,[2] Kln. F. M. S. XIII (2 & 8), 1926, p. 193,[3] I. F. R. XIII (VI), 1929, p. 237.[4] — Sylhet.[4] C. Almora Div.[4] Kumaon.[4], Pahang.[3] Perak., Selangor.[3]

frontalis C. O. Wat. Tr. E. S. L., 1878, p. 111;[5] Ill. Typ. Sp. 1879, p. 63, t. XV, f. 4,[6] Bourg. A. M. C. G. XVIII, 1883, p. 645; A. S. E. Fr. 1890, p. 166; Pic. F. I. Fr., 1923, No. 1, p. 12.[7] — Perak.[3, 4] Cochin-China.[7] Sumatra.[5, 6, 7] Borneo.[2, 7]

tonkineus Pic. F. I. Fr., 1923, No. 1, p. 12.[1] — Tonkin.[1]

22. **Metanoeus** C. O. Waterhouse (1879).

Type: *dispar* C. O. Wat.

C. O. Wat. Ill. Typ. Sp. 1879, p. 73.

aestimatus Kln. I. F. R. XIII (VI), 1929, p. 238.[1] Pl. II, F. 49-50. — Assam.[1]

angustatus Pic. M. L. 1927, p. 35.[1] — Tonkin.[1]

dispar C. O. Wat. Tr. E. S. L. 1878, p. 115;[1] Ill. Typ. Sp. 1879, p. 74, t. XVIII, f. 1;[2] Kln. F. M. S. XIII, (2 & 3), 1926, p. 194;[3] I. F. R., XIII (VI), 1929, p. 238.[4] — Malacca.[1, 2], Assam.[1] Burma.[4], Selangor.[3] Pahang.[3], Perak.[3]

fulvus C. O. Wat. Ill. Typ. Sp. 1879, p. t. XVIII, f. 2.[1] — Penang.[1]

inapicalis Pic. H. T. E. No. 419, 1925, p. 8.[1] — Tonkin.[1]

quadraticollis Kln. I. F. R. XIII (VI), 1929,[1] p. 239, Pl. II, F. 51. — Assam.[1] Burma.[1]

scutellaris Pic. H. T. E. No. 406, 1921, p. 11.[1] — Malacca.[1]

23. **Prometanoeus** Kleine (1925).

Type: *ochraceus* Kln.

Kln. S. E. Z. LXXXV, 1925, p. 133.

ochraceus Kln. S. E. Z. LXXXV, 1925, p. 133, f. 1, 2.

V. ATELINAE Kleine.

I. F. R. XIII (VI), 1929, p. 222.

24. Atelius C. O. Waterhouse (1878). Ceylon.[1]

Type: *expansicornis* F. Wlk.

C. O. Wat. Tr. E. S. L. 1878, p. 104; Ill. Typ. Sp. 1879, p. 76.

expansicornis F. Wlk. A. M. N. H. 1858, II, p. 282,[1] C. O. Wat. Ill, Typ. Sp. 1870, p. 76, t. XVIII, f. 5.[2] Ceylon.[1,2]

25. Scarelus C. O. Waterhouse (1878).

Type: *longicornis* C. O. Wat.

C. O. Wat. Tr. E. S. L. 1878, p. 104; Ill. Type. Sp., 1879, p. 76.

inapicalis Pic. Shornik III, 1925, p. 161.[1] India.[1]

orbatus C. O. Wat. Tr. E. S. L. 1878, p. 117:[1] Ill. Typ. Sp. 1879, p. 77, t. XVIII, f. 7;[2] Kln. F. M. S. XIII, (2 & 3), 1926, p. 195.[3] Singapore.[1,2] Perak.[3] Pahang.[3]

VI. TRICHALINAE Kleine.

I. F. R. XIII (VI), 1929, p. 222.

26. Trichalus C. O. Waterhouse (1877).

Type: *flavopictus* C. O. Wat.

C. O. Wat. Tr. E. S. L. 1877, p. 82; Ill. Typ. Sp. 1879, p. 67.

chapaensis Pic. F. I. Fr. 1923, Nos. 1 and 2, p. 54.[1] Tonkin.[1]

communis C. O. Wat. Ill. Typ. Sp. 1879, p. 71, t. XVII, f. 9;[1] Bourg. A. S. E. Fr. 1890, p. 166;[2] Pic. F. I. Fr. 1923, No. 1, p. 12;[3] Kln. F. M. S. XIII (2 & 3) 1926, p. 193;[4] I. F. R. XIII (VI) 1929, p. 239.[5] Johore.[5], Selangor.[4] Pahang.[4,5] Cochin-China.[2,3] Cambodia.[2,3] Krakataua. Sumatra., Java.[1]

inconditus Kln. T. IX (4) 1926, p. 312, f. 69, 111.[1] Perak., Java.[1]

niger C. O. Wat. Ill. Typ. Sp. 1879, p. 71, t. XVII, f. 8.[1] Perak., Banda.[1] Java.[1]

27. Leptotrichalus Kleine (1925).

Type: *cyaniventris* Kirsch.

Kln. P. J. S. XXVIII, No. 2, 1925, p. 296.

cyaniventris Kirsch, M. Z. M. D. I., 1875, p. 35;[1] Kln. P. J. S. XXVIII, No. 2, 1925, p. 303, t. 2, f. 10;[2] F. M. S. XIII (2 & 3) 1926, p. 194.[3] *Metriorrhynchus cyaniventris* Kirsch, 1. c.	Perak. Malacca.[1, 2] Sumatra.[2] Mentawei.[2] Borneo.[2], Java.[2]
rubrosignatus Kln. F. M. S., XVI, 1930, p. 170.[1]	Perak.

VII. METRIORRHYNCHINAE Kleine.

I. F. R. XIII (VI), 1929, p. 222.

28. Metriorrhynchus Guérin Méréville (1838).

Type: *parallelus* Guér.

Guér Voy. Coquille 1838, Ent. p. 72.

inaequalis Fabr. Syst. El. II, 112; Kln. F. M. S. XIII (2 & 3), 1926, p. 192;[1] I. F. R. XIII (VI), 1929, p. 239.[2] *kirschi* C. O. Wat. Ill. Typ. Sp., 1879, p. 50.[3] *notaticollis* Pic. H. T. E. No. 406, 1921, p. 11. *enganoensis* Pic. l. c., p. 11.[4] *Dictyopterus lineatus* Kirsch, M. Z. M. D. I, 1875, p. 35.[5]	Malacca.[1, 2, 5] Selangor.[1] Negri Sembilan.[1, 2] Perak.[1, 2], Siam.[1, 2] Penang.[2, 3], Pahang.[1] Johore.[2] Singapore.[2, 3] Sumatra.[3], Engano.[4] Borneo.[3]
orientalis C. O. Wat. Ill. Typ. Sp., 1879, p. 58, t. XIV, f. 8.[1]	Siam.[1]
rubicundus C. O. Wat. Ill. Typ. Sp., 1879, p. 56, t. XIV, f. 2;[1] Gorh. Tr. E. S. L. 1890, p. 545.	Sylhet.[1]
sericans C. O. Wat. Ill. Typ. Sp., 1879, p. 53, t. XIII, f. 7;[1] Gorh. Tr. E. S. L., 1890, p. 545; Kln. I. F. R. XIII (VI), 1929, p. 239.[2]	India.[1, 2], Assam.[2] Burma.[2] Tenasserim.[2]
sericeus C. O. Wat. Ill. Typ. Sp., 1879, p. 56, t. XIV, f. 2;[1] Gorh. Tr. E. S. L., 1890, p. 545.	Sylhet.[1]

VIII. PLATERODINAE Kleine.

I. F. R. XIII (VI), 1929, p. 222.

29. Plateros Bourgeois (1879).

Type: *alatus* Newm.

Bourg. C. R. B. XXII, 1879, p. XIX; C. O. Wat. Tr. E. S. L., 1878, p. 100; Ill. Typ. Sp., 1879, p. 25.

Microditoneces Pic. H. T. E. No. 404, 1921, p. 5.

amplipennis Pic. H. T. E., No. 405, 1921, p. 7:[1] F. I. Fr., 1923, No. 1, p. 10.[2] — Saigon.[1, 2]

annamitus Pic. H. T. E. No. 405, 1921, p. 7;[1] F. I. Fr. 1923, No. 1, p. 10.[2] — Annam.[1, 2]

binhanus Pic. H. T. E. No. 420, 1925, p. 10.[1] — Tonkin.[1]

birmanicus Kln. I. F. R. XIII (VI), 1929, p. 245, Pl. II, F. 72-73, Pl. III, F. 74. — Burma.[1], Assam.[1] Tenasserim.[1], Perak.[1]

brevehumeralis Pic. M. L., 1927, p. 35.[1] — Tonkin.[1]

brunnescens Kln. I. F. R. XIII (VI), 1929,[1] p. 246, Pl. III, F. 79-81. — Tenasserim.[1]

callimorpha Kln. I.F.R., l. c., p. 241, Pl. II,[1] Fig. 54-56. — Ceylon.[1]

ceylonicus Kln. I. F. R., l. c., p. 244, Pl. II, Fig. 66-68. — Ceylon.[1]

chapaensis Pic. F. I. Fr., 1923, No. 1, p. 11.[1] — Tonkin.[1]

cimicoides Kln. I. F. R. XIII (VI), 1929,[1] p. 242, Pl. II, F. 62-64. — Kumaon.[1] Mussoorie, U. P. Bengal.

confrater Kln. I. F. R. XIII (VI), 1929, p. 241,[1] Pl. II, F. 57-58. — Ranikhet.[1] Mussoorie 6-7,000'. Lambatach 7,600, U.P.

coomani Pic. A. S. L. L. LXXII, 1925, p. 74.[1] — Tonkin.[1]

cribripennis Bourg. A. S. E. B. XXXVI, 1892, p. 235;[1] Kln. I. F. R. XIII (VI), 1929, p. 240.[2] — Mandar.[1] Central India.[2]

dispellens F. Wlk. A. M. N. H. 1858, II, p. 282,[1] C. O. Wat. Ill. Typ. Sp., 1879, p. 26, t. VI, f. 8;[2] Gorh, Tr. E. S. L., 1890, p. 544;[3] Kln. I. F. R. XIII (VI), 1929, p. 240.[4] — Ceylon.[1, 2, 4] South India.[2, 3, 4]
Lycus dispellens F. Wlk., l. c.
Lycus gracilis C. O. Wat. Tr. E.S.L., 1878, p. 99, 107.

distortus Kln. F.M.S. XIII (2 & 3), 1926, p. 183, f. 11-13.[1] — Selangor.[1], Penang.[1] Perak.[1]

elisus Pic. H. T. E. No. 405, 1921, p. 7:[1] F. I. Fr., 1923, No. 1, p. 10.[2] — Tonkin.[1, 2]

flavonotatus Kln. I. F. R. XIII (VI), 1929,[1] p. 242, Pl. II, F. 59-61.	Sylhet.[1] E. Ramganga. Kumaon 5,000'. Ranikhet 5,000'. W. Almora.
fletcheri Kln. S. E. Z. LXXXXI, 1930, p. 97.[1]	Nilgiris.[1] Shillong 5,000'.
forcipatus Kln. I. F. R. XIII (VI), 1929, p. 246, Pl. III, F. 75-77.[1]	Manipur.[1]
fulgidus Kln. I. F. R., l. c., p. 243, Pl. II, F. 65.[1]	Punjab[1], Assam.[1] Tenasserim.[1]
fuscicornis Gorh. Tr. E. S. L. 1890, p. 544.[1]	Andamans.[1]
fuscipennis C. O. Wat. Ill. Typ. Sp. 1879, p. 27, t. VII, f. 3;[1] Gorh. Tr. E. S. L., 1890, p. 544:[2] Pic. F. I. Fr., 1923, No. 1, p. 10.[3]	Sylhet.[1, 2] Cambodia.[3]
gardneri Kln. I. F. R. XIV (XIV), 1930, p. 288.[1]	Rangirum 6,000'. Debrepani 6,000'. Darjeeling.
harmandi Bourg. Bull. Mus. Paris 1902, VIII, p. 92.[1]	Himalaya.[1]
imitator Kln. I.F.R. XIII (VI), 1929, p. 240, Pl. II, F. 53.[1]	Haldwani Div.[1] Malabar.[1], Bengal.[1]
indianus Gorh. A.S.E.B. XLVII, 1903, p. 1903, p. 324.[1]	Nilgiris 6,000'.[1]
lacosus Pic. H. T. E. No. 423, 1926, p. 24.[1]	Tonkin.[1]
languidus C. O. Wat. Ill. Typ. Sp., 1879, p. 27, t. VII, f. 2:[1] Gorh. Tr. E. S. L. 1890, p. 544;[2] Kln. I. F. R. XIII (VI), 1929, p. 240.[3]	Ceylon.[1, 2] Andamans.[3]
lateauriantiacus Pic. M. LI, 1928, p. 16.[1]	India.[1]
lautus Kln. I.F.R. XIII (VI), 1929, p. 245,[1] Pl. II, F. 71-72.	Ceylon.[1]
limbaticollis Bourg. A.S.E. Fr. LXXVIII, 1910, p. 426.[1]	Ceylon.[1]
minutissimus Kln. I. F. R. XIII (VI), 1929, p. 246[1], Pl. III, F. 78.	Assam.[1] Tenasserim.[1]
minutus Pic. H. T. E. No. 405, 1921, p. 6.[1]	Ceylon.[1]
multiimpressus Pic. H. T. E. No. 426, 1926, p. 33.[1]	Tonkin.[1]
mussooriensis Kln. S. E. Z. LXXXXI, 1930, p. 97.[1]	Mussoorie.[1]

nietneri Kln. I. F. R. XIII (VI), 1929, p. 244,[1] Pl. II, F. 69-70. — Ceylon.[1]

nilgirensis Kln. S. E. Z. LXXXXI, 1930, p. 98.[1] — Nilgiris.[1]

ochraceipennis Kln. I. F. R. XIII (VI), 1929, p. 243, Pl. V, F. 143. — India.[1] Tenasserim.[1] Burma.[1], Perak.

peguensis Pic. H. T. E. No. 405, 1921, p. 6.[1] — India.[1]

planatus C. O. Wat. Ill. Typ. Sp. 1879, p. 27, t. VII, f. 1;[1] Kln. F. M. S. XIII (2 & 3), 1926, 1. 183.[2] — Penang.[1], Siam.[2] Shanghai.[1]

proximus Bourg. Bull. Mus. Paris VIII, 1902, p. 93.[1] — Himalaya.[1]

punctatus Pic. M. LI, p. 16.[1] — Tonkin.[1]

robustithorax Pic. F. I. Fr. 1923, No. 1, p. 10.[1] — Tonkin.[1]

robustus I.F.R. XIII (VI), 1929,[1] p. 247, Pl. IV, F. 133-134. — Perak[1], Pahang.

rubripennis Pic. M. XXX, 1919, p. 21;[1] Kln. I. F. R. XIII (VI), 1929, p. 240.[2] — Madura.[1], Shillong.[1] Gory valley 4,000'.[1, 2]

sikkimanus Kln. I. F. R., l. c., p. 240, Pl. II, F. 52.[1] — Sikkim.[1]

sincerus Kln. I. F. R., l. c., p. 247, Pl. V, F. 144-146. — Dehra Dun.[1]

suturalis Pic. M. XXX, 1919, p. 21;[1] Kln. I. F. R. XIII (VI), 1929, p. 240.[2] — India.[1] Assam.[2]

sycophanta Fairm. A. S. E. Fr. 1888, (6), VIII, p. 352;[1] Bourg. A. S. E. Fr. 1890, p. 166;[2] Pic. F. I. Fr., 1923, No. 1, p. 10.[3] — Indo-China.[1] Cochin-China.[3] Annam.[2, 3] Tonkin.[2, 3]

tonkineus Pic. H. T. E. No. 425, 1926, p. 31.[1] — Tonkin.[1]

30. Lyponia C. O. Waterhouse (1878).

Type: *debilis* C. O. Wat.

C. O. Wat. Tr. S. E. L., 1878, p. 99; Ill. Typ. Sp. 1870, p. 24; Kln. E. B. XX, 1924, No. 3, p. 170.

alternata Pic. M. IL, 1927, p. 5. — Indo-China.[1]

aurantiaca Pic. M. IL, 1927, p. 5.[1] — Indo-China.[1]

debilis C. O. Wat. Tr. E.S.L., 1878, p. 107;[1] Ill. Typ. Sp., 1879, p. 25, t. VI, f. 9;[2] Pic. F.I.Fr., 1923, No. 1, p. 9.[3] — Tonkin.[3] China.[1, 2]

dolosa Kln. E. B. XX, 1924, No. 3, p. 173, f. 5, 9;[1] Kln. I. F. R. XIII (VI), 1929, p. 248.[2] — Sikkim.[1] Tsuntang 6,000.[2]

himalajica Bourg. A. S. E. Fr., 1885, (6) V, p. 79;[1] Kln. E. B. XX, 1924, No. 3, p. 172;[2] I. F. R. XIII (VI), 1929, p. 248.[3] — Darjeeling.[1, 2] Burma.[2, 3] Laos.[2]

ochraceicollis Pic. F. I. Fr., 1923, No. 1, p. 9.[1] — Laos.[1]

waterhousei Gorh. Tr. E. S. L., 1890, p. 543.[1] — India.[1], Assam. Formosa.[1]

31. **Ditoneces** C. O. Waterhouse (1879).

Type: *pubicornis* F. Wlk.

C. O. Wat. Ill. Typ. Sp., 1879, p. 31.

Subg. *Cautirodes* Pic. H. T. E. No. 404, 1921, p. 3.

aculeatus Kln. I. F. R. XIII (VI), 1929, p. 249, Pl. III, F. 87-89. — Colombo.[1], Dikoya.[1] Bogawantalawa.[1]

assamenis Pic. M. LI, 1928, p. 15.[1] — Assam.[1]

assamicus Kln. I. F. R. XIII (VI), 1929, p. 252[1], Pl. V, F. 147-148. — Assam, Shillong.[1] Pahang.

atripennis Pic. H. T. E. No. 425, 1926, p. 32.[1] — Tonkin.[1]

bifoveiceps Pic. H. T. E. No. 404, 1921, p. 4;[1] Pic. F. I. Fr. 1923, No. 1, p. 11.[2] — Cochin-China.[1, 2]

binhanus Pic. S. L. L. LXXII, 1925, p. 73.[1] — Tonkin.[1]

breviramosus Kln. I. F. R. XIII (VI), 1929, p. 251, Pl. III, F. 99-101. — Burma.[1]

curvicollis Kln. F. M. S., XVI, 1930, p. 169.[1] — Perak.[1]

disconiger Pic. H. T. E. No. 425, 1926, p. 32.[1] — Tonkin.[1]

distortus I.F.R. XIII (VI), 1929, p. 250, Pl. III, F. 90-91. — Travancore.[1] Nilgiris. Anaimalais.

donckieri Pic. F. I. Fr., 1923, No. 1, add. 1, p. 49.[1] — Tonkin.[1]

elongatissimus Pic. F.I. Fr., 1923 No. 1, add. 1, p. 49.[1] — Tonkin.[1]

fabrei Bourg. B. S. E. Fr., 1888, (6) III, p. XXV.[1] — Hindostan.[1]

gorhami Pic. H. T. E. No. 404, 1921, p. 3.[1] — Ceylon.[1]

hoanus Pic. H. T. E. No. 425, 1926, p. 32.[1] — Tonkin.[1]

inconstans Kln. F.M.S., XIII, (2 & 3), 1926, p. 182, f. 9.[1] — Perak.[1], Borneo.[1]

indistinctus Kln. F. M. S. XVI, 1930, p. 169.[1] — Perak.[1], Borneo.[1]

lanatus Kln. I. F. R. XIII (VI), 1929, p. 248,[1] Pl. III, F. 87-89. — Dikoya.[1], Kandy.[1] Halaphani.[1] Haldumulle.[1]

laoensis Pic. M. 1916, XVIII, p. 16;[1] F. I. Fr., 1923, No. 1, p. 11.[2] — Laos.[1, 2]

laticornis Pic. M., 1916, XVIII, p. 16.[1] — Laos.[1]

limbatus Pic. H. T. E. No. 425, 1926, p. 31.[1] — Tonkin.[1]

malaccanus Pic. H. T. E. No. 404, 1921, p. 3.[1] — Perak.[1]

maindroni Pic. H. T. E. No. 404, 1921, p. 3.[1] — India.[1]

malayensis Kln. F.M.S. XIII, (2 & 3), 1926, p. 182, f. 10.[1] — Lubok Kedondong.[1] Lalang., Beluka.[1]

nigriceps Pic. H. T. E. No. 404, 1921, p. 3.[1] — Ceylon.[1]

nigroscutellatus Kln. I. F. R. XIII (VI), 1929,[1] p. 248, Pl. III, F. 82-83. — Burma.[1]

nigrosulcatus Kln. I. F. R., l. c., p. 250,[1] Pl. III, F. 95-96. — Assam.[1], Burma.[1] Tenasserim.[1] Sikkim.[1]

nilgirensis Gorh. A. S. E. B., 1903, XLVII, p. 323.[1] — Nilgiris., Punjab.[1]

obscurus C. O. Wat. A. M. N. H., 1880 (5) V, p. 213.[1] — Malabar.[1]

pallidicolor Pic. F. I. Fr., 1923, No. 1, p. 11.[1] — Cambodia.[1]

phungi Pic. F. I. Fr., 1923, No. 1, add. 2, p. 53.[1] — Tonkin.[1]

pubicornis F. Wlk. A. M. N. H., 1858, II, p. 281, ♂;[1] C. O. Wat. III, Typ. Sp. 1879, p. 32, t. VII, f. 9.[2] — Ceylon.[1, 2]

Lycus pubicornis F. Wlk., l. c., p. 281♂.

Lycus melanopterus F. Wlk., l. c., p. 281♀.

Lycus duplex F. Wlk., l. c., p. 281 ♀.

Lycus revocans F. Wlk., l. c., p. 282 ♂.

pubipennis F. Wlk. A. M. N. H., 1859, III, p. 259.[1] Kln. I. F. R. XIII (VI), 1929, p. 248.[2] — Ceylon.[1] Colombo.[2]

Lycus pubipennis F. Wlk., l. c.

punctipennis C. O. Wat. Tr. E. S. L., 1878, p. 100 & 108; Ill. Typ. Sp. 1879, p. 31, t. VII, f. 10;[1] Bourg. A. M. C. G. XVIII, 1883, p. 646; A. S. E. Fr., 1890, p. 166;[2] Pic. F. I. Fr., 1923, No. 1, p. 11.[3] — Cambodia.[2, 3] Java.[1]

Lycus punctipennis C. O. Wat. Tr. E. S. L., l. c.

purpureopubens Pic. H. T. E. No. 404, 1921, p. 4.[1] — Perak.[1]

ramosus Kln. I. F. R. XIII (VI), 1929, p. 251, Pl. III, F. 97-98.[1] — India.[1]

simulator Kln. I. F. R. XIII (VI), 1929,[1] p. 250, Pl. III, F. 92-94. — Ceylon.[1]

sobrinus C. O. Wat. Ill. Typ. Sp., 1879, p. 32, t. VIII, f. 2.[1] — Burma.[1]

soror Kln. F. M. S., XVI, 1930, p. 170.[1] — Perak.[1]

subrecticollis Pic. H. T. E. No. 405, 1921, p. 5.[1] — Ceylon.[1]

terminalis C.O. Wat. Ill. Typ. Sp., 1879, p. 33, t. VIII, f. 3.[1] — Ceylon.[1]

32. **Melampyrus** C. O. Waterhouse (1879).

Type: *alternans* C. O. Wat.

C. O. Wat. Ill. Typ. Sp., 1879, p. 30.

alternans C. O. Wat. Tr. E. S. L., 1878, p. 100 & 108,[1] Ill. Typ. Sp. 1879, p. 30, t. VIII, f. 7;[2] Kln. F. M. S. XIII, (2 & 3), 1926, p. 185;[3] I. F. R. XIII (VI), 1929, p. 252.[4] — Singapore.[4] Perak.[3], Pahang.[3] Sarawak.[1, 2]

Lycus alternans C. O. Wat., l. c.

miscellus C. O. Wat. Tr. E. S. L., 1878, p. 100 & 108;[1] Ill. Typ., Sp., 1879, p. 30, t. VIII, f. 6.[2] — Penang.[1, 2]

Lycus misellus C. O. Wat., l. c.

33. **Dihammatus** C. O. Waterhouse (1879)

Type: *cribripennis* C. O. Wat.

C. O. Wat. Ill. Typ. Sp., 1879, p. 29.

malayensis Kln. I. F. R. XIII (VI), 1929,[1] p. 253, Pl. III, F. 103-104. — Pahang.[1]

persimilis Kln. I. F. R., l. c., p. 252, Pl. III, F. 102.[1] — Burma.[1], Assam.[1]

piceus Kln. F. M. S. XIII (2 & 3), 1926, p. 184, f. 14.[1] — Pehang.[1] Tenasserim.[1]

34. **Planeteros** Gorham (1883).

Type: *ochropterus* Gorh.

Gorh. A. M. C. G. XVIII, 1883, p. 591.

exutus Bourg. A. S. F. Fr. LXXVIII, 1910, p. 427.[1] — Cylon.[1]

IX. LYGISTOPTERINAE Kleine.

I. F. R. XIII (VI), 1929, p. 222.

35. Calochromus Guérin Ménéville (1833).

Type: *glaucopterus* Guér. Ménév.

Guér. A. S. E. Fr., 1833, p. 158.

Micronychus Motsch. B. M. XXXIV, 1861, I, p. 138.

aemulus C. O. Wat. Cist. Ent. II, 1877, p. 198;[1] Ill. Typ. Sp., 1879, p. 3, t. I, f. 11;[2] Kln. F. M. S. XIII, (2 & 3), 1926, p. 178.[3]	Perak.[3] Sarawak.[1, 2]
andrewesi Kln. S. E. Z. LXXXXI, 1930, p. 100.[1]	Nilgiris.[1]
apicalis Hope, Gray's Zool. Miscell., 1831, p. 24;[1] C. O. Wat. Ill. Typ. Sp., 1879, p. 3, t. I, f. 10;[2] Gorh. Tr. E. S. L., 1890, p. 542.[3]	Nepal.[1, 2, 3]
atricollis Pic. H. T. E. No. 420, 1925, p. 12.[1]	India.[1]
assamicus Kln. I. F. R. XIII (VI), 1929, p. 262[1], Pl. IV, F. 132.	Assam.[1] Tenasserim.[1]
bimaculatus Motsch. B. M. XXXIV, 1861, I, p. 139, t. IX, f. 14.[1]	Ceylon.[1]
Micronychus bimaculatus Motsch., l. c.	
birmanicus kln. I. F. R. XIII (VI), 1929, p. 262,[1] Pl. IV, F. 126-129.	Tenasserim.[1]
coomani Pic. H. T. E. No. 425, 1926, p. 30.[1]	Tonkin.[1]
costulatus kln. I. F. R. XIII (VI), 1929,[1] p. 260, Pl. IV, F. 120-122.	Kashmir.[1] Assam.[1]
crinitus Kln. I. F. R., l. c., p. 261, Pl. IV,[1] F. 123-124.	Nilgiris.[1]
cyanescens Kln. S. E. Z. LXXXXI, 1930, p. 101.[1]	Malaya.[1] Cameron's Highlands.
darjeelinensis Bourg. A. M. C. C. XVIII, 1883, p. 622.[1]	Himalaya.[1] Rangirum 6,000'[1] Darjeeling.

densecostatus Kln. I. F. R. XIV (XIV), 1930, p. 288.[1]	Siam.[1]
fracticornis Kln. I. F. R. XIII (VI), 1929,[1] p. 260, Pl. IV, F. 119.	Nilgiris.[1]
fruhstorferi Pic. E. Nr. 343, 1913, p. 147.[1]	Perak., Java.[1]
giganteus Pic. M. VII, 1913, p. 8;[1] Kln. I. F. R. XII (VI), 1929, p. 259.[2]	India.[1, 2] Burma.[2]
gorhami Pic. H. VII, 1913, p. 9.[1]	Khasi Hills.[1]
impressicornis Pic. M. XVIII, 1916, p. 15;[1] F. I. Fr. 1923, No. 1, p. 7.[2]	Indo-China.[1] Laos.[2]
impressipes Pic. M. XVIII, 1916, p. 14;[1] F. I. Fr. 1923, No. 1, p. 7;[2] Kln. I. F. R. XIII (VI), 1929, p. 259.[2]	India.[3] Nilgiris (Waterfalls). Assam.[3], Sikkim.[3] Indo-China.[1] Tonkin.[2]
kaschmirensis Kln. I. F. R. XIII (VI) 1929, p. 259.[1]	Kashmir.[1]
lepidus C. O. Wat. Cist. Ent. II, 1877, p. 201;[1] Ill. Typ. Sp. 1879, p. 7, t. II, f. 3.[2]	Penang.[1, 2] Java.[1, 2]
longipennis C. O. Wat. Cist. Ent. II, 1877, p. 201,[1] Ill. Typ. Sp. 1879, p. 7, t. II, f. 4;[2] Kln. I. F. R. XIII (VI) 1929, p. 258.[3]	Malacca.[3] Sumatra.[1, 2]
madurensis Pic. M. VII, 1913, p. 9;[1] Kln. I. F. R. XIII (VI), 1929, p. 259.[2]	Madura.[1] Nilgiris.[2]
melanurus C. O. Wat. Cist. Ent. II,[1] 877, p. 196;[1] Ill. Typ. Sp. 1879, p. 2, t. I, f. 5;[2] Kln. F. M. S. XIII (2 and 3), 1926, p. 177;[3] I. F. R. XIII (VI), 1929, p. 258.[4]	Johore.[4], Penang.[1] Pahang.[3], Selangor.[3] Singapore.[4] Sumatra.[1, 2], Java.[1, 2]
miniatus Kln. S. E. Z. LXXXXI, 1930, p. 99.[1]	Nilgiris.[1]
minutus Pic. F. I. Fr. 1923, No. 1, add. 1, p. 48.[1]	Tonkin.[1]
monticola Kln. I. F. R. XIV (XIV), 1930, p. 287.	S. E. Thibet: Tsangpo Gorge 8000'-9000'
morestratus Kln. F. M. S. XIII (2 and 3), 1926, p. 178, f. 1-3.[1]	Pahang.[1]
mordax Kln. F. M. S. XIII (2 & 3), 1926, p. 179, f. 4-5.[1]	Pahang.[1]
multifidus Kln. F.M.S. XIII (2 and 3), 1926, p. 179, f. 6.[1]	Pahang.[1], Siam.[1]
mysticus Kln. F.M.S. XIII (2 & 3), 1926, p. 180, f. 7-8.[1]	Perak.[1]

nanus Kln. S. E. Z. LXXXXI, 1930, p. 101.[1]	South India.[1] Anamalais.[1]
nilgirensis Kln. S. E. Z. LXXXXI, 1930, p. 99.[1]	Nilgiris.[1]
nigronotatus Pic. M. VII, 1913, p. 8;[1] Kln. I. F. R. XIII (VI), 1929, p. 258.[2]	Nilgiris.[1], Manipur.[2] Haldwani.[1, 2]
nigrovittatus Pic. H. T. E. No. 420, 1925, p. 12;[1] Kln. I. F. R. XIII (VI), 1929, p. 258.[2]	India.[1, 2], Manipur.[2] Dudhatoli. 10000'.[8] Sunderdhunga.[1, 2] 8-9000.'[2] Debrepani 6000.' Darjeeling.
orbatus C. O. Wat. Cist. Ent. II, 1877, p. 197;[1] Ill. Typ. Sp. 1879, p. 3, t. I. f. 3;[2] Gorh. Tr. E. S. L. 1890, p. 541;[3] Kln. F. M. S. XIII (2 & 3), 1926, p. 177[4] I. F. R. XIII (VI), 1929, p. 258.[5]	Allahabad.[5] Madras.[5] Darjeeling.[1, 2, 5] Pahang.[4], Assam.[5] Philippines.[1, 2] Andamans.[3, 5]
pecularius Kln. I. F. R. XIII (VI), 1929, p. 261, Pl. IV, F. 125.[1]	Ceylon.[1]
perakensis Kln. F. M. S., XVI, 1930, p. 172[1]	Perak.[1]
reverens Kln. S. E. Z. LXXXXI, 1930, p. 98[1]	Nilgiris.[1]
ruber C. O. Wat. Cist. Ent. II, 1877, p. 199;[1] Ill. Typ. Sp. 1879, p. 4, t. I, f. 6;[2] Gorh. Tr. E. S. L. 1890, p. 542:[3] Kln. I. F. R. XIII (VI), 1929, p. 258.[4]	Allahabad.[1, 2, 3, 4] Nilgiris.[4], Madras.[4] Singapore.[4] Malabar.
rugatus C. O. Wat. Cist. Ent. II, 1877, p. 199;[1] Ill. Typ. Sp. 1879, p. 4, t. I, f. 4.[2]	Allahabad.[1, 2]
sanguinolentus Kln. I. F. R. XIII (VI), 1929,[1] p. 262, Pl. IV, F. 129-131.[1]	Burma.[1]
serraticornis Kln. I. F. R. l. c. p. 263,[1] Pl. V. F. 149-150.	India.[1]
siamensis Kln. F. M. S., XVI, 1930, p. 172[1].	Siam.[1]
simplicicornis Pic. F. I. Fr. 1923, No. 1, add. p. 48.[1]	Tonkin. (?)
singaporensis Pic. H. T. E. No. 420, 1925, p. 12;[1] Kln. I. F. R., XIII, 1929, p. 259.[2]	Nilgiris.[2] Singapore.[1] Dehra Dun.
substrictus Kln. S. E. Z. LXXXXI, 1930, p. 99.[1]	Nilgiris.[1]

sulcatipennis Pic. M. XVIII, 1916, p. 15;[1] F. I. Fr. 1923, No. 1, p. 7.[2]	Indo-China.[1] Laos.[2]
tarsalis C. O. Wat. Ill. Typ. Sp. 1879, p. 6, t. I, f. 9,[1] Gorh. Tr. E. S. L. 1890, p. 542.[2]	India.[1, 2]
terminatus C. O. Wat. A. M. N. H. 1886 (5) XVII, p. 498,[1] Kln. I. F. R. XIII (VI), 1929, p. 259.[2]	Ceylon.[1, 2]
testaceipes Pic. E. No. 343, 1913, p. 148.[1]	India.[1]
thoracicus Kln. S. E. Z. LXXXXI, 1930, p. 100.[1]	Nilgiris.[1]
tonkinensis Pic. M. XVII, 1916, p. 20;[1] F. I. Fr. 1923, p. 7.[2]	Tonkin.[1, 2]
triguttatus Kln. I. F. R. XIII (VI), 1929, p. 260, Pl. IV, F. 118.[1]	Sunderhunga. 8-12000'[1] W. Almora.[1] Kumaon.[1] Gori Valley. 7000'[1]
velutinus C. O. Wat. Cist. Ent. II, 1877, p. 199,[1] Ill. Typ. Sp. 1879, p. 5, t. I, f. 7,[2] Gorh. Tr. E. S. L. 1890, p. 542.[3]	Burma.[1, 2, 3]
vestitus C. O. Wat. Cist. Ent. II, 1877, p. 200;[1] Ill. Typ. Sp. 1879, p. 6, t. II, f. 2.[2]	Penang.[1, 2] Perak.
vesticollis Gorh. A. S. E. B. XXXIX, 1895, p. 301;[1] Kln. I. F. R. XIII, (VI), 1929, p. 259.[2]	Nilgiris.[1], Haldwani.[2] Chamba.[1], Laos.
vitalisi Pic. M. XVII, 1916, p. 15;[1] F.I.Fr. 1923, No. 1, p. 7.[2]	Indo-China.[1], Laos.[2]

36. **Lygistopterus** Mulsant (1838).

Type: *acuticollis* Fairm.

Mulsant, A. Soc. Agr. Lyon, I, 1838, p. 79 (Opusc. Ent. XI. 1860, p. 165).

aurantiacus Pic. M. IL. 1927, p. 5.[1]	Tonkin.[1]

X. DILOPHOTINAE Kleine.

I. F. R. XIII (VI), 1929, p. 222.

37. **Dilophotes** C.O. Waterhouse (1879). Type: exilis C.O. Wat.

C. O. Wat. Ill. Typ. Sp. 1879, p. 75.

atricollis Pic. E. M. XV, 1926, No. 1, p. 68;[1] Kln. I. F. R. XIII (VI), 1929, p. 253.[2]	Sylhet.[2], Sikkim.[2] Assam.[2], Tenasserim.[2] Formosa.[1]

commendatus Kln. I. F. R. XIII (VI), 1929, p. 253,[1] Pl. III, F. 105-106.	Perak.[1]
costatus Kln. F. M. S., XVI, 1930, p. 170.[1]	Perak.[1]
depressicornis Pic. F. J. Fr. 1923, No. 1 add. 1, p. 49.[1] Kln. F.M.S. XIII (2 and 3), 1926, p. 191, f. 38-40.[2]	Pahang.[2] Tonkin., Perak.
dilaticornis Pic. H. T. E. No. 429, 1925, p. 6.[1]	Tonkin.[1]
discoidalis Pic. M. XXXIII, 1921, p. 15.[1]	Perak., Borneo.[1]
dohertyi Pic. M. III, 1912, p. 9.[1]	Malacca.[1]
exilis C. O. Wat. Tr. E. S. L. 1878, p. 116.[1] Kln. I. F. R. XIII (VI), 1929, p. 253.[2]	Assam.[2] Borneo.[1]
grossus Kln. F. M. S., XVI, 1930, p. 172.[1]	Perak.
major Kln. I. F. R. XIII (VI), 1929,[1] p. 254, Pl. IV, F. 135-136.	Perak.[1]
pulchellus Kln. S. M. J. III, (3) Nr. 10, 1926, p. 360.[1]	Selangor., Kina Balu.[1]

38. **Flabellodilophotes** Pic. (1912).

Type: *obscurus* Pic.

Pic. Mel. III. 1912, p. 9.

nietneri Kln. I. F. R. XIII (VI), 1929, p. 254, Pl. IV, F. 107-108.	Ceylon.[1]
niger Kln. I. F. R. l. c. p. 255, Pl. IV,[1] F. 109.	Assam.[1]

39. **Libnetis** C. O. Waterhouse (1878).

Type: *pumilio* C. O. Wat.

C. O. Wat. Tr. E. S. L. 1878, p. 104; Ill. Typ. Sp. 1879, p. 77.

birmanensis Kln. I. F. R. XIII (VI), 1929, p. 255, IV, F. 110-111.[1]	Burma.[1] Assam.[1]
dohertyi Pic. E. 1911, p. 174.[1]	Malacca.[1]
indicus Pic. M. XXIII, 1217, p. 6.[1]	East India.[1]
kannegieteri Pic. E. Nr. 3211, 911, p. 166.[1]	Perak., Sumatra.
niger Pic. H. T. E. No. 426, 1926, p. 34.[1]	Tonkin.[1]
pumilio C. O. Wat. Tr. E. S. L. 1878, p. 177;[1] Ill. Typ. Sp. 1879, p. 77, t. XVIII, f. 10.[2]	Ceylon.[1,2]

rescidus Kln. S. E. Z. LXXXXI, 1930, p. 102.[1] — Nilgiris.[1]

similis Kln. I. F. R. XIII (VI), 1929, p. 255.[1] — Landsdowne.[1]

testaceus Pic. M. XXXIII, 1921, p. 14.[1] — India.[1] Nilgiris, South India.

40. **Libnetisia** Pic. (1912).

Type: *atricornis* Pic.

Pic. H. T. E. No. 404, 1921, p. 1.

atricornis Pic. H. T. E. No. 404, 1921, p. 1;[1] Kln. I. F. R. XIII (VI), 1929, p. 256.[2] — India.[1] Nilgiris.[2]

atronotata Pic. F. I. Fr. 1923, No. 1, p. 9.[1] — Tonkin.[1]

curticollis Pic. H. T. E. No. 404, 1921, p. 1.[1] — Malacca.[1]

flavohumeralis Kln. I. F. R. XIII (VI), 1929,[1] p. 256. — Assam.[1]

laticollis Pic. H. T. E. No. 404, 1921, p. 1;[1] Kln. I. F. R. XIII (VI), 1929, p. 256. — India.[1] Shambaganur.[2] Madura.[2]

testacea Pic. H. T. E. No. 404, 1921, p. 1.[1] — India.[1]

41. **Lyropaeus** C. O. Waterhouse (1878).

Type: *fallax* F. Wlk.

C. O. Wat. Tr. E. S. L. 1878, p. 105; 111. Lype. Sp. 1879, p. 78.

aurantiacus Bourg. A. S. E. Fr. LXXVII 1908, p. 503.[1] — Travancore.[1]

biguttatus Gorh. A. M. N. H. 1880 (5) V, p. 213;[1] Kln. I. F. R. (VI), 1929, p. 256.[2] — India.[1, 2] Cochin.

cinnamomi Kln. S. E. Z. LXXXI, 1930, p. 102.[1] — Nilgiris.[1]

coccineicollis Pic. M. VII 1913, p. 6.[1] — Shambaganur.[1]

contrarius Kln. I. F. R. XIII (VI), 1929,[1] p. 257. — Nilgiris.[1]

dohertyi Pic. E. No. 320, p. 166;[1] Kln. F. M. S. XIII (2 and 3) 1926, p. 196, f. 41-42;[2] I. F. R. XIII (VI), 1929, p. 257. — Malacca.[1] Perak.[2] Assam.[3]

densepilosus Kln. I. F. R. XIII (VI), 1929,[1] p. 257. — Ceylon.[1] Travancore.

fallax F. Wlk. A. M. N. H. 1858, II, p. 281;[1] C. O. Wat. Ill. Typ. Sp. 1879, p. 78, t. XVIII, f. 4.[2] — Ceylon.[1]

gorhami Pic. M. I, 1911, p. 16.[1]	India.[1]
grandissimus Kln. I. F. R. XIII (VI), 1929,[1] p. 257.	Chamenkaiam hill.[1]
granulosus Kln. S. E. Z. LXXXXI, 1930, p. 102.[1]	South India.[1] Kanara.
laticollis Pic. Echange No. 423, 1926, p. 4.[1] I. F. R. XIII (VI), 1929, p. 257.[2]	Ceylon.[1,2]
optabilis Kln. F.M.S., XIII (2 and 3) 1926, p. 196, f. 43.[1]	Perak.[1], Pahang.[1]

GENERA SEDIS INCERTAE

42. **Pristolycus** Gorham (1883).

Type: *sagulatus* Gorh.

Gorh. Tr. E. S. L. 1883, p. 407; E. Olivier, Rev. Sc. Bourb. XXIV, 1911, p. 68.

annamitus Pic. M. XVIII 1916, p. 26; F. I. Fr. 1923, No. 1, p. 20.[2]	Annam.[1,2] Laos.[2]
laboissierei Pic. M. XVI, 1915, p. 6;[1] F. I. Fr. 1923, No. 1, p. 20.[2]	Tonkin.[1,2]
tricolor Pic. E. N. II No. 1, 1928, p. 4.[1]	Tonkin.[1]

PART XXII – PHALONIADAE AND CHLIDANOTIDAE

PHALONIADAE

Phaloniadae: Meyr., Handb., pp. 543-544 (1895) [charact.]; Meyr., Proc. Linn. Soc. N. S. W. XXXV 158 (1910) [charact.]; Meyr., Rev. Handb., p. 483 (1928) [charact.]

Phaloniidae: Forbes, Lep. New York, pp. 499-501, ff. 288-290 (1924) [charact.]

Phalonianae: Wlsm., T. E. S. 1895. 498 (1895).

Phaloniinae: Busck, Jl. N. Y. Ent. Soc. XV 19-21 (1907) [charact.]; Kennel, Pal. Tortric., pp. 47, 59-60 (1908) [charact.]; Kennel; Spuler's Schmett. Eur. II 254 (1910) [charact.]; Kennel, Pal. Tortric., p. 233 (1913) [charact.]

Lozoperidae [part]: Stainton, Manual II 265 (III. 1859) [charact.]; Wilk., Brit. Tortric., pp. 9, 278 (V. 1859) [charact.]; Barrett, Lep. Brit. Isds. X 278 (1905) [charact.].

Conchylidae: Meyr., P. Linn. Soc. N. S. W., V 413 (1881) [charact.]; Meyr., P. Linn. Soc. N. S. W. VI 693 (1881) [charact.]; Meyr., Trans. N. Z. Inst. XV 66 (1883) [charact.].

1. Phalonia, Hübner 1826

Phalonia, Hb., Verz. Schmett., p. 393 (1826); Meyr., Handb., p. 545, fig. [neur. *tesserana*] (1895) [redescr.]; Fernald, U. S. Nat. Mus. Bull. 52. p. 486 (1903); Dyar, Proc. E. S. Wash. V 231-232 (1903); Kearfott, Smith's Check List Lep. Bor. Amer., p. 106 (1903); Busck, Jl. N. Y. Ent. Soc. XV 21 (1907) [redescr.]; Kennel, Spuler's Schmett. Eur. II 255 (1907) [redescr.]; Walsm., P. Z. S. 1907. 491 (1908); Fernald, Genera Tortric., pp. 17, 54 (1908); Meyr., P. Linn. Soc. N. S.

Type: *tesserana*, Schiff. [Europe]

W. XXXVI 297 (1911) [redescr.]; Kennel, Pal. Tortric., pp. 241-242 (1913) [redescr.]; Walsm., Biol. Centr. Amer., Het. IV 294 (1914): Pierce and Metcalfe, Genit. Brit. Tortric., p. 29 (1922) [genit. charact.]; Forbes, Lep. N. York, p. 503 (1924) [redescr.]; Meyr., Rev. Handb., p. 484, fig. [neur. *tesserana*] (1928) [redescr.]

Dapsilia, Hb., Verz. p. 394 (1826); Stainton, Manual II 277 (1859); Wilk., Brit. Tortric., pp. 317-318 (1859) [redescr.]; Barrett, Brit. Lep. X 338 (1908) [redescr.] — Type: *rutilana,* Hb. [Europe]

Cochylis, Treitschke, Schmett. Eur. VIII 272 (1830); Curtis., Brit. Entom. XI 491 (1. III. 1834) [redescr.; type cited as *rubellana*]; Steph., Ill. Brit. Ent., Haust. IV 185 (30. IX. 1834) [redescr.]; Duponchel, Ann. S. E. France III 448 (1834) [redescr.]; Westwood, Synops. Brit. Ins., p. 109 (1840); Westw., Brit. Moths II 175 (1845) [redescr.]; Guenée, Ann. S. E. France (2) III 303 (1845); Gn., Eur. Micr. Index, p. 65 (1846); Duponchel, Cat. Méth. Lep. Eur., p. 308 (1846) [redescr.]; Herrich-Schäffer, Schmett. Eur. IV 181 (1851) [redescr.]; Pierce and Metcalfe, Genit. Brit. Tortr., p. 36 (1922).

Conchylis, Treits., Schmett. Eur. VII 233 (1829) [non-descr.]; Sodoffsky, Bull. Mosc. X. No. 6, p. 93 (1837) [more correct reading of *Cochylis*']; Zett., Ins. Lapp., p. 990 (1840) [redescr.]; Led., Wien. Ent. Monats. III 273-274, t. 1 f. 12 (1859) [redescr.]; Hein., Kleinschmett. Deutschl. I. i. 68-69 (1863) [redescr.]; Zell., Verh. z.-b. Ges. Wien. XXV 242 (1875) [remarks on N. Amer. spp.]; Snell., Vlind. Ned. Microlep. pp. 232-233, ff. 84, 89 (1882) [redescr.]; Rebel, Cat. Pal. Lep. II 94 (1901). — Type: *roseana,* Hw.] *rubellana,* Hb. [Europe]

Stenodes, Guenée, Ann. S. E. France (2) III 300 (1845) [non-descr.]; Gn., Eur. Micr. Index, p. 62 (1846) [non-descr.] [Invalid, as never described; *elongana,* F. R., is the sole species included.]

Chrosis, Stainton, Manual II 268 (III. 1859); Wilk., Brit. Tortr., p. 285 (V. 1859) [redescr.; type fixed]; Guenée, Ann. S. E. France (2) III 300 (1845) [non-descr.]; Gn., Eur. Micr. Index, p. 62 (1846) [non-descr.]. — Type: *tesserana,* Schiff. [Europe]

Argyridia, Stainton, Manual II 277 (5. III. 1859); Wilk., Brit. Tortric., pp. 318-319 (1. V. 1859) [redescr.]; Steph., List Brit. Anim. B. M. X 83 (1852) [non-descr.]. — Type: *dipoltella* Hb. [Europe]

Thyralia, Walsingham, P. Z. S. 1897. 138-139 (1897).	Type: *bunteana*, Robinson [N. America: W. Indies]
Aethes, Pierce, Genit. Brit. Tortric., p. 32 (1922); Billberg, Enum. Ins. (1820) [non-descr.]; Zett., Ins. Lapp., p. 990 (1840) [non-descr.; sunk as syn. of *Conchylis*, Tr.]	Type: *smeathmanniana*, Fb. [Europe]
Diceratura, Djakonov, Rev. Russe Ent. XXIII 155-156, ff. 6-18 (1929).	Type: *purpuratana*, H. S. [S. Europe: N. Syria]

definita, Meyrick 1928

Phalonia definita, Meyr., Exot. Micr. III 438 (1928) ♀[1]	Shillong[1]

geniculata, Meyrick 1930

Phalonia geniculata, Meyr., Exot. Micr. III 5 91 (1930) ♂[1]	Shillong[1]

glycitis, Meyrick 1928

Phalonia glycitis, Meyr., Exot. Micr. III 438 (1928) ♂[1]	Khasis[1]

mellita, Meyrick 1916

Phalonia mellita, Meyr., E. M. M. LII 277-278 (1916) ♂ ♀[1]	Ceylon Puttalam[1] Trincomali[1] Madulsima[1]
Phalonia manniana [nec F. R.], Meyr., Proc. Linn. Soc. N. S. W. XXXVI 297 (1911)[2]	India N. Coorg.[1] Ganges Delta[1] Khulna distr[2] Chapra (Bihar) Pusa ——— N. Australia Port Darwin[1] Tonkin.

thermoconis, Meyrick 1925

Phalonia thermoconis, Meyr., Exot. Micr. III 139 (1925) ♂ ♀[1]	Kashmir, Srinagar[1] Kumaon Muktesar (7,000 ft.)[1]

2. Clysia, Hübner 1826

Clysia, Hb., Verz. Schmett., p. 409 (1826); Meyr., Handb., p. 556 (1895) [redescr.]; Kennel, Spuler's	Type: *ambiguella*, Hb. [Europe: Asia Minor]

Schmett. Eur. II 255 (1907) [redescr.]; Fernald, Genera Tortric., pp. 18, 58 (1908) [type fixation]; Kennel, Pal. Tortric., p. 240 (1913) [redescr.]; Walsm., Biol., Centr. Am., Het. IV 298 (1914); Pierce, Genit. Brit. Tortric., p. 26 (1922); Meyr., Rev. Handb., p. 494 (1928) [redescr.]

charixantha, Meyrick 1928

Clysia charixantha, Meyr., Exot. Micr. III 435 (1928) ♂ ♀[1]	Ceylon, Maskeli[1] Madulsima[1] India Dibidi[1] (Coorg; 3,000 ft.)

eucalypta, Meyrick 1928

Clysia eucalypta, Meyr., Exot. Micr. III 436 (1928) ♀[1]	Ceylon, Bentœa[1]

Larva, in case, on *Scaevola koenigii*[1]

turbinaris, Meyrick 1928

Clysia turbinaris, Meyr., Exot. Micr. III 435 (1928) ♂ ♀[1]	Assam, Khasis[1] Cherrapunji[2] Shillong
Clysia ambiguella [nec Hb.], Wlsm., A. M. N. H. (7) V 488 (1900)[2]; Fletcher, Ind. Agric Ent. Mem. VI 34 (1921)[3]	Naga Hills[2] Mao[2] (N. Manipur: 5,000-7,000 ft.) Burma Pegu: Karen Hills[2]

3. Euxanthis, Hübner 1826

Euxanthis, Hb., Verz. Schmett., p. 391 (1826); Meyr., Handb., p. 557, fig. [neur. *angustana*] (1895) [redescr.]; Rebel, Cat. Pal. Lep. II 100 (1901); Kennel, Spuler's Schmett. Eur. II 260 (1907) [redescr.]; Fernald, Genera Tortric., p. 16 (1908) [type fixation]; Kennel, Pal. Tortric., p. 306 (1913) [redescr.]; Meyr., Rev. Handb., p. 496, fig. [neur. *angustana*] (1928) [redescr.]	Type: *hamana,* L. [Europe]
Pharmacis [nec Hb. 1822 - Hepialidae], Hb., Zütr. Exot. Schmett. II 10 (1823); Fernald, U. S. Nat. Mus. Bull. 52, p. 488 (1903); Busck, Jl. N. Y. Ent. Soc. XV 29 (1907) [redescr.]; Walsm., P. Z. S. 1907. 992 (1908); Pierce, Genit. Brit. Tortric., p. 30 (1922); Forbes, Lep. New York, p. 502 (1924) [redescr.]	Type: *sartana,* Hb. [U. S. America]

Commophila, Hb., Verz. Schmett., p. 392 (1826); Fernald, U. S. Nat. Mus. Bull. 52, p. 488 (1903); Busck, Jl. N. Y. Ent Soc. XV 30 (1907) [redescr.]; Walsm., Biol. C. Am., Het. IV 297 (1917), p. 458 (1915); Pierce, Genit. Brit. Tortric., p. 30 (1922); Forbes, Lep. N. York, p. 503 (1924) [redescr.] — Type: *aeneana,* Hb. [Europe]

Eugnosta, Hb., Verz. Schmett., p. 394 (1826) — Type: *lathoniana,* Hb. [Europe; Asia Minor]

Argyrolepia, Stephens, Ill. Brit. Ent. Haust. IV 175-176 (1834); Steph., Cat. Brit. Ins. II 190 (1829) [non-descr.]; Duponchel, Ann. S. E. France III 448 (1834) [redescr.]; Westw., Synops. Brit. Ins., p. 109 (1840) [type fixation]; Westw., Brit. Moths II 169 (1845) [part.]; Guenéee, Ann. S. E. Fr. (2) III 301 (1845); Gn., Eur. Micr. Index, p. 63 (1846); Dup., Cat. Méth. Lep. Eur., p. 309 (1846) [redescr.]; Stainton, Manual II 268 (1859) [redescr.]; Wilk., Brit. Tortric. p. 287 (1859) [redescr.]; Hein., Kleinschmett. Deutschl. I. i. 75 (1863) [redescr.]; Barrett, Lep. Brit. Isds. X 326 (1905) [redescr.; part.] — Type: *lathoniana,* Hb.

Eupoecilia, Stephens, Ill. Brit. Ent., Haust. IV 181 (1834); Steph., Cat. Brit. Ins. II 190 (1829) [non-descr.]; Westw., Synops. Brit. Ins., p. 109 (1840) [type fixation]; Guenée, Ann. S. E. Fr. (2) III 298 (1845); Gn., Eur. Micr. Index, p. 60 (1846); Walsm., A. M. N. H. (7) VI 446 (1900); Barrett, Lep. Brit. Isds. X 280 (1905) [redescr.; part.] — Type: *angustana,* Hb. [Europe]

Xanthosetia, Stephens, Ill. Brit. Ent., Haust. IV 190-191 (1834); Steph., Cat. Brit. Ins. II 192 (1829) [non-descr.]; Dup., Ann. S. E. Fr. III 447-448 (1834) [redescr.]; Westw., Synops. Brit. Ins., p. 109 (1840); Westw., Brit. Moths II 178 (1845) [redescr.] Guenée, Ann. S. E. Fr. (2) III 300 (1845); Gn., Eur. Micr. Index, p. 62 (1846); Dup., Cat. Méth. Lep. Eur., p. 289 (1846) [redescr.]; Herrich-Schäffer, Schmett. Eur. IV 175 (1851) [redescr.]; Stainton, Manual II 276 (1859) [redescr.]; Wilk., Brit. Tortric., p. 315 (1859) [redescr.]; Barrett, Lep. Brit. Isds. X 339-340 (1905) [redescr.] — Type: *hamanu,* L. [Europe]

Eupecillia, Herrich-Schäffer, Schmett. Eur IV 179 (1851) [Probably emendation of *Eupœcilia,* Steph.]. — Type: *lathoniana,* Hb.

Hypostromatia, Zeller, Stett. Ent. Stg. XXVII 141-142 (1866). — Type: *versicolorana,* Z. [Colombia]

Agapete, Pierce, Genit. Brit. Tortric., p. 31 (1922); Hb., Cat. Lep. Coll. Franck, p. 98 (1825) [non-descr.]; Walsm., Biol. Centr. Am., Het. IV 458 (1915) [non-descr.] — Type: *zoegana,* L. [Europe]

amphimnesta, Meyrick 1928

Euxanthis amphimnesta, Meyr., Exot. Micr. III 436-437 (1928). ♂ [1] — Bhim-Tal[1] (Kumaon; 5,000 ft.) Muktesar (Kumaon; 7,000 ft.)

innotatana, Warren 1888

Xanthosetia innotatana, Warren, P. Z. S. 1888. 337 (1888)[1]; Cotes and Swinhoe, Cat. Moths India, p. 697, No. 4754 (1889)[2] — Kala Pani.[1-3] (Punjab Hills) Dras Ladak[3] (Kashmir; 7,000 ft.)

Pharmacis innotatana, Walsm., A. M. N. H. (7) V. 488 (1900).[3]

10 species.

Species omitted.

Phalonia hybridella, Hübner 1811-1817

Tinea hybridella, Hb., Samml. Eur. Schmett., Tin., t. 51 f. 351 (1811-1817)[1] — Dharmsala[2]

C. & S. E. Europe
Asia Minor

Phalonia hybridella, Walsm., A. M. N. H. (7) V 486 (1900)[2]; Fletcher, Ind. Agric. Ent. Mem. VI 34 (1921)[3]

[It seems probable that Lord Walsingham's determination of the Dharmsala specimen (possibly in poor condition) was not correct, and *P. hybridella* is better omitted from the Indian List pending further evidence. T. B. F.].

Phalonia manniana, Fischer von Röslerstamm 1839

Cochylis manniana, F. R., Abb. Microlep., p. 134, t. 51 ff. 2 *a-b* (1839).

[*See* under *P. mellita,* Meyr.]

Clysia ambiguella, Hübner 1796

Tinca ambiguella, Hb. Samml. Eur. Schmett., Tin., t. 22 f. 153 (1796). ♀.

[*See* under *C. turbinaris,* Meyr.]

CHLIDANOTIDAE.

Chlidanotidae: Meyr., B. J. XVII 412 (1906); Meyr., Proc. Linn. Soc. N. S. W. XXXVI 293 (1911) ["Chilanotidae" - error typogr.].

1. Metrenis, Meyrick 1906

Metrernis, Meyr., B. J. XVII 414 (1906). — Type: *ochrolina,* Meyr.

ochrolina, Meyrick 1906

Metrernis ochrolina, Meyr., B. J. XVII 414 (1906). ♂ ♀[1] — Ceylon: Maskeliay.[1]

2. Electracma, Meyrick 1906

Electracma, Meyr., B. J. XVII 413 (1906) — Type: *hemichroa,* Meyr.

hemichroa, Meyrick 1906

Electracma hemichroa, Meyr., B. J. XVII 413-414 (1906). ♂ ♀[1] — Ceylon: Maskeliya.[1]

3. Chlidanota, Meyrick 1906

Chlidanota, Meyr. B. J. XVII 412-413 (1906). — Type: *thriambis.* Meyr.

thriambis, Meyrick 1906

Chlidanota thriambis, Meyr., B. J. XVII 413 (1906). ♀[1] — Ceylon.[1] [*e.r.* J. Pole Coll., without exact locality.]

4. Trymalitis, Meyrick 1905

Trymalitis, Meyr., B. J. XVI 590 (1905); Meyr., B. J. XVII 412 (1906); Meyr., Proc. Linn. Soc. N. S. W. XXXVI 294 (1911) [redescr.] — Type: *margarias,* Meyr.

cataracta, Meyrick 1907

Trymalitis cataracta, Meyr., B. J. XVIII 153 (1907). ♂[1] — Ceylon: Maskeliya.[1]

margarias, Meyrick 1905

Trymalitis margarias, Meyr., B. J. XVI 590 (1905). ♂[1]	Ceylon: Puttalam.[1] Colombo. Trincomali. Kandy.[1]

optima, Meyrick 1911

Trymalitis optima, Meyr., Proc. Linn. Soc. N. S. W. XXXVI 294-295 (1911). ♂ ♀[1]	Andamans: (Mt. Harriet; 1,200 ft.). —— Queensland: Brisbane.[1]

5. Archimaga, Meyrick 1905

Archimaga, Meyr., B. J. XVI 608 (1905)	Type: *pyractis,* Meyr.

euplocamis, Meyrick 1918

Archimaga euplocamis, Meyr., Exot. Micr. II 171-172 (1918). ♀[1]	Ceylon: Pattipola.[1]

philomima, Meyrick 1918

Archimaga philomima,, Meyr., Exot. Micr. II 171 (1918) ♂ ♀[1]	Assam: Khasis.[1] Cherrapunji.

pyractis, Meyrick 1905

Archimaga pyraetis, Meyr., B.J. XVI 609 (1905).	Ceylon: Maskeliya.[1]

3 Species.

PART XXIII - CHALCIDOIDEA*

INTRODUCTION

Among the various major groups of the parasitic Hymenoptera, the superfamily Chalcidoidea is the largest and economically the most important. It comprises several thousands of species contained in two thousand genera. Its members are characterized by elbowed antennae, pronotum not extending behind to the tegulae and two-jointed trochanters. Fore-wings have but few veins, are devoid of large stigma and closed cells. Abdominal sterna are hard, chitinised and devoid of folds. Ovipositor issues some distance before the anal extremity of the abdomen.

The species vary in size from about 0.2 mm. to 25 mm. They are generally sombre-coloured, sometimes black, red or yellow, and in some families brilliantly iridescent blue or green. The head is large and the thorax is compact, thus giving them great powers of flight. The legs are generally well developed; sometimes the coxa and femur are greatly swollen, serrate or dentate; or are provided with powerful, long, apical tibial spurs.

Except in one or two phytophagous groups, they are either parasites or hyperparasites of other insects. The orders of insects most commonly patasitized by them are the Lepidoptera, Diptera and Homoptera. The Coleoptera, Hymenoptera and Orthoptera are not, however, exempted, and numerous species attack beetles, Braconids, Ichneumonids, bees, Chalcids and cockroaches. Most of the insects parasitized by them are either directly or indirectly injurious to man. Only in a relatively small number of cases beneficial or useful insects, like Braconid parasites or lac-insects, are attacked. The phytophagous species are contained in the families Agaonidae, which are "fig insects", and the Torymidae and Eurytomidae, some of which are gall-formers, while others are pests of cultivated plants.

The study of Chalcids may be said to have begun with Linuaeus who described several species under his genera *Ichneumon, Cynips.* etc. Later, the names of Swederus, Fabricius, Dalman, Dahlbom, Thomson, Foörster, Mayr, Motschulsky, Haliday, Westwood, Walker, Saunders, Kirby, Riley, Howard, Ashmead and Cameron stand out prominent as students of this group. In 1904, Ashmead published his monumental work on the classification of the Chalcidoidea and in 1909, Schmiedexnecht monographed the group in the *Genera Insectorum.* Since then, our knowledge of the Chalcids has been greatly augmented by the studies of Howard, Ashmead, Gahan and Rohwer from America; Silvestri, Masi, Grandi, Mercet and Kryger from Europe; and Girault from Australia.

Very little work, either systematic or bionomic, has been done on these insects in India. In the last century, Saunders and Westwood described several fig-inhabiting Chalcids from Ceylon and Calcutta. Numerous other Ceylonese species were also described by Walker, Motschulsky and Howard. Kirby and Ashmead also described a few Indian forms. In 1898, Dalla Torre listed thirty species from India in his *Catalogus Hymenopterorum.* Several new forms were later added by Cameron, mainly from the North-Western India. In 1912, Cameron also described several species associated with the lac-insect in Dehra Dun. Dr. Ramakrishna Ayyar made extensive collections, which were described by Gahan, Rohwer, Weld and Ferriere. He also published, from time to time, annotated lists of Indian Chalcids. Recently Waterston and Ferriere have considerably increased our knowledge of this group in India. In more recent years Ramakrishna Ayyar and myself have described a few species of economic importance. Altogether about six hundred species, comprised in about one hundred and eighty genera, have so far been described from India; and considerably many more forms await discovery.

CLASSIFICATION

The superfamily Chalcidoidea is divided into fourteen families, falling into two natural groups. The single family Mymaridae constitutes one group and is characterized by the extremely long and pedunculated hind wings. The rest of the families have normal hind wings and form the second group. The families Elasmidae, Eulophidae and Trichogrammidae form a sub-group, recognised by the axillae advanced forwards into the region of the seapulae. The families Agaonidæ, Torymidæ, Chalcididæ, Eurytomidæ, Perilampidæ, Eucharidæ, Miscogasteridæ, Cleonymidæ, Encyrtidæ and Pteromalidæ are comprised in another sub-group distinguished by the normal axillæ.

The Agaonidae, first given the family rank by Walker in 1846, include some of the most peculiar-looking Chalcids. Members of this family are recognised by the oblong head, serrate palpi and mandibles, stout fore and hind legs and slender or aborted mid legs. The sexes differ very much, the males being almost always wingless. The species are entirely phytophagous and live in the "figs" of different Urticaceæ.

The family is divided into two well-marked subfamilies : Agaoninae and Sycophaginae. The family was extensively studied by Westwood,

Saunders, Gustav Mayr and Paul Mayer. In recent years our knowledge of the family has been very greatly augmented by the studies of Grandi. Over five hundred species are known from different parts of the world; and fourteen species comprised in seven genera are known from India.

The family Tormidae, first recognised by Walker in 1833, is a connecting link between the Agaonidae on the one hand and the Chalcididae on the other hand, but several of its species approach the former more than any other group. It is distinguished by the normal head, 3-4-dentate mandibles, axillae separate internally, by the large, subtriquetrous hind coxa and swollen and generally simple hind femur. Six well-defined subfamilies are recognised : Idarninae, Toryminae, Monodontomerinae, Podagrioninae, Megastigminae and Ormyrinae. The bulk of the species are parasites of gall insects, while others are phytophagous and attack seeds of plants. Members of the Idarninae are either inquilines or parasites in figs. The Podagrioninae are mostly parasites of mantid oothecae. Extensive studies of this family were made by Walker, Gustav Mayr, Forster and Ashmead. About thirty-two species, distributed over twenty-four genera, are known from India.

The family Chalcididae differs from the Torymidae in the elongated, subcylindrical hind coxae, greatly swollen, usually dentate or serrate hind femur and in the strongly curved hind tibiae. Two subfamilies, Leucospidinae and Chalcidinae, are recognised. The Chalcidinae comprise five tribes : Brachymerariae, Halticellariae, Chalcidariae (= *Smicrarice*) Chalcitellariae and Dirhinariae. Ashmead and Howard made extensive studies of the family. Schletterer in 1890 and Weld in 1922 have monographed the subfamily Leucospidinae. The species are mostly primary or secondary parasites of Lepidoptera; though several of the Leucospidinae are parasites of bees and wasps. About ninety species, comprised in about twenty-five genera, have been recorded so far from India.

The family Eurytomidae differs from the Torymidae and the Chalcididae mainly in the normal hind coxae. The family is divided into five tribes : Aximaciae, Isosomariae, Eurytomariae, Rileyarise and Decatomariae. Gustav Mayr, Ashmead and Hedicke have paid considerable attention to the family. The members comprising this family have diverse habits : some are phytophagous or gall-makers; others inhabit the nests of bees and wasps or are parasites of gall insects; while several of them parasitize the Coleoptera. Twenty-three species, comprised in six genera, are known from India.

The small family Perilampidae, first recognised by Forster in 1856, is related to the Eucharidae. It is easily recognised by the large thorax and small triangulated abdomen. The species are mainly parasitic on Lepidoptera. Smith has recently studied the biology of *Perilampus.* Six species distributed over three genera have been recorded so far from India.

The family Eucharidae include very singular-looking species, often brilliantly coloured iridescent green. The scutellum is generally greatly produced behind or armed with long forked spinous processes. The species are parasitic on ants. Kirby monographed the family in 1887 and more recently Ruschka has revised the European forms of the family. Eight species, comprised in three genera, have been recorded so far from India.

The family Miscogasteridae is closely related to the Pteromalidae from which it is easily distinguished by the two apical tibial spurs of the hind legs. Five distinct subfamilies are recognised : Pireninae, Tridyminae, Miscogaterinae, Lelapinae and Colotrechninae. Several of the genera originally included in the Pteromalidae have since been transferred to this family. Numerous species have been described by Girault, and a few by Waterston and Masi. Seven species, comprised in four genera, are known from India.

Cleonymidae, related to the Encyrtidae, comprise three subfamilies. Chalcedectinae, Cleonyminae and Pelecinellinae. In this family are included some of the largest and the most striking of all Chalcids. The species described so far have mainly been bred from Coleoptera. No member of the family has yet been recorded from India.

The very large family Encyrtidae is recognised by the greatly elongated apical tibial spur of the mid leg and absence of a femoral impression on the mesopleura. Three subfamilies, Eupelminae, Encyrtinae and Signiphorinae, are recognised. The family was monographed by Ashmead in 1900 and 1904. In recent years Howard, Girault and Mercet have studied the American, Australian and the European forms respectively. Smith and Compere have made special studies of the coccidiphagous forms. The species parasitize all orders of insects, though the Homoptera and the Lepidoptera are more generally attacked. Certain genera show marked restricted specificity of hosts. Fifty-seven species, comprised in twenty-eight genera, are known from India.

The largest family, Pteromalidae, is distinguished from the Miscogasteridae by the single apical tibial spur of hind legs. It is divided into six subfamilies : Pteromalinae, Merisinae, Eunotinae, Sphegigasterinae, Spalanginae and Diparinae. Ashmead and Rubsaamen, and in recent years Giraut, have extensively studied the family. Masi has recently monographed the subfamily Eunotinae. The species attack almost all orders of insects. Twenty-four species, comprised in seventeen genera, are known from India.

The Elasmidae are a small family resembling the Eulophidae but distinguished by the enlarged hind coxae, compressed femur and differently shaped abdomen. The species mostly attack Lepidopterous insects, though several are known to parasitize Braconids. Ferriere has recently described several Indian forms of this family. The twelve Indian species of this family are comprised in the single genus *Elasmus*.

The Eulophidae are a very large family, mostly consisting of small species recognised by the normal hind coxae and by thoracic and venational characters. Five subfamilies are recognised : Entedoninae, Aphelininae, Tetrastichinae, Elachertinae and Eulophinae. The species parasitize almost all orders of insects. The Aphelininae are mostly parasitic on Aphids and Diaspine Coccids. Howard, Smith and Compere have extensively studied the family. The sixty-six species, known from India, are comprised in twenty-eight genera.

The Trichogrammidae, first recognised by Forster in 1856, are readily recognised by the trimercus tarsi. Two subfamilies, Oligositinae and Trichogramminae, are recognised. The family has been very extensively studied by Kurdinmov, Kryger, Girault and others, while in India relatively little is known about them. Over one hundred species have been described from different parts of the world but only seven species comprised in five genera are known so far from India The species are among the minutest of Chalcids and all are egg parasites, generally of the Lepidoptera.

The Mymaridae, first recognised by Haliday in 1833, are recognised by their long, petiolated hind wings. There are two subfamilies : Gonatocerinae and Mymarinae. The species are very minute like those of the Trichogrammidae and parasitize the eggs of various insects. One or two species of the family are aquatic and are capable of swimming under the water by means of their wings. Girault and Dozier have studied the family in recent years. Four species contained in four genera have been recorded so far from India.

KEY TO FAMILIES, SUBFAMILIES AND TRIBES*

1. Hind wing exceedingly narrow and penduncculate; ovipositor issuing just anterior to the tip of abdomen; antennae without ring joints; scape small, short and compressed.	17 (p. 24)
Hind wing neither exceedingly narrow nor pedunculate; ovipositor issuing far anterior to the tip of abdomen; antennae elbowed, with one, two or three ring joints; scape large and long.	2.
2. Axillae triangularly produced into the region of the scapulae; their anterior margins on or in advance	15 (p. 21)

* This key is adapted from Ashmead (1904) and Schmiedeknech (1909). In this form it has been found most useful by the writer in his work. The couplets dealing with families are numbered in Arabic numerals and those of subfamilies and tribes, as A, B, C and *a, b, c* respectively. Thus, although inserted directly into the family keys, the subfamily keys remain entirely independent so far as the couplet numbers are concerned. The subfamily key for Eulophidæ is more or less modified from Girault (*Mem. Queensland Mus.*, II, pp. 294-295, 1913).

	of an imaginary line drawn from tegula to tegula; anterior tibial spurs generally small; tarsi three to four jointed, rarely five-jointed or hete romerous.	
	Axillae not triangularly produced into the region of the scapulae, their anterior margins always behind an imaginary line drawn from tegula to tegula; anterior tibial spurs large; tarsi five-jointed, rarely three- or four-jointed in apterous males.	3.
3.	Head in female oblong, with a deep, broad, longitudinal furrow above; palpi or mandibles most frequently with serrate appendages; fore and hind legs very stout, their tibiae very much shorter than their femora; mind legs very slender, or sometimes aborted; hypopygium very prominent; males always apterous, with the abdomen long and tubular; inhabitants of the figs.	Family I, AGAONIDÆ
A.	Mandibles with an appendage, abdomen narrowed towards apex.	Subfamily I. *AGAONINÆ*
B.	Mandibles without an appendage, palpi sometimes with an appendage; abdomen broadened towards the apex.	Subfamily II. *SYCOPHAGINÆ*
	Head rarely oblong, generally of a different shape; never with a deep, broad, longitudinal furrow above; palpi and mandibles without a serrate appendage; fore and hind legs often stout but their tibiae are never shorter than their femora; mind legs not especially slender; hypopygium rarely prominent; males only rarely apterous and if apterous the abdomen is not tubular.	4
4.	Mesopleura large, entire, without a femoral furrow or impression, except occasionally in some males; mesepisternum large, triangular, not extending to the base of the fore coxae; tibial spur of mid leg long, stout or dilated at base.	13 (p. 16)
	Mesopleura always with a femoral impression or furrow; mesepister-num never very large, except in some *Cleonymidae*, and extending to the base of the fore coxae; tibial spur of the mid legs normal.	5
5.	Hind tibiae with *two* apical spurs; rarely with one only, in which case the radius has a large rounded stigma, ovipositor long.	6

Hind tibiae with only *one* apical spur; ovipositor rarely long, if long the stigma is small.	14 (p. 18)
6. Mandibles falcate, with one to two teeth within; thorax gibbous; scutellum often large, abnormally developed, elevated and produced posteriorly; axillae connate or indistinguishable from the surrounding area.	11 (p. 13)
Mandibles not falcate, usually with three to four teeth within, only rarely falcate; thorax not gibbous; scutellum normal; axillae separate.	7
7. Hind coxae rarely very much larger than the fore coxae, generally smaller or at the most equal; if larger the pronotum is elongate, mesepisternum large, hind legs long, post marginal vein long; ovipositor rarely prominent.	9 (p. 12)
Hind coxae much larger than the fore coxae.	8
8. Hind coxae subtriquetrous, at least compressed into a sharp ridge above; hind femur rarely very much swollen, generally simple, occasionally with one or two teeth; hind tibiae STRaight; abdomen subcompressed, rarely depressed; ovipositor not exseerted; abdomen conical with a peculiar sculpture; stigmal vein very short.	Family II. TORYMIDÆ
A. Parapsidal furrows not well defined; abdomen conically pointed or produced in female, oblong in male, peculiary sculptured in both sexes; ovipositor not exserted.	Subfamily I. *ORMYRINÆ*
Parapsidal furrows well defined; abdomen without the peculiar sculpture; ovipositor always prominen-tly exserted and often long.	B
B. Hind tibiae with on apical spur; radius with a large, rounded stigma, the basal nervure distinct and STRaight; post marginal vein as long as or longer than the marginal vein.	Subfamily II. *MEGASTIGMINÆ*
Hind tibiae with two apical spurs; radius short.	C.
C. Hind femur neither serrated nor armed with teeth or spines; posterior margin of the mesepisternum incised beyond the middle.	D.

Hind femur more or less swollen, either serrate or armed with teeth beneath.	E.
D. Stigmal vein long; abdomen usually depressed, if sub-compressed the hypopygium is large and prominent; males frequently apterous, head oblong with a triangular fovea in which the antennae lie.	Subfamily III. *IDARNINÆ*
Stigmal vein very short; abdomen compressed, hypopygium not prominent; males winged, if apterous head not oblong and without the triangular fovea.	Subfamily IV. *TORYAMINÆ*
E. Hind femur not much swollen, as long as the tibia, feebly serrate beneath or armed with one or sometimes two teeth beyond the middle towards the apex, tibiae STRaight; radius short but distinctly longer than in the Toryminae.	Subfamily V. *MONODONTO-MERINÆ*
Hind femur much swollen, with four to eight teeth beneath; hind tibia STRongly arched; radius very short, the knob almost sessile.	Subfamily VI. *PODAGRIONINÆ*
Hind coxae usually elongated, subcylindrical; hind femur very much swollen, serrate or dentate; hind tibiae STRongly arched; abdomen generally conical, rarely globose or oblong-ovate; ovipositor rarely prominent; radius rarely short.	Family III. CHALCIDIDÆ
A. Fore wings longitudinally folded; ovipositor curving upwards and backwards over the dorsum of abdomen.	Subfamily I. *LEUCOSPIDINÆ*
B. Fore wings not longitudinally folded; ovipositor when prominent STRaight and not curving back over the abdomen.	Subfamily II. *CHALCIDINÆ*
(a) Abdomen sessile.	b.
Abdomen petiolate.	d.
(b) Post marginal vein distinctly developed, stigmal vein rarely very short.	c.
Post marginal vein wanting or only slightly developed, stigmal vein very	e.

short or absent; antennae inserted close to the mouth border, always below an imaginary line from the lower borders of the eyes.

(c) Antennae inserted in the middle of the face, at least above an imaginary line from the lower borders of the eyes. — Tribe 1. BRACHYMERARIÆ (=Chalcidariæ)*

Antennae inserted close to the mouth border, at least below an imaginary line drawn from the lower borders of the eyes. — Tribe 4. HALTICELLARIÆ

(d) Antennae inserted in the middle of the face; post marginal vein very long; ovipositor if prominent not very slender; eighth segment of abdomen often produced into a long, compressed style. — Tribe 2. CHALCIDARIÆ (=Smicrariae)*

Antenmae inserted near the mouth border. — *e.*

(e) Head not cornuted. — Tribe 3. CHALCITELLARIÆ

Head cornuted. — Tribe 5. DIRHINARIÆ

9. Pronotum rarely transverse quadrate, either conical or conically produced anteriorly, or very short, transverse liners and very much narrowed medially, rarely as wide as the mesonotum. — 12. (p. 13)

Pronotum large, quadrate, at least as wide as the mesonotum. — 10.

10. Pronotum quadrate or subquadrate; abdomen in female not triangular, second dorsal segment never very large, hypopygium plough-share-shaped and prominent. — Family IV. EURYTOMIDÆ

A. Metathorax viewed from above longer than the scutellum, gradually sloping and — B.

* This change in the tribe names is rendered necessary by the fact that the genus *Chalcis* of Fabricius is no louger contained in the Chalcidariæ of authors; it is used in the sense of *Smicra* of authors. The *chalcis* of authors (not of Fabricus) is a synonym of *Brachymeria.* Also *vide* foot-notes under the genera *Brachymeria* and *Chalcis.*

	not abruptly declivous (in one case quadrate or squarely truncate); marginal vein at least twice as long as the stigmal.	
	Metathorax viewed from above shorter than the scutellum, abruptly declivous; marginal vein sometimes short and stout.	C.
B.	Head broader than the thorax, cornuted; eyes rounded.	Tribe 1. AXIMARIÆ
	Head not broader than the thorax, not cornuted; eyes oval.	Tribe 2. ISOSOMARIÆ
C.	Marginal vein short, stout, stigmated; antennae similarly segmented in both sexes, segments 11; abdomen short, round or oval; wing with a dusky spot in the midmost.	Tribe 5. DECATOMARIÆ
	Marginal vein more or less long and not as described above.	D.
D.	Antennae with ten to twelve segments and only *one* ring joint, segments dissimilar in the sexes, in female filiform or subclavate, in male excised or petiolated at apex, with whorls of long hairs or with sparse long hairs.	Tribe 3. EURYTOMARIÆ
	Antennae with 13 segments and two or three ring joints, segments similar in both sexes, neither petiolate nor with whorls of hairs.	Tribe 4. RILEYARIÆ.
	Pronotum short, as wide as mesonotum; abdomen in female generally triangular; second and third abdominal segments occupying most of the dorsal surface; hypopygium not prominent.	Family V. PERILAMPIDÆ
11.	Second abdominal segment very large and often enclosing the rest; coxae globose and nearly equal; stigmal vein scarcely developed; scutelluin often produced backwards into a spine or a forked process.	Family VI. EUCHARIDÆ
12.	Mesepisternum not large triangular; fore femur never much swollen; marginal vein in hind wings usually long, the costal cell not reaching the hooklet of spines and frequently narrow; stigmal vein well developed.	Family VII. MISCO-GASTERIDÆ
A.	Metathorax at apex produced beyond the insertion of the hind coxae; abdomen	Subfamily VI. *LELAPINÆ*

	petiolate or subpetiolate; antennae with 13-14 segments, subclavate, inserted below the middle of the face; marginal vein in the fore wings very long; second abdominal segment much lengthened; ovipositor exserted.	
	Metathorax normal.	B.
B.	Abdomen petiolate, if subsessile it is elongate and STRongly carinate beneath; antennae with 12-13 segments; marginal vein shorter than submarginal; costal cell normal.	Subfamily III. *MISCOGASTERINÆ.*
	(*a*) parapsidal furrows incomplete, clypeus transverse.	Tribe 1. HALTICOPTERARIÆ
	(*b*) Parapsidal furrows complete, clypeus not transverse.	Tribe 2. MISCOGASTERARIÆ
	Abdomen sessile or subsessile.	C.
C.	Antennae with 8-11 segments, inserted close to the mouth border.	Subfamily I. *PIPENINÆ*
	Antennae with 12-13 segments, inserted well above the mouth border, rarely below.	D.
D.	Mesonotum with complete or incomplete parapsidal furrows.	Subfamily II *TRIDYMTNÆ.*
	(*a*) Parapsidal furrows complete.	Tribe 1. TRIDYMARIÆ
	(*b*) Parapsidal furrows incomplete.	Tribe 2. METASTINARIÆ
	Mesonotum without parapsidal furrows.	Subfamily V. *COLOTRECHNINÆ*
	Mesepisternum large, triangular, either the anterior or the posterior femur much swollen, sometimes toothed or the posterior femur swollen and toothed; if with slender legs, the hind ones very long, stigmal vein in fore wings very short, post marginal vein extends to the apex of the wing.	Family VIII. CLEONYMIDÆ
A.	Abdomen with a long petiole.	C.
	Abdomen sessile or subpetiolate.	B.

B. Hind femur much swollen and toothed as in the genera *Brachymeria* and *Chalcis**; abdomen usually depressed; ovipositor rarely exserted; antennae at the most with 11 segments.	Subfamily II. *CHAL-CEDECTINÆ*
Hind femur neither much swollen nor toothed; fore femur swollen, excised or dentate beneath towards the apex; ovipositor often exserted; antenmae with 11-13 segments.	Subfamily III. *CLEO- NYMINÆ*
C. Abdomen STRongly compressed; ovipositor long; marginal and post marginal veins very long, the latter extending to the tip of the wing; stigmal vein very small stigma subsessile; antennae with 11 segments, inserted close to the mouth border; hind legs long, their coxae long and cylindrical.	Subfamily I. *PELE-CINELLINÆ.*
13. Mesonotum either depressed, with more or less distinct parapsidal furrows or convex and entirely without parapsidal furrows, rarely convex with distinct parapsidal furrows; axillae most frequently meeting at their inner basal angles, rarely separated by a wide gap.	Family IX. ENCYRTIDÆ
A. Mesonotum entire, convex or subconvex; parapsidal furrows entirely absent.	B.
Mesonotum entire, usually depressed or impressed, rarely convex, parapsidal furrows distinct, at least never entirely wanting; marginal vein usually long.	Subfamily I. *EUPELMINÆ*
(a) Mesonotum in female always depressed or impressed, concave or subconcave, medially usually with a triangular elevation anteriorly; parapsidal furrows not or rarely sharply defined, never short; mesonotum in male subconvex, with the furrows rarely complete, neither short nor curved off laterally.	Tribe 1. *EUPEL-MARIÆ*

* *Chalcis.* Fabr. = *Smicra* spinola. *Brachymeria* is the *Chalcis* of authors. *Vide* foot-note under *Chalcis.*

(b)	Mesonotum convex in both the sexes; parapsidal furrows delicate but always complete, short and STRongly curved off laterally.	Tribe 2. TANAOST-IGMABLÆ.

B.	Marginal vein rarely very long, often punctiform and always much shorter than the submarginal; stigmal vein usually short, rarely long; scutellum never short or transversely linear; middle tibiae without lateral spurs.	Subfamily II. *ENCYTRINÆ*

(a)	Mandibles 2-3-dentate or broadly truncate apically, never acute; tarsi always five-jointed.	b.
	Mandibles without teeth, acute or truncate.	e.
(b)	Mandibles bidentate or truncate.	c.
	Mandibles tridentate, teeth sometimes minute, labrum usually conspicuous.	d.
(c)	Mandibles not stout or only rarely moderately so, rather long and always bidentate at apex; hypopygium in female large, plough-share-shaped, enclos-ing ovipositor and extending beyond the anus; second abdominal segment most frequently large; antennae usually inserted just above the clypeus, rarely just below :	
	Mandibles short, stout, polished, broadly truncate apically, unarmed; hypopy-gium; not very prominent; body stout, non-metalic; hind tibiae with two apical spurs; antennae inserted below the middle of the face.	Tribe 2. ENCYRTARIÆ
(d)	Hind tibiae with one apical spur; hypopygium not prominent; body most frequently metallic.	Tribe 3. MIRINARIÆ
(e)	Tarsi 4-5-jointed; marginal vein punctiform or subobsolete.	Tribe 4. ARRHEN-OPHAGARIÆ

Marginal vein long, as long as the submarginal; scutellum very short, transverse-linear; middle tibiae with lateral spurs, the apical spur lobed.	Subfamily III. *SIGNIPHORINÆ*
14. Mesonotum subconvex, parapsidal furrows complete or incomplete; hind coxae rarely much larger than the fore coxae; axillae separated; mesepisternum small; hind wings with long marginal veins.	Family X. PTEROMALIDÆ
A. Abdomen sessile or nearly so.	B.
Abdomen distinctly petiolate.	D.
B. Head very wide, much wider than thorax; antennae with 6-10 segments, inserted near the mouth border; scutellum large; hind angles of metanotum acute.	Subfamily I. *EUNOTINÆ*
Head not specially wide; antennae with 12-13 segments; scutellum normal.	C.
C. Metanotum with spiracular sulcilateral folds also present, though sometimes incomplete; median carina usually more or less distinct, antennae with 12-13 segments.	Subfamily II. *PTEROMALINÆ.*
(a) Both mandibles tridentate.	Tribe 1. METOPONARIÆ.
Both mandibles not tridentate.	b.
(b) Both mandibles 4-dentate.	Tribe 2. PTEROMALARIÆ.
The right mandible 4- and the left mandible 3-denate.	c.
(c) Antennae inserted on or near the middle of the face.	Tribe 3. RHAPHITELARIÆ.
Antennae inserted far below the middle of the face.	Tribe 4. EUTELARIÆ.
Metanotum without spiracular sulci, lateral folds absent.	Subfamily III. *MERISINÆ.*
(a) Metanotum with the median carina more or less distinct or at least indicated basally; parapsidal furrows incomplete.	Tribe 1. ROPTROCERARIÆ.

	Metanotum without a trace of the median carina.	*b.*
(b)	Parapsidal furrows incomplete.	Tribe 2. MERISINARIÆ.
	Parapsidal furrows complete.	Tribe 3. ISOPLATI-NARIÆ.
D.	Marginal vein very long, stigmal vein short, stigma minute; antennae inserted in the middle of the face.	Subfamily IV. *DIPARINÆ.*
	Marginal vein not very long, often short and thick; costal cells rarely very narrow; head oblong; sometimes apterous.	E.
E.	Head viewed in front short and rounded, the occipital line complete; antennae 9-13-jointed.	Subfamily V. *SPHEGI GASTERINÆ.*
(a)	Antennae inserted close to the mouth border; head usually lenticular, much wider than thorax, deeply concave behind, rarely normal; parapsidal furrows complete.	Tribe 1. ASAPHESARIÆ.
	Antennae inserted near the middle of face.	b.
(b)	Fore wings with a thick, usually short marginal vein.	Tribe 2. PACHY-NEURARIÆ.
	Fore wings with a slender marginal vein.	c.
(c)	Head transverse.	Tribe 3. SPHEGI-GASTERARIÆ.
	Head very large, frons sometimes cornuted.	Tribe 4. CERATO-MARIÆ.
	Head viewed in front oblong, occipital line complete, antanae 8-12-jointed.	Subfamily VI. *SPALANGINÆ.*
15.	Hind coxae normal, mesopleura impressed.	16.
	Hind coxae abnormally large, dilated; hind femur flat, compressed; tarsi very long; mesopleura entire; marginal vein in fore wings most frequently extraordinarily lengthened; stigmal vein short, scarcely dilated; mesonotum without parapsidal furrows.	Family XI. ELASMIDÆ

16. Tarsi 4- 5-jointed, rarely heteromerous; fore wings not short and broad, pubescence normal; marginal and stigmal veins normal; post marginal vein often wanting; mesonotum with complete or incomplete parapsidal furrows.	Family XII. EULOPHIDÆ.
A. Submarginal vein of fore wing usually entire, not distinctly broken distad just before curving up to the marginal or not with all of the long basal portion abruptly more slender than the short curved apical portion; usually as long as or longer than the marginal; stigmal vein rarely very short or sessile; the post marginal vein always present, usually as long as or longer than the stigmal.	B.
Submarginal vein of fore wings broken distad just before curving up to the marginal or the long basal portion abruptly more slender than the short, curved apical portion; usually distinctly shorter than the marginal, frequently much shorter; stigmal vein usually of moderate length, frequently short, rarely long and slender; post marginal vein most frequently absent or only slightly developed, rarely as long as or longer than the stigmal.	C.
B. Parapsidal furrows distinct and well developed.	Subfamily I. *ELACHERTINÆ.*
(a) Hind tibiae with only *one* apical spur, which is *rarely* very long.	Tribe 1. ELACHERTARIÆ.
Hind tibiae with *two* apical spurs.	b.
(b) Hind tibial spurs very long, pronotum anteriorly acute.	Tribe 2. EUPLECTRARIÆ.
Hind tibial spurs not very long, pronotum anteriorly rounded.	Tribe 3. OPHELINARIÆ.
Parapsidal furrows wanting or incomplete or at the most slightly indicated only anteriorly.	Subfamily II. *EULOPHINÆ.*
(a) Hind tibiae with two apical spurs.	Tribe 1. EULOPHARIÆ.
Hind tibiae with one apical spur.	Tribe 2. HEMIPTASENARIÆ.

C.	Submarginal vein very short; marginal vein very long; postmarginal vein variable, often very short or only slightly developed; metapleura very small; abdomen often petiolate, but sometimes sessile or subsessile.	Subfamily III. *ENTEDONINÆ.*
(a)	Tarsi 4-articulate in both sexes; antennae 10-articulate or less; funicle 5-articulate.	b.
	Tarsi 5-articulate in female; antennae with 11 or more segments, never 10 or less, funicle 6-articulate.	Tribe 1. TETRA-CAMPARIÆ.
(b)	Parapsidal furrows complete.	c.
	Parapsidal furrows incomplete, at the most indicated only anteriorly.	Tribe 4. PEDIOBARIÆ.
(c)	Abdomen sessile or subsessile, never distinctly petiolate.	Tribe 2. OMPHALARIÆ.
	Abdomen distinctly petiolate.	Tribe 3. ENTEDONARIÆ.
	Submarginal vein not very short, usually longer than the marginal; postmarginal vein always wanting; abdomen usually sessile, rarely petiolate.	D.
D.	Stigmal vein very short or nearly obsolete; stigma sessile or nearly so; mesopleura usually without a femoral impression.	Subfamily IV. *APHELININÆ.*
(a)	Tarsi 5-articulate.	Tribe 1. APHELINARIÆ.
(b)	Tarsi 4-articulate.	Tribe 2. PTEROPTRICARÆ
	Stigmal vein distinct, stigma never subsessile; mesopleura with a distinct femoral impression.	Subfamily V. *TETRASTICHINÆ.*
(a)	Abdomen distinctly petiolate; hind wings with a long clavate marginal vein.	Trible I. CERATO-NEURARIÆ
(b)	Abdomen sessile or subsessible; hind wings with a slender marginal vein.	Trible 2. TETRASTI-CHARIÆ.

Tarsi 3-jointed; fore wings short and broad, broadly rounded at apex, with the pubescence most frequently arranged in rows, more rarely normally pubescent; marginal and stigmal veins united to form a Ω-shaped Strongly curved line.	Family XIII. TRICHO-GRAMMIDÆ.
Wings without regular rows of hairs.	Subfamily I. *OLIGOSITINÆ.*
Wings with regular rows of hairs.	Subfamily II. *TRICHOGRAMMINÆ.*
17. Pronotum usually large, rounded, conically produced anteriorly; wings always with a long marginal fringe, nearly veinless and usually without the stigmal vein, marginal vein most frequently reduced to a mere dot; antennae in female most frequently terminating in a distinct fusiform or egg-shaped solid club, more rarely biarticulate club; tarsi 4- 5-jointed.	Family XIV. MYMARIDÆ.
A. Tarsi 5-articulate.	Subfimily I. *GONA-TOCERINÆ*
(a) Abdomen distinctly petiolate.	Tribe 1. OCTOÆRIE
B. Tarsi 4-articulate.	Subfamily II. *MYMARINÆ.*
(a) Abdomen distinctly petiolate.	Tribe 1. MYMARARIÆ.
(b) Abdomen sessile or subsessile.	Tribe 2. ANAPHESARIÆ.

SUPERFAMILY CHALCIDOIDEA

Linn., *Syst. Nat.*, (ed. 10), I p. 343, (1758). Dalm., *Kongl. Svensk. Vet. Akad. Handl.*, p. 132, (1820). Westw., *Intr. Mod. Class. Ins.*, II, p. 154, (1840). Walk., *List. Chalc. Brit. Mus.*, I, p. 1, (1846). New-port, *Trans. Linn. Soc. London.* pp. 61-77, (1855), {comparative anatomy and development]. Foörst., *Hymen. Stud.*, II, pp. 1-90, (1856). Thoms., *Hymen. Skandin*, IV, p. 3, (1875). How, *Amer, Nat.*, XVI, pp. 60-62, 149-151, (1882), [pupation]; *Ent. Amer.*, I, p. 197, (1886); *Trans. Amer. Ent. Soc.*, I, pp. 197-219; II, pp. 33-38, 97-101, (1887); *Proc. Ent. Soc. Washington*, I, pp. 65-74, (1888), [morphology]; *Ins. Life*, IV, pp. 193-196, (1891), [methods of pupation among the Chalcididae]; *Proc. U.S. N. Mus.*, (*Smith. Inst. Pub.*), XIV, pp. 567-588, (1892), [biology]. Ashm., *Proc. Ent. Soc. Washington*, IV, p. 243, (1897). Dalla

TORRE, *Wien. Ent. Zeit.*, XVI, pp. 83-88, (1897), [generic nomenclature]; *Cat. Hymen.*, V, pp. 1-431, (1898) [world list]. Marchal, *Ann. Soc. Ent. France*, LXIX, pp. 102-112, (1900) [biological notes]; ASHM., *Mem. Carnegie Mus.*, I, (4), pp. 228-365, (1904), [classification]. SilveSTR., Martel. & MASI, *Boll. Lab. Zool. Portici*, II, pp. 1-82 (1907), [habits of parasitic forms]. Vassiliew, *Zs. Wiss. Insektenbiol.*, III, p. 386, (1908), [parthenogenesis]. SCHMIED., *Gen. Ins.*, *fas.* 97, pp. I-550, pl. 8, (1909), [tables and generic descriptions]. Lefroy and Howlett, *Ind. Ins. Life*, pp. 172-177, (1909) [general account]. SilveSTR., *Boll. Lab. Zool. Portici*, X, pp. 66-68, pl. vi, (1915), [embryology]. Gatenby, *Quart. Journ. Micr. Sc.*, LXIII, pp. 175-196, pls. 2, (1918), [polyembryony]. SilveSTR., *Boll. Lab. Zool. Portici.*, XIII, pp. 88192, (1919), [biology]. RAMAKR., *Rep. Proc. Third Ent. Meet. Pusa*, III, pp. 931-936, (1920), [notes]; *Rep. Proc. fourth Ent. Meet. Pusa*, pp. 29-40, (1921), [notes]. Fahringer, *Zs. Wiss. Insektenbiol.*, XVI, pp. 228-235, (1921), [life-history notes]; *ibid.*, XVII, pp. 7-13, 41-47, (1922), [life-history notes.], [homonymy and synonymy]. GAHAN, *Proc. Ent. Soc. Washington*, XXIV, pp. 33-58, (1922), [list of phytophagous forms]. GAHAN and Fagan. *Bull. U.S.N. Mus.*, No. 124, pp. 1-173, (1923), [list of type, pp. 759-772, (1924) [general account, classification and literature]. Husain and Mathur, *Rep. Proc. Fifth Ent. Meet. Pusa*, pp. 119-121], (1924), [list of Chalcids of economic importance from the Punjab]. Parker, *Ann. Soc. Ent. France*, XCIII, pp. 261-379, pl. 38, (1924), [researches on post embryonic forms]. Parker and Thomson, *Ann. Soc. Ent. Amer.*, XVIII, pp. 384-395, pl. 3, (1925), [notes on larvae]. RAMAKR., *Spol. Zeyl.*, XIII, pp. 235-254, (1925), [lists of Indo-Ceylonese species]; *Bull. Ent. Res.*, XVIII, pp. 73-78, (1927), [annotated list of species of economic importance from South India]. GRANDI, *Boll. Lab. Ent. Bologna*, I, pp. 211-233, (1928), [catalogue of world species of Agaonidae]. Parker and Thomson, *Ann. Soc. Ent. France*, XCVII, pp. 425-465, (1928), [biology of entomophagous species]. RAMAKR., *Madras Agric. Journ.*, XX, (4), pp. 1-12, (in reprint), (1932), [general notes on economic importance]. BRUES and melander, *Class. Ins.*, pp. 482-488, 519-520, (1932), [tales of families]. Handl., *Handb. Zool.*, IV, pp. 893-1036, figs. 211, (1933) [general treatise]. RAMAKR. & MARGAB., *Journ. Bombay Nat. hist. Soc.*, XXXVII, pp. 193-196, (1934), [additional list of Indo-Ceylonese species]. IMMS, *Text Book Ent.*, pp. 573-579, (1934), [general acccunt]. GRANDI, *Boll. Lab. Ent. Bologna*, VII, pp. 214-240, (1934), [catalogue of world species of fig insects]. Mathur, R.N., *Ind. For. Rec.*, (N.S.), *Ent.*, I, (2), pp. 67-69, (1935), [Notes on forms parasitizing *Psyllids*].

FAMILY I. AGAONIDAE

WESTW.,*Trans. Ent. Soc. London*, p. 214, (1840). WALK., *List Chalc. Brit. Mus.*, I, p 23, (1846). FÖRST., *Hymen. Stud.*, II, p. 29, (1856). MOTSCH., *Bull. Soc. Imp. Nat. Moscou*, XXXVI, p. 47, (1863). WALK., *Notes Chalc.*, IV, p. 58, (1871). Saund., *Trans. Ent. Soc. London*, p. 20, (1882); *Ent. Month., Mag.*, XIX, p. 163, (1882). WESTW., *Trans. Ent. Soc. London*, pp. 47-60, (1882). Paul Mayer, *MT. Zool. St. Biol. Centralbl.*, II, pp. 545-550; *Bull. Ent. Ital.*, XIV, p. 242, (1882). WESTW., *Trans. Ent. Soc. London*, pp. 378-380, (1883). GUSTAV MAYR, *Verh. Zool.-Bot. Ges. Wien*,

XXXV, pp. 147-250, (1885). Ashm., *Proc Ent. Soc. Washington,* IV, p. 243, (1897). Dalla Torre, *Cat. Hymen.,* V, p. 315, (1898). Ashm., *Mem. Carnegie Mus.,* I, (4), pp. 231, 394, (1904). Schmied., *Gen., Ins., fas.* 97, pp. 5-12, (1909); Gir., *Mem. Queensland Mus.,* IV, p. 310, (1915). Grandi, *Boll. Lab. Zool. Portici,* X, pp. 121-296, (1915); XI, pp. 183-234, (1916); XV, pp. 183-234, (1921); XVI, pp. 1-57, (1922); *Ann. Mus. Civ. Stor. Nat. Genoa,* XLIX, pp. 304-316, (1922); *Bull. Ent. Res. London,* XIII, pp. 295-299, (1923); *Ann. Mus. Civ. Stor. Nat. Genoa,* LI, pp. 101-108, (1923); *Boll. Lab. Zool. Portici,* XVIII, pp. 3-31, (1924); *Att. Soc. Ital. Sc. Nat., Milan,* LXIII, pp. 288-311, (1925); *Boll. Lab. Zool. Portici,* XX, pp. 169-188, (1927); *Boll, Lab. Ent. Bologna,* I, pp. 107-235, (1928); *Mem, R. Accad. Bologna,* el. sc. fis., (8), VI, pp. 15-31, (1930); *Boll, Lab. Ent. Bologna,* VII, pp. 214-240, (1935).

SUBFAMILY I. AGAONINAE

Ashm., *Mem. Carnegie Mus.,* I, (4), p. 232, (1904), Schmied., *Gen. ins. fas.* 97, p. 6, (1909).

1. Eupristina Saunders

Eupristina, Saund., *Trans. Ent. Soc. London,* p. 5, pl. i, (1883), [Genotype *E. masoni* Saund.] Ashm., *mem. Carnegie Mus.,* J, (4), pp. 233, 234, 375, (1904). Schmied., *Gen. Ins., fas.* 97, p. 8 (1909). Grandi, *Boll. Lab. Zool. Portici,* XI, pp. 207-218, (1916); *Boll. Lab. Ent. Bologna,* VII, p. 230, (1935).

Grassii Grandi

Eupristina grassii, Grandi, *Boll, Lab. Zool. Portici,* XI pp. 218-225, (1916)[1]. Ramakr., *Spol. Zeyl.,* XIII, p. 236, (1925). Grandi, *Boll. Lab. Ent. Bologna,* VII, p. 230, (1935). (Probably identical with *E. Masoni* Saund.). [1]Peradeniya: Ceylon.

Masoni Saunders.

Kupristina masoni, Saund., *Trnas. Ent. Soc. London,* p. 5, pl. i, (1883)[1]; *Ent. Month. Mag.,* XIX, p. 163, (1883)[2]. Schmied., *Gen. Ins., fas.* 97, p.8, (1909). Ramakr. *Spol. Zeyl.,* XIII, p. 236, (1925). Grandi, *Hall Soc. Zool. France,* LIII, p. 81, (1828); *Boll. Lab. Ent. Bologna,* VII, p. 230, (1935). [From fig. of *Ficus indica* L.] [1, 2]Calcutta.

Saundersi Grandi

Epristina saundersi, Grandi, *Boll. Lab. Zool Portici,* XI p. 225, (1616)[1]. Ramakr., *Spol Zeyl.,* XIII, p. 236, (1925). [1]Calcutta.

GRANDI, *Boll. Lab. Ent. Bologna,* VII, p. 231, (1935) [From fig. of *F. religiosa* L.]

2. Blastophaga Gravenhorst

Blastophaga, Graven., *Ubers. Arbeit. schles. Ges. vaterl. Cultur,* I, p. 23, (1826), [Genotype *B. psens* (Linn.)]. ASHM., *Mem. Carnegie Mus.,* I, (4), pp. 233, 234, 394, (1904). SCHMIED., *Gen. Ins., fas.* 97, p. 9, (1909). GRANDI, *Boll. Lab. Zool. Portici,* X, pp. 122-129, (1915); *Boll. Lab. Ent. Balogna,* VII, p. 214, (1935).

Kradibia, Saund., *Trans. Ent. Soc. London,* pp. 23-24, (1883).

breviventris MAYR.

Blastophaga breviventris, MAYR, *Verh. Zool.-Bot. Ges. Wien,* XXXV, p. 181, (1885)[1]. SCHMIED., *Gen. Ins., fas.* 97, p. 9, (1909). RAMAKR., *Spol. Zeyl.,* XIII, p. 236, (1925). GRANDI, *Boll. Lab. Ent. Bologna,* I, pp. 130-132, (1928); VII, p. 215, (1935).

[1]India.

geSTRoi GRANDI

Blastophaga geSTRoi, GRANDI, *Boll. Lab. Zool. Portici,* X, pp. 126, 127, 128, (1915); XI, p. 184, (1916)[1]. RAMAKR., *Spol. Zeyl.,* XIII, p. 236, (1925). GRANDI, *Boll. Lab. Ent. Bologna,* VII, p. 216, (1935). [From fig. of *F. asperrima* Roxb.].

[1]Peradeniya Ceylon.

Gomberti GRANDI.

Blastophaga gomberti, GRANDI, *Bull. Soc. Zool. France,* LIII, p. 70, pl. i, figs. 1-15, (1928)[1]. RAMAKR. & MARGAB., *Journ. Bombay Nat. Hist. Soc.,* XXXVII, p. 193, (1934). GRANDI, *Boll. Lab. Ent. Bologna,* VII, p. 216, (1935). [From fig. of *F. guttata* Blum.].

[1] India.

quadraticeps Westwood

Blastophaga guadraticeps, WESTW., *Trans. Ent. Soc. London,* p. 43, (1882)[1]. MAYR, *Verh. Zool. -Bot. Gen. Wien,* XXXV, pp. 176-177, (1885)[2]. SCHMIED., *Gen. Ins., fas.* 97, p. 9, (1909). GRANDI, *Bull. Ent. Res.,* XIII, p. 296, (1923)[3]; *Boll. Lab. Ent. Bologna,* I pp. 123-126, (1928); VII, p. 219, (1935). [From fig. of *F. religiosa*].

[1]Singapore.
[2] Malacca.
[3]Ceylon (also Pusa).

3. **Ceratosolen** Mayr.

Ceratosolen, Mayr, *Verh. Zool.-Bot. Ges. Wien*, XXXV, p. 164, (1885), [Genotype *C. appendiculatus* Mayr]. Schmied., *Gen. Ins., fas.* 97, p. 9, (1909). Grandi, *Boll. Lab. Zool. Portici*, X, pp. 133-150, (1916); *Boll. Lab. Ent. Bologna*, VII, p. 225, (1935). *Ceratosolens*, Ashm., *Mem. Carnegie Mus.*, I, (4), pp. 233, 234, (1904).

fusciceps Mayr.

Ceratosolen fusciceps, Mayr, *Verh. Zool.-Bot. Ges. Wien*, XXXV, p. 167, (1885).[1] Schmied., *Gen. Ins., fas.* 97, p. 10, (1909). Ramakr., *Spol. Zeyl.*, XIII, p. 236, (1925). Grandi, *Bull. Soc. Zool. France*, LIII, pp. 79-80, (1928); *Boll. Lab. Ent. Bologna*, I, p. 171, (1928); VII, p. 227, (1925). [From figs. of *F. glomerata & F. covelli* (?)].

Ceratosolen fuscipes, Grandi, *Bull. Ent. Res.*, XIII, p. 299, (1923)[2].

[1]Java.
[2]Colombo: Ceylon (also Pusa, & Bangalore).

gravelyi Grandi.

Ceratosolen gravelyi, Grandi, *Boll. Lab. Zool. Portici*, X, pp. 150, 152, (1915); XI, pp. 201-207, (1916)[1]. Ramakr., *Spol. Zeyl.*, XIII, p. 237, (1925). Grandi, *Boll. Lab. Est. Bologna*, VII, 227, (1935). [From fig. of *F. cunia* Hann.].

[1]Chota Nagpur.

marchali Mayr.

Ceratosolen marchali, Mayr, *Wien. Ent. Zeit.*, XXV, pp. 155-156, (1906).

Ceratosolen berlandi, Grandi, *Bull. Soc. Ent. France*, LIII, pp. 74-79, (1928)[1]. Ramakr. & Margab., *Journ. Bombay Nat. Hist. Soc.*, XXXVII, p. 193, (1934). Grandi, *Boll. Lab. Ent. Bologna*, VII, p. 228, (1935).

Ceratosolen marchali, Grandi, *Boll. Lab. Ent. Bologna*, I, pp. 175, 176, (1928); VII, p. 228, (1935). [From fig. of *F. hispida* L.].

[1]India

4. **Liporrhopalum** Waterston

Liporrhopalum, Waterst., *Trans. Ent. Soc. London*, p. 129, (1920), [Genotyupe *L. rutherfordi* Waterst.]. Grandi, *Boll. Lab. Ent. Bologna*, VII, p. 224, (1935).

rutherfordi WATERSTON

Liporrhopalum rutherfordi, WATERST., *Trans. Ent. Soc. London*, p. 130, (1920)[1]. RAMAKR., *Spol. Zeyl.*, XIII, p. 237, (1925). GRANDI, *Boll. Lab. Ent. Bologna*, VII, p. 224, (1935). [1]Paradeniya: Ceylon.

SUBFAMILY II. SYCOPHAGINÆ

ASHM., *Mem. Carnegie Mus.*, I, (4), p. 234, (1904). SCHMIED., *Gen. Ins.*, *fas.* 97, p. 10, (1909).

5. **Sycophaga** Westwood

Cynips, Linne, (*partim*), *Syst. Nat.*, (10), I, p. 554, (1758).

Sycophaga, Westws., *Trans. Ent. Soc. London*, pp. 214, 222, (1840), [Genotype *S. sycomeri* (Linn.)]. ASHM, *Mem. Carnegic Mus.*, I, (4), p. 235, (1904). SCHMIED., *Gen. Ins.*, *fas.* 97, p. 12, (1909). GRANDI, *Boll. Lab. Zool. Portici*, X, pp. 221-227, (1916); XVII, p. 113, (1923); *Boll. Lab. Ent. Bologna*, WII, p. 234, (1935).

Apocrypta, COQUEREL, (*partim*), *Rev. Mag. Zool, Guerin-Meneville*, (2), VII, p. 367 (185).

brevitarsus GRANDI.

Sycophaga brevitarsus, GRANDI, *Boll. Lab. Zool. Portici*, X, p. 236, (1915)[1]; XI, p. 229, (1916)[1]. RAMAKR., *Spol. Zeyl.*, XIII, p. 237, (1925). [1]Colombo: Ceylon.

6. **Apocrypta** COQUEREL.*

Apocrypta, Coq., *Rev. Mag. Zool. Guerin-Meneville*, (2), VII, p. 367, (1855), [Genotype *A. perplexa* Coq.]. ASHM., *Mem. Carnegie Mus.*, I, (4), p. 235, (1904). SCHMIED., *Gen. Ins.*, *fas.* 97, p. 12 (1909). GRANDI, *Boll. Lab. Zool. Portici*, X, pp. 264-273, (1915).

Sycophaga, MAYR. (*partim*), *Verh. Zool. Bot. Ges. Wien*, XXXV, p. 193, (1885).

westwoodi GRANDI

Apocrypta westwoodi Randi, Boll. *Lab. Zool. Portici*, X, p.

* Dalla Torre (*Cat. Elymen.*, V. p. 322, 1808) makes this a synonym of *Sycophag* westw. but the later authers do not agree with him.

273, (1915); XI, p. 232, (1916). RAMAKR. *Spol. Zeyl.*, XIII; p. 237, (1925).

7. Platyscapa MOTSCHULSKY.

Platyscapa, MOTSCH., *Bull. Soc. Imp. Nat. Moscou,* XXXVI, (2), p. 48, (1863), [Genotype *P. frontalis* MOTSCH.]. ASHM., *Mem. Carnegie Mus.*, I, (4), p. 23, (1904). SCHMIED., *Gen. Ins., fas.* 97, p. 12, (1909). GRANDI, *Boll. Lab., Ent. Bologna,* VII, p. 240, (1935).

frontalis MOTSCHULSKY.

Platyscapa frontalis, MOTSCH., *Bull. Soc. Imp. Nat. Moscou,* XXXVI, (2), p. 48, (1863)[1]. SCHMIED., *Gen. Ins., fas.* 97, p. 12, (1909). RAMAKR., *Spol. Zeyl.*, XIII, p. 237, (1925). GRANDI, *Boll. Lab. Ent. Bologna,* VII, p. 240, (1935).

[1]Mt. Patannas: Ceylon.

Family II. TORYMIDÆ

WALK., *Ent. Month. May.*, I, p. 115, (1833); *List Chalc. Brit. Mus.*, I, p. 14, (1846); II, p. 100, (1848). FÖRST., *Hymen. Stud.*, II, pp. 19, 23, 43, (1856). WALK. *Notes Chalc.*, II, pp. 19-36, (1871). GUSTAV MAYR, *Verh. Zool,-Bot. Ges. Wien,* XXIV, pp. 53-142, (1874), [European forms]. THOMS., *Hymen. Skandin.*, IV, pp. 11, 59, (1875). How., *Ent. Amer.*, I, p. 198, (1886). ASHM., *Proc. Ent. Soc. Washington,* IV, p. 243, (1897); *Mem. Carnegie Mus.*, I, (4), p. 235, (1904). SCHMIED., *Gen. Ins. fas.* 97, p. 85, (1909), GIR., *Mem. Queensland Mus.*, IV, p. 275, (1915), [AuSTRalian forms]. Rusch., *Zeits. angew. Ent.*, IX, pp. 395-408, (1923) [European *Monodontomerus*].

SUBFAMILY I. IDARNINÆ

ASHM., *Proc. Ent. Soc. Washington,* IV, p. 235, (1897); *Mem. Carnegie Mus.*, I, (4), p. 237, (1904). SCHMIED., *Gen. Ins., fas.* 97, p. 86, (1909).

8. Polanisa WALKER.

Polanisa, WALK., *Entomologist,* VIII, p. 17, (1875), [Genotype *P. transiens* (WALK.)]. ASHM., *Mem. Carnegie Mus.*, I, (4), p. 237, (1904). SCHMIED., *Gen. Ins., fas.* 97, p. 90, (1909).

Idarnella, WALK., *Trans. Ent. Soc. London,* p. 37, (1883).

transiens (WALKER).

Idarnes transiens, WALK., *Notes Chalc.*, IV, p. 62, (1871)[1].

[1]Ceylon.

Polanisa lutea, WALK., *Entomologist*, VIII, p. 18, (1875).

Idarnella transiens, WALK., *Trans, Ent. Soc. London*, p 37, (1883).

Polanisa transiens, SCHMIED., *Gen. Ins., fas.* 97, p. 90, (1909). RAMAKR., *Spol. Zeyl*, XIII, p. 241, (1925). HOFFMEYER. *Ent. Medd.*, XVIII, p. 248, (1933).

9. **Sycoscaptella** Westwood

Sycoscaptella, WESTW., *Trans. Ent. Soc. London*, p. 36, (1883),[Genotype *S. affiris* WESTW.] ASHM., *Mem. Carnegie Mus.*, I, (4), p. 237, (1904). SCHMIED., *Gen. Ins., fas.* 97, p. 90, (1909).

affinis Westwood

Sycoscaptella affinis, WESTW., *Trans. Ent. Soc. London*, p. 42, pl. vi, figs. 30-35, (1883)[1]. SCHMIED., *Gen. Ins., fas.* 97, p. 91, (1909). RAMAKR., *Spol, Zeyl.*, XIII, p. 241, (1925).

[1]Calcutta.

10. **Idarnes** WALKER

Idarnes, WALK., *Ann. Mag. Nat. Hist.*, XII, p. 47, (1843), [Genotype *I. carme* WALK.]. ASHM., *Mem. Carnegie Mus.*, I, (4), pp. 238, 239, (1904). SCHMIED., *Gen. Ins., fas.* 97, p. 91, (1909).

Tetragonaspis, MAYR, *Verh. Zool.-Bot. Ges, Wien*, XXXV, p. 157, (1885).

Stabilis WALKER.

Idarnes stabilis, WALK., *Notes Chalc.*, IV, p. 62, pl. IV, (1871).[1] WESTW., *Trans. Ent. Soc. London*, p. 338, (1883)[1]. SCHMIED. *Gen. Ins., fas.* 97, p. 91, (1909)[2]. RAMAKR., *Spol., Zeyl.*, XIII, p. 241, (1925). (=*Sycopcaptes stabilis* (WALK.). det. TIMBERLAKE)[3].

[1]Ceylon.
[2]India.
[3]Pusa.

11. **Goniogaster** MAYR.

Goniogaster, MAYR, *Verh. Zool., bot. Ges. Wien*, XXXV, pp. 157, 240, (1885), [Genotype *G. varricolor* MAYR]. ASHM., *Mem. Carnegie Mus.*, I, (4), pp. 238, 240, (1904). SCHMIED., *Gen. Ins., fas.* 97, p. 92 (1909).

Varricolor MAYR.

Goniogaster varricolor, MAYR, *Verk. Zool.-Bot. Ges. Wien,* XXXV, p. 241, (1885)[1]. SCHMIED., *BGen. Ins., fas.* 97, p. 92, (1909). — [1]Java.

Goniogaster varricornis. ASHM., *Mem. Carnegie Mus.,* I, (4), p. 238, (1904). RAMAKR., *Spol. Zeyl.,* XIII, p. 242, (1925)[2]. [From *Picus* (Cystogyne) *sub-opposita* Meig.] — [2]Coimbatore.

12. Platneura MOTSCHULSKY.

Platyneura, MOTSCH., *Bull. Soc. Imp. Nat. Moscou,* XXXVI, (3), p. 49, (1863), [Genotype *P. testacea* MOTSCH]. ASHM., *Mem. Carnegie Mus.,* I, (4), pp. 238, 240, (1904). SCHMIED., *Gen. Ins., fas.* 97, p. 93, (1909).

testacea MOTSCHULSKY.

Platyneura testacea, MOTSCH., *Bull. Soc., Imp. Nat.,* XXXVI, (3), p. 50, (1863)[1], SCHMIED., *Gen. Ins., fas.* 97, p. 93, (1909). RAMAKR., *Spol. Zeyl.,* XIII, p. 241, (1925). — [1]Mt. Patannas: Ceylon.

13. Sycophila WALKER.

Sycophila, WALK., *Notes Chalc.,* IV, p. 63, (1871), [Genotype *S. decatomoides* WALK.]. ASHM., *Mem. Carnegie Mus.,* I, (4), pp. 238, 240, (1904). SCHMIED., *Gen. Ins., fas.* 97, p. 95, (1909).

Isanisa, WALK., *Entomologist,* VIII, p. 15, (1875).

Pseudisa, WALK., *Entomologist,* VIII, p. 15, (1875).

decatomoides WALKER

Sycophila decatomoides, WALK., *Notes Chalc.,* IV, p. 64, (1871)[1], SCHMIED., *Gen. Ins., fas.* 97, p. 95, (1909). RAMAKR., *Spol. Zeyl.,* XIII, p. 241, (1925). — [1]India.

Megastigmoides WALKER

Sycophila megastigmoides, WALK., *Notes, Chalc.,* IV, p. 64, (1871)[1]. SCHMIED., *Gen. Ins., fas.* 97. p. 95, (1909). RAMAKR., *Spol. Zeyl.,* XIII, p. 241, (1925). — [1]India.

Pseudisa smicroides, WALK., *Entomologist,* VIII, p. 15, (1875)[1]. — [I]India.

14. **Tetranemopteryx** Ashmead.

Sycoscaptella, WESTW. *(partim)*, *Trans. Ent. Soc. London*, p. 43, (1883).

Tetranemopteryx, ASHM, *Mem. Carnegie Mus.*, I, (4), p. 239, (1904), [Genotype *T. quadrisetosa* (WESTW.)]. SCHMIED., *Gen. Ins.*, *fas.* 97, p. 95, (1909).

Quadrisetosa (Westwood).

Sycoscaptella quadrisetosa, WESTW., *Trans. Ent. Soc. London*, pp. 43, 375, pl. xvi, (1883)[1]. [1]Ceylon.

Tetranemopteryx quadrisetosa, ASHM. *Mem. Carnegie Mus.*, I, (4), p. 239, (1904). SCHMIED., *Gen. Ins.*, *fas.* 97, p. 96, (1909). RAMAKR., *Spol. Zeyl.*, XIII, p. 242, (1925).

15. **Sycoscapterella** Ashmead.

Sycoscapter, WESTW., *(partim)*, *Trans. Ent. Soc. London*, p. 42, (1883).

Sycoscapterella, ASHM. *Mem. Carnegie Mus.*, I, (4), p. 239, (1904), [Genotype *S. anguliceps* (WESTW.)]. SCHMIED., *Gen, ins.*, *fas.* 97, p. 96, (1909).

angulicepus (Westwood).

Sycoscapter anguliceps, WESTW., *Trans. Ent. Soc. London*, p. 42, (1883)[1]. [1]Ceylon.

Sycoscapterella anguliceps, ASHM., Mem. Carnegie Mus., I, (4), p. 239, (1904). SCHMIED., *Gen. Ins.*, *fas.* 97, p. 96, (1909). RAMAKR., *Spol. Zeyl.*, XIII, p. 242, (1925).

16. **Sycoscapter** (Saunders) Westwood.

Sycoscapter, WESTW., *Trans. Ent. Soc. London*, p. 34, (1883), [Genotype *S. insignis* (Saund.)]. ASHM., *Mem. Carnegie Mus.*, I, (4), p. 239, (1904). SCHMIED., *Gen. Ins.*, *fas.* 97, p. 96, (1909).

gracilipes Westwood

Sycoscapter gracilipes, *Trans. Ent. Soc. London*, p. 41, pl. viii, (1883)[1]. SCHMIED., *Gen. Ins.*, *fas.* 97, p. 96, (1909). RAMAKR., *Spol. Zeyl*, XIII, p. 242, (1925). [1]Ceylon.

insignis (Saunders) Westwood

Sycoscapter insignis, Saund., *Trans. Ent,. Soc. London,* pp. 34, 351, (1883).[1] SCHMIED., *Gen. Ins., fas.* 97, p. 96, (1909). RAMAKR., *Spol. Zeyl.,* XIII, p. 242, (1925). [1]Calcutta.

17. Sycoscapteridea Ashmead

Sycoscapter, WESTW., (*partim*), *Trans. Ent. Soc. London,* p. 41, (1883).

Sycoscapiteridea, ASHM., *Mem. Carnegie Mus.,* I, (4), p. 239, (1904). [Genotype *S. monilifera* (WESTW.)]. SCHMIED., *Gen. Ins., fas.* 97, p. 96, (1909).

monilifera (Westwood).

Sycosapter monolifera, WESTW., *Trans. Ent. Soc. London,* p. 41, (1883)[1]. [1]Ceylon.

Sycoscapteridea monilifera, ASHM., *Mem. Carnegie Mus.,* I(4), p. 239, (1904). SCHMIED., *Gen. Ins., fas.* 97, p. 96, (1909). RAMAKR., *Spol. Zeyl.,* XIII, p. 242, (1825).

Sycobiella, WESTW., *Trans. Ent. Soc. London,* p. 33, (1883), [Genotype *S. saundersi* WESTW.]. ASHM., *Mem. Carnegie Mus.,* I, (4), p. 240, (1904). SCHMIED., *Gen. Ins., fas.* 97, p. 97, (1909).

Pteromaloides (WALKER).

Idarnes pteromaloides, WALK., *Notes Chalc.,* IV, p. 63, (1871)[1]. [1]India (also Pusa).

Micranisa pteromaloides, WALK., *Entomologist,* VIII, p. 15, (1875); *Trans. Ent. Soc. London,* (*Proc.*), p. xvii, (1884). DALLA TORRE, *Cat. Hymen.,* V. p. 317, (1898). SCHMIED., *Gen. Ins., fas.* 97, p. 94, (1909). RAMAKR., *Spol. Zeyl.,* XIII, p. 241, (1925).

Saundersi Westwood

Sycobiella saundersi, WESTW., *Trans. Ent. Soc. London,* p. 34, (1883)[1]. SCHMIED., *Gen. Ins., fas* 97, p. 97, (1909). RAMAKR., *Spol. Zeyl.,* XIII, p. 242, (1925). [1]Calcutta.

19. Walkerella Westwood

Walkerella, WESTW., *Trans. Ent. Soc. London,* p. 32, (1883), [Genotype *W. temeraria,* WESTW.]. ASHM., *Mem.*

Carnegie Mus., I, (4), p. 240, (1904). SCHMIED. , *Gen. Ins., fas.* 97, p. 97, (1909).

temeraria Westwood

Walkerella temeraria, WESTW., *Trans. Ent. Soc. London*, p. 32, p.. iv, figs. 9-12, (1883)[1]. SCHMIED., *Gen. Ins., fas.* 97, p. 97, (1909). RAMAKR., *Spol. Zeyl.*, XIII, p. 242, (1925). [1]Madras.

20. Sycobia WALKER

Sycobia, WALK., *Notes Chalc.*, IV, p. 60, (1871), [Genotype *S. bethyloides* WALK.]' *Trans. Ent. Soc. London, p. 31, pl. iv, (1871).* ASHM., *Mem. Carnegie Mus.*, I, (4), p. 241, (1904). SCHMIED., *Gen. Ins., fas.* 97, p. 97, (1909).

Agrianisa, WALK., *Entomologist*, VIII, p. 16, (1875).

bethyloides WALKER

Sycobia bethyloides, WALK., *Notes Chalc.*, IV, p. 61, (1871)[1]; *Trans, Ent. Soc. London*, (also Pusa). p. 31, (1871). SCHMIED., *Gen. Ins., fas.* 97, p. 97, (1909). RAMAKR., *Spol. Zeyl.*, XIII, p. 242, (1925). [1]Calcutta (also Pusa).

21. Otitesella Westwood.

Otitesella, WESTW., *Trans. Ent. Soc. London*, p. 39, (1883), [Genotype *O. digitata* WESTW.]. ASHM., *Mem. Carnegie Mus.*, I, (4), p. 240, (1904). SCHMIED., *Gen. Ins., fas.* 97, p. 97, (1909). GRANDI, *Boll. Lab. Zool. Portici*, XVI, p. 16, (1922).

digitata Westwood

Otitesella digitata, WESTW., *Trans. Ent. Soc. London*, p. 39, pl. vii, (1883)[1]. SCHMIED., *Gen. Ins., fas.* 97, p. 98, (1909). RAMAKR., *Spol. Zeyl.*, XIII, p. 242, (1925). [1]Ceylon (also Pusa).

religiosa Westwood

Otitesella religiosa, WESTW., Trans. Ent. Soc. London, p. 40, (1883)[1]. SCHMIED., *Gen. Ins., fas.* 97. p. 98, (1909). RAMAKR., *Spol. Zeyl.*, XIII, p. 242, (1925). [1]Ceylon (also Pusa).

SUBFAMILY II. TORYMINAE

ASHM., *Mem. Carnegie Mus.*, I, (4), p. 241, (1904). *Ins., fas.* 97, p. 98, (1909). SCHMIED., *Gen.*

22. Ecdamua WALKER

Ecdamua, WALK., *Trans. Ent. Soc. London,* I, (4), p. 387, (1862), [Genotyupe *E. macrotelus* WALK.] ASHM., *Mem. Carnegie Mus.,* I, (4), p. 241, (1904). SCHMIED., *Gen. Ins., fas.* 97, p. 101, (1909).

indica WALKER

Ecdamua indica, WALK., *Notes Chalc.,* II, p. 35, (1871)[1]. SCHMIED., *Gen. Ins., fas.* 97, p. 101, (1909). RAMAKR., *Spol. Zeyt.,* XIII, p. 242, (1925). [1]India

23. Torymus (DALMAN) Thomson

Ichneumon, Linne, (*partim*), *Syst. Nat.,* (10), p. 567, (1758).

Torymus, Dal., (*partim*), *Svensk. Vet.-Akad. Handl.,* XLI, pp. 135, 178, (1820), [Genotype *T. bedegauris* (Linn.)]. THOMS., *Hymen. Skandin.,* IV, (1), p. 80, (1875). ASHM., *Mem. Carnegie Mus.,* I, (4), p. 242, (1904). SCHMIED., *Gen, Ins., fas.* 97, p. 102, (1909).

Callimome, Spinola, (*partim.*). *Ann. Mus. Hist. Nat.,* XVII, pp. 146, 148, (1871).

Misocampus, LATREILLE, *Nouv. Dict. Hist. Nat.,* (ed. 2), (1817).

ceylonicus (MOTSCHULSKY).

Callimome ceylonicus, MOTSCH., *Bull. Soc. Imp. Nat. Moscou,* XXXVI, (3), p. 47, (1863)[1]. [1]Mt. Nura Ellia: Ceylon.

Torymus ceylonicus, SCHMIED., *Gen. Ins., fas.* 97, p. 104, (1909). RAMAKR., *Spol. Zeyl.,* XIII, p. 242, (1925).

24. Torymoides WALKER

Torymoides, WALK., *Notes Chalc.,* III, p. 37, (1871), [Genotype *T. amabilis* WALK.]. ASHM., *Mem. Carnegie Mus.,* I (4), p. 242, (1904). SCHMIED., *Gen. Ins., fas.* 97, p. 109, (1909).

amabilis WALKER

Torymoides amabilis, WALK., *Notes Chalc.,* III, p. 38, (1871)[1]. SCHMIED., *Gen. Ins., fas.* 97, p. 109, (1909). RAMAKR., *Spol. Zeyl.,* XIII, p. 243, (1925). [1]Ceylon.

SUBFAMILY III. MONODONTOMERINAE

ASHM., *Proc. Ent. Soc. Washington*, IV, p. 247, (1899); *Mem. Carnegie Mus.*, I, (4), p. 242, (1904). SCHMIED., *Gen. Ins., fas.* 97, p. 109, (1909). Rusch., *Zeits. angew. Ent.*, IX, pp. 395-408, (1923), [European *Monodontomerus*].

25. **Monodontomerus** Westwood

Monodontomerus, WESTW., *Philos. Mag.*, (3), II, p. 443, (1833), [Genotype *M. obscurus* WESTW.]. ASHM., *Mem. Carnegie Mus.*, I, (4), p. 243, (1904). SCHMIED., *Gen. Ins., fas.* 97, p. 112, (1909). WATERST., *Ind. For. Rec.*, IX, p.2, (1921). Rusch., *Zeits. angew. Ent.* IX, pp. 395-408, (1923), [European forms].

Paroligosthenus, CAM., *Ind. For. Rec.*, IV p. 94, (1912).

sp. RAMAKRISHNA AYYAR.

Monodontomerus ap., RAMAKR., *Spol. Zeyl.*, XIII, p. 243, (1925)[1].

[1]Coimbatore.

trichiophthalmus (CAMERON)

Paroligosthenus trichiophthalmus, CAM., *Ind. For. Rec.*, IV, p. 94, (1912).[1]

Monodontomerus trichiophthalmus, Westerst., *Ind. For. rec.*, IX, p.2, (1921). RAMAKR., *Spol. Zeyl.*, XIII, p. 243, (1925). MANI, *Rec. Ind., Mus.*, XXXVIII, p. 125, (1936)[2]. [Host: a leaf-cutting bee.]

[1]Dehra Dun´

[2]Lyallpur

SUBFAMILY IV. PODAGRIONINÆ

ASHM., *Mem. Carnegie Mus.*, I, (4), p. 224, (1904). SCHMIED., *Gen.* Ins., *fas.* 97, p. 116, (1909). GIR., *Mem. Queensland Mus.*, IV, p. 286, (1915). Rabaud, *Bull. Soc. Ent. France*, p. 178, (1917). KIEFF., *Bull. Soc. Ent France*, pp. 357-359, (1919). GAHAN, *Philip. Journ, Sc.*, XXVII, p. 89, (1925).

26. **Podagrion** Spinola

Podagrion, Spin., *Ann. Mus. Hist. Nat.*, XVII, p. 147, (1811), [Genotype *P. slpendens* Spin.]. ASHM., *Mem. Carnegie Mus.*, I, (4), p. 244, (1904). SCHMIED., *Gen. Ins., fas.* 97, p. 116, (1909). MASI, *Navit. Zool.*, XXIV, p. 126, (1917). GAHAN, *Proc. U.S.N. Mus.*, LXI, (24), p. 265, (1922). RAMASWAMAYYA, *Madras Dept. Agric.*

Year Book, 1922, pp. 55-57, (1923). GAHAN, *Philip. Journ. Sc.,* XXVII, p. 89, (1925). MASI, *Konowia,* V, pp. 9-19, 170, (1926).

Palmon, DALMAN, *Svensk. Vet.-Akad. Handl.,* XLVI, p. 388, (1825). WESTW., *Trans. Ent. Soc. London,* p. 259, (1847). WALK., *Notes Chalc.,* II, p. 19, (1871).

Priomerus, WALK., *Ent. Month. Mag.,* I, p. 116, (1833).

Blephonira, Holmgren, *Eugen. Resa. Ins.,* p. 438, (1856).

Bactryischion, A. COSTA, *Men. Accad. Sc. Napoli,* II, p. 223, (1857).

Greeni CRAWFORD

Podagrion greeni, CRAWFORD, *Proc. U.S.N. Mus.,* XLII, p. 3, (1912)[1]. RAMAKR., *Spol. Zeyl.,* XIII, p. 243, (1925). [1]Punduloya: Ceylon.

Indicum Girault (?)*

Podagrion indicum GIR. (?), RAMAKR., *Rep. Proc. Third Ent. Meet. Pusa,* III, p. 935, (1920)[1]. [1] Coimbatore.

Mantoidæ (MOTSCHULSKY).+

Anastatus mantoidae, MOTSCH., *Etud. Ent.,* VIII, p. 116, pl. II, fig. 11, (1859); *Bull. Soc. Imp. Mat. Mascou,* XXXVI, (3), p. 57, pl. II, fig. 11, (1863).[1] [1]Ceylon.

Podagrion mantoidae, DALLA TORRE, *Cat. Hyme.,* V, p. 280, (1898). SCHMIED., *Gen., Ins., fas.* 97, p. 117, (1909). RAMAKR., *Spol. Zeyl.* XIII, p. 243, (1925). [Host : Mantids].

micans STRand

Podagrion micans, STRand, *Arch. Naturg. Berlin,* LXXVII, Bd. 1, H. 2, pp. 154-158, (1911).[1] [1]Ceylon.

---

* The original description of this species could not be traced in the literature. In a series of privately published pamphlets is said to have proposed a number of generic and specific names. These pamphlets unfortunately, are not available in this country. It is probable that this and a few other Indian species were described in these pamphlets. These names are not found in the *Zool. Rec.* up to the year 1935.

\+ This species wasfirst described by Moteshulsky as type of the genus *Anastatus* Metsch. Though he accepted the genus *Anastuits* as valid, Dalla Torre transferred the species to *Pedagrion* Spin. Gahan and Fagan *(Bell. U. S. N. Mus.,* No. 124, p. 12, 1923) do not agree with Dalla Torre and the inclusion of the species under *Podagrion* in this catalogue is only provisional.

pachymerum WALKER

Plamon pachymerus, WESTW., *Trans. Ent. Soc. London*, IV, (4), p. 259, (1847).

Bactryischion bicoloratum, A. COSTA, *Mem. Accad. Sc. Napoli*, II, p. 226, (1857).

Palmon splendidum, WALK., *Notes Chalc.*, II, p. 19, (1871).

Priomerus pachymerus, WALK., *Ent. Month. Mag.*, I, p. 118, (1883).

Podagrion pachymerum, DALLA TORRE, *Cat. Hymen.*, V., p. 370, (1898). SCHMIED., *Gen. Ins.*, *fas.* 97, p. 117, (1909). GAHAN, *Proc. U.S.N. Mus.*, LXI, (24), p. 265, (1922)[1]. RAMASWAMAYYA, *Madras Dept. Agric. Year Book*, 1922, pp. 55-57, (1923). RAMAKR. & MARGAB., *Journ. Bombay Nat. Hist. Soc.*, XXXVII, p. 194, (1934). [Host : *Mantis religiosa*]. [1]Coimbatore.

repens (MOTSCHULSKY)

Priomerus repens, MOTSCH., *Etud. Ent.*, VIII, p. 118, (1859)[1]. [1]Ceylon.

Podagrion repens, WALK., *Notes Chalc.*, II, p. 28, (1871). SCHMIED., *Gen. Ins.*, *fas.* 97, p. 117. (1909). RAMAKR., *Spol. Zeyl.*, XIII, p. 243, (1925).

Sp. RAMAKRISHNA AYYAR.

Podagrion sp., RAMAKR., *Bull. Ent. Res.*, XVIII, p. 77, (1927).[1] [Host : Mantid]. [1]Coimbatore.

SUBFAMILY V. MEGASTIGMINÆ

THOMS., *Hymen. Skandin.*, IV, p. 59, (1875). ASHM., *Proc. Ent. Soc. Washington*, IV, p. 246, (1899); *Mem. Carnegie Mus.*, I, (4), p. 244, (1904). SCHMIED., *Gen. Ins.*, *fas.* 97, p. 118, (1909).

27. Megastigmus DALMAN

Ichneumon, FABR., (*partim*), *Suppl. Ent. Syst.*, p. 231, (1798).

Megastigmus, DALMAN, *Svensk. Vet.-Akax. Handl.*, XLI, p. 178, (1820), [Genotype *M. dorsalis* (FABR.)].* ASHM., *Mem. Carnegie Mus.*, I (4), p. 245, (1904).

---

* Genotype *M. bipunctatus* (Swed.) (=*Pteromalus bipunctatus* Swed.) according to Gahan and Fagan. *(Bull. U. S. N. Mus.*, No. 124, p. 84, 1923}.

Schmied., *Gen. Ins., fas.* 97, p. 118, (1909). Gir., *Mem. Queensland Mus.,* IV, p. 297, (1915).

Trogocarpus, Rond., *Bull. Soc. Ent. Ital.,* IX, p. 204, (1877).

Megalostigmus, Schulz, (emend.), *Spol. Hymen.,* p. 147, (1906).

indi Girault (?)†

Megastigmus indi, Gir. (?), Ramkr., *Rep. Porc. Third Ent. Meet. Pusa,* I, p. 37; III, p. 935, (1920).[1] [In pods of *agathi* and *daincha* as parasite of *Eurytoma indi*]. [1]Combatore.

Sp. Ramakrishna Ayyar.

Megastigmus sp., Ramakr., *Spol. Zeyl.,* XIII, p. 243, (1925)[1]. [1]Combatore. (Also Pusa.)

28. Spilomegastigmus Cameron

Spilomegastigmus, Cam., *Spol. Zeyl.,* III, p. 73, (1903), [Genotype *S. ruficeps* Cam.]. Schmied., *Gen. Ins., fas.* 97, p. 120, (1909).

ruficeps Cameron

Spilomegastigmus ruficeps, Cam., *Spol. Zeyl.,* III, p. 74, (1903)[1]. Schmied., *Gen. Ins., fas.* 97, p. 120, (1909). Ramakr., *Spol. Zeyl.,* XIII, p. 243, (1925). [1]Kandy: Ceylon.

29. Bootania Dalla Torre

Metamorpha, Walk., *Trans. Ent. Soc. London,* I, (4), p. 347, (1862)..

Bootania, Dalla Torre, *Wien. Ent. Zeit.,* XVI, p. 86, (1897), [Genotype *B. leucospoides* (Walk.)]. Ashm., *Mem. Carnegie Mus.,* I, (4), p. 245, (1904). Schmied., *Gen. Ins., fas.* 97, p. 120, (1909).

leucospoides (Walker)

Metamorpha leucospoides, Walk., *Trans. Ent. Soc. London,* I, (4), p. 347, (1862)[1]. Westw., *Thesaur. Ent. Oxon.,* p. 136, pl. XXV, fig. 8, (1874). [1]Bhutan.

Bootania leucospoides, Schmied., *Gen. Ins., fas.* 97, p. 120, (1909). Ramakr., *Spol. Zeyl.,* XIII, p. 243, (1925).

---

†The original description of this species could not be traced in the literature.

30. **Odopoia** WALKER.

Odopoia, WALK., *Notes Chalc.*, II, p. 36, (1871), [Genotype *O. atra* WALK.]. ASHM., *Mem. Carnegie Mus.*, I, (4), p. 245, (1904). SCHMIED., *Gen. Ins.*, *fas.* 97, p. 120, (1909).

atra WALKER

Odopoia atra, WALK., *Notes Chalc.*, II, p. 36, (1871)[1]. SCHMIED., *Gen. Ins.*, *fas.* 97, p. 120, (1909). RAMAKR., *Spol. Zeyl.*, XIII, p. 243, (1925).

[1]Ceylon.

SUBFAMILY VI. ORMRINAE

FÖRST., *Hymen. Stud.*, II, pp. 19-22, 24, (1856). THOMS., *Hymen. Skand.*, IV, (1), p. 100, (1875). ASHM., *Proc. Ent. Soc. Washington*, IV, p. 247, (1899); *Mem. Carnegie Mus.*, I, (4), p. 245, (1904). SCHMIED., *Gen. Ins.*, *fas.* 97, p. 121, (1909).

31. **Ormyrus** Westwood

Ormyrus, WESTW., *Philos. Mag.*, I, p. 127, (1832), [Genotype *O. punctiger* WESTW.]. ASHM., *Mem. Carnegie Mus*,. I, (4), p. 245, (1904). SCHMIED., *Gen. Ins.*, *fas.* 97, p. 122, (1909).

Periglyphus, Boheman, *Svensk. Vet.-Akad. Handl.*, LIV, p. 378, (1833).

Siphonura, Nees, *Hymen, Ichneum. Affin. Mon.*, II, p. 81, (1834).

Cyrtosoma, Perris, *Ann. Soc. Ent. France*, IX, p. 96, (1840).

orientalis WALKER

Ormyrus orientalis, WALK., *Notes Chalc.*, IV, p. 68, (1971)[1]. SCHMIED., *Gen. Ins.*, *fas.* 97, p. 123, (1909). RAMAKR., *Spol. Zeyl.*, XIII, p. 243, (1925).

[1]Ceylon.

FAMILY III. CHALCIDIDÆ

Leach, *Edinb. Encyc.*, IX, p. 144, (1830). WESTW., *Introd. Mod. Class. Ins.*, I, p. 166, (1840). WALK., *List Chalc. Brit. Mus.*, I, p. 2, (1846). FÖRST., *Hymen. Stud.*, II, pp. 18, 21, 29, (1856). THOMS., *Hymen. Skandin.*, IV, pp. 11, 12, (1875). KIRBY, *Journ. Linn., Soc. London, (Zool.)*, XVII, pp. 53-78, (1884). How., *Ent. Amer.*, I, p. 197, (1886). KIRBY, *Journ. Linn. Soc. London*, (*Zool.*). XX, pp. 28-37, (1887). How., *Trans. Amer. Ent. Soc.*, I, pp. 197-219; II, pp. 33-38, 97-101, (1887); *Proc. Ent. Soc. Washington*, I, pp. 65-74, (1888). ASHM., *Proc. Ent. Soc.*

Washington, IV, p. 245, (1899); *Proc. U.S.N. Mus.*, XXIII, p. 202, (1900); *Mem, Carnegie Mus.*, I, (4), p. 246, (1904). SCHMIED., *Gen. Ins.*, *fas.* 97, p. 17, (1909). MASI, *Eos*, III, pp. 29-48, (1927); *Boll. Lab. Ent. Bologna*, II, pp. 155-188, (1927).

WALK., *Ent. Month, Mag.*, II, p. 13, (1833). HALIDAY, *Hymen. Synop.*, p. ii, (1939). WALK. *List Chalc. Brit. Mus.*, I. p. 1, (1946). FÖRST., *Hymen. Stud.*, II, pp. 18, 20, 29, (1856). How, *Ent. Amer.*, I, p. 197, (1886). Schletterer, *Berlin. Ent. Zeitschr.*, XXXV, pp. 141-302, (1890), [Revision]. ASHM., *Proc. Ent. Soc. Washington*, IV, p. 235, (1899); *Mem Carnegie* Mus., I, (4), p. 246, (1909). SCHMIED., *Gen. Ins.*, *fas.* 97, p. 13, (1909). WELD, *Proc. U.S.N. Mus.*, LXI, art. 6, pp. 1-43, (1922), [Revision : biological, morphological and systematic]. MASI, *Boll, soc. Ent. Bologna*, LXVII, pp. 36-43, (1935), [Systematic notes].

32. **Polistomorpha** Westwood.*

Polistomorpha, WESTW., *Zeit. Ent.*, I, p. 265, (1839), [Genotype *P. surinamensis* WESTW.].** WALK., *Thesaur. Ent. Oxon.*, p. 133, (1874). Schlett., *Berlin,. Ent. Zeitschr.* XXXV, p. 292, (1890). ASHM., *Mem. Carnegie Mus.*, I, (4), p. 247, (1904). Ducke, *Bull. soc. Ent. France*, pp. 163-166, (1906). SCHMIED., *Gen. Ins.*, *fas.* 97, p. 14, (1909). WELD, *Proc. U.S.N. Mus.*, LXI, art. 6, 9. 4, (1922). MANI, *Rec. Ind. Mus.*, XXXVII, p. 242, (1935).

indica MANI

Polistomorpha indica, MANI, *Rec. Ind. Mus.*, XXXVII, p. 243, (1935)[1].

[1]Yercaud.

33. **Leucospis** Fabricius.†

Leucospis, FABR., *Syst. Ent.*, I, p. 361, (1775), [Genotype *L. dorsigera* FABR.]. WESTW., *Thesaur. Ent. Oxon.*, p. 135, (1874). Schlett., *Berlin. Ent. Zeitschr.* XXXV, pp. 141-302, (1890). ASHM., *Mem. Carnegic Mus.*, I, (4), p. 247, (1904). SCHMIED., *Gen. Ins.*, *fas.* 97, p. 14, (1909). WELD, *Proc. U.S.N. Mus.*, LXI, art. 6, p. 14, (1922) MANI, *Rec. Ind. Mus.*, XXXVII, p. 244, (1935).

---

* For additional records of species *vide* addendum.

† In a recent letter Mr. B. D. Burks of the State Natural History Survey Division Illinois, U. S. America, writes that Gahan of U. S. N. Museum is of opinion that *Polictomorpha* (ass determined by an examination of the genotype species in the Oxford and British Museums) does not being to *Leucospidinae*, but is semewhat allied to *Brachymeria*. This needs confirmation by further studies.

Exochloenus, Shipp, *Entomologist*, XXVII, p. 16, (1894). ASHM., *Mem. Carnegie Mus.*, I, (4), [. 247, (1904). SCHMIED., *Gen. Ins.*, *fas.* 97, p. 17, (1909). MANI, *Rec. Ind-Mus.*,, XXXVII, p. 244, (1935).

atra Fabricius.*

Leucospis atra, FABR., *Suppl. Ent. Syst.*, p. 259, (1798)[1]. SCHMIED., *Gen. Ins.*, *fas.* 97, p. 15, (1909). RAMAKR., *Spol. Zeyl.*, XIII, p. 237, (1925). — [1]India.

bengalensis WELD

Leucospis bengalensis, WELD, *Proc. U.S.N. Mus.*, LXI, art, 6, p. 32, (1922)[1]. RAMAKR., *Spol. Zeyl.*, XIII, p. 237, (1925). MANI, *Rec. Ind. Mus.*, XXXVIII, p. 339, (1936), [Notes and description of the male]. — [1]Bengal.

bombayensis MANI

Leucospis bombayensis, MANI, *Rec. Ind. Mus.*, XXXVII, p. 246, (1935)[1]. — [1]Satara : Bombay.

guzeratensis WESTWOOD

Leucospis guzeratensis, WESTW., *Zeit. Ent.*, I, p. 253, (1839)[1]. Schlett., *Berlin. Ent. Zeitschr.*, XXXV, p. 224, (1890). SCHMIED., *Gen. Ins.*, *fas.* 97, p. 16, (1909). WELD, *Proc. U.S.N. Mus.*, LXI, p. 33, (1922)[2]. RAMAKR., *Spol. Zeyl.*, XIII, p. 237, (1925). — [1]Gujarat Tranquebar. [2] Coimbatore.

indiensis WELD

Leucospis indiensis, WELD, *Proc. U.S.N. Mus.*, LXI, art. 6, p. 20, (1922)[1]. RAMAKR., *Spol. Zeyl.*, XIII, p. 237, (1925). MANI, *Rec. Ind.*, *Mus.*, XXXVIII, p. 339, (1936)[2], [Notes]. — [1]Coimbatore. [2]Kankondigee: Sunderbans.

malabarensis BRUES.

Leucospis malabarensis, BRUES, *Psyche*, XXXII, p. 27, (1925)[1]. RAMAKR. and MARGAB., *Journ. Bombay Nat. Hist. Soc.*, XXXVII, p. 193, (1934). — [1]Malabar.

meenakshiæ MANI

Leucospis meenakshiæ, MANI, *Rec. Ind. Mus.*, XXXVII, p. 244, (1935)[1]. — [1]Tanjore.

---

* According to Schletterer and other authors, this is a Synonym of *L. guzeratensis*.

nursei CAMERON

Leucospis nursei, CAM., *Journn. Bombay Nat. Hist. Soc.*, XVII, p. 92, (1906)[1]. RAMAKR., *Spol. Zeyl.*, XIII, p. 237, (1925). [1]Quetta.

Petiolata FABRICIUS.*

Leucospis petiolata, FABR., *Mant. Ins.*, I, p. 285, (1787)[1]. WESTW., *Zeit. Ent.*, I, p. 253, (1839). SCHMIED., *Gen. Ins., fas.* 97, p. 16, (1909). RAMAKR., *Spol. Zeyl.*, XIII, p. 237, (1925). [1]India.

quettaensis CAMERON

Leucospis quettaensis, CAM., *Journ. Bombay Nat. Hist. Soc.*, XVII, p. 91, (1906)[1]. RAMAKR., *Spol. Zeyl.*, XIII, p. 237, (1925). [1]Quetta.

ramakrishnai MANI

Leucospis ramakrishnai, MANI, *Rec. Ind. Mus.*, XXXVII, p. 248, (1935)[1]. [1]Pusa.

viridissima ENDERLEIN.

Leucospis viridissima, Ender., *Ent. Mitteil.*, I, p. 144, (1912)[1]. RAMAKR., *Spol. Zeyl.*, XIII, p. 237, (1925). [1]Ceylon.

34. Epexochlaenoides GIRAULT

Epexochlaenoides, GIR., *Mem. Queensland Mus.*, IV, p. 357, (1915), [Genotype *E. bicinctus* Fig.]. WELD, *Proc. U.S.N. Mus.*, LXI, art. 6, p. 35, (1922).

pyriformis WELD.

Epexochlaenoides pyriformis, WELD, *Proc. U.S.N. Mus.*, LXI, art. 6, p. 35, (1922)[1] RAMAKR., *Spol. Zeyl.*, XIII, p. 237, (1925). MANI, *Rec. Ind. Mus.*, XXXVIII, p. 338, (1936)[2], [Notes]. [Host : *Rhynchium nitidulum*]. [1, 2]Pusa. (Also Coimbatore.)

SUBFAMILY II. CHALCIDINAE

WALK., *Ent. Month, mag.*, II, p. 20, (1835). HALIDAY, *Hymen. Synop.*, p. ii, (1839) FÖRST., *Hymen. Stud.*, II, p. 29, (1856). KIRB., *Journ. Linn. Soc.*

---

* According to Schletterer and authors, this is a synonyme of *L. gazeratensis*.

London, XVII, pp. 53-78, (1884). ASHM., *Proc. Ent. Soc. Washington*, pp. 87-88, (1888), iv, p. 247, (1897); *Mem. Carnegie Mus.*, I, (4), p. 247, (1904). SCHMIED., *Gen. Ins., fas.* 97, p. 17, (1909). GIR., *Mem. Queensland Mus.*, Iv, p. 314, (1915). MASI, *Boll. Lab. Ent. Bologna*, II, pp. 156-188, figs. 9, (1929), [Oriental forms].

TRIBE 1. BRACHYMERARIÆ

ASHM., *Mem. Carnegie Mus.*, I, (4), p. 248, (1904), SCHMIED., *Gen. Ins., fas.* 97, p. 18, (1909).

35. Oncochalcis CAMERON

Oncochalcis, CAM., *Entomologist*, XXXVII, p. 161, (1904). SCHMIED., *Gen. Ins., fas.* 97, p. 20, (1909) [Genotype *O. marginata* CAM.].

deesae CAMERON

Oncochalcis deesoe, CAM., *Journ. Bombay Nat. Hist. Soc.*, XVII, p. 580, (1907)[1]. RAMAKR., *Spol. Zeyl.*, XIII, p. 238, (1925).

[1]Bombay.

deesensis CAMERON

Oncochalcis deesensis, CAM., *Zeit. Hymen. Dipt.*, V, p. 285, (1905)[1]. RAMAKR., *Spol. Zeyl.*, XIII, p. 237, (1925).

[1]Deesa: Bombay.

marginiscutis CAMERON

Oncochalcis marginiscutis, CAM., *Journ. Bombay Nat. Hist. Soc.*, XVII, p. 168, (1907)[1]. RAMAKR., *Spol. Zeyl.*, XIII, p. 238, (1925).

[1]India.

nursei CAMERON

Oncochalcis nursei, CAM., *Journ. Bombay Nat. Hist. Soc.*, XVII, p. 580, (1907)[1]. RAMAKR., *Spol. Zeyl.*, XIII, p. 238, (1925).

[1]Bombay. (Also Punjab?)

quettaensis CAMERON

Oncochalcis quettaensis, CAM., *Journ. Bombay Nat. Hist. Soc.*, XVII, p. 194, (1906)[1]. RAMAKR., *Spol. Zeyl.*, XIII, p. 238, (1925).

[1]Quetta.

rotundata CAMERON

Oncochalcis rotundata, CAM., *Zeit. Hymen. Dipt.*, V, p. 231, (1905).[1] RAMAKR., *Spol. Zeyl.*, XIII, p. 237, (1925).

[1]Quetta.

rufescens CAMERON

Oncochalcis rufescens, CAM., *Journn. Bombay Nat. Hist. Soc.,* XVII, p. 195, (1906)[1]. RAMAKR., *Spol. Zeyl.,* XIII, p. 238, (1925). [1]Quetta.

Chalcis rufescens, Husain, *Rep. Proc. Fifth ent. Meet. Pusa,* p. 120, (1924)[1] (?). [Host : Cotton *Earias*]. [1]Punjab.

36. Trigoneura Sichel.

Trigoneura, Sich., *Ann. Soc. Ent. France,* V, p. 376, (1865), [Genotype *T. crassicauda* Sich.]. ASHM., *Mem. Carnegie Mus.,* I, (4), p. 349, (1904). SCHMIED., *Gen. Ins. fas.* 97, p. 22, (1909). WATERST., *Ind. For. Rec.,* IX, p. 60, (1923).

Centrochalcis, CAM., (*partim*), *Ind. For. Rec.,* IV, p. 92, (1913).

ruficaudis (CAMERON)

Centrochalcis ruficaudis, CAM., *Ind, for. Rec.,* IV, p. 92, (1913)[1]. [1]Dehra Dan.

Trigoneura ruficaudis, WATERST., *Ind. for. Rec.,* IX, p. 60, (1922)[1]. RAMAKR., *Spol. Zeyl.,* XIII, p. 238, (1925). [Bred out of *Heritiera fomes* attacked by *Chrysobothris sp.* in company with *Glenea Derolus dissicollis* and *Diorthus simplex*]. [1]Dehra Dan. (Also Sunderbans Bengal.)

tenuicaudis WATERSTON

Trigoneura tenuicaudis, WATERST., *Ind. for. Rec.,* IX, p. 62, (1992)[1]. RAMAKR., *Spol. Zeyl.,* XIII, p. 238, (1925)[2]. (Host : a Buprestid *Chrysobothris* (?)]. [1]Dehra Dun. [2]Bengal.

37. Stenochalcis MASI

Stenochalcis, MASI, *Boll. lab. Ent. Bologna,* II, p. 155, (1929), [Genotype *S. quadridentata* MASI].

quadridentata MASI

Stenochalcis quadridentata, MASI, *Boll. Lab. Ent. Bologna,* II, p. 157, (1929)[1]. RAMAKR. & MARGAB., *Journ. Bombay Nat. Hist. Soc.,* XXXVII, p. 194, (1934). [1]Assam.

38. Pseudochalcis KIRBY

Halticella, WALK., (*partim*), *Trans. Ent. Soc. London,* I, p. 360, (1862).

Pseudochalcis, KIRB., *Journ. Linn. Soc. London*, (*zool.*), XVII, pp. 54, 62, (1883), [Genotype *P. declarator* (WALK.)]. ASHM., *Mem Carnegie Mus.*, I, (4), p. 249, (1904). SCHMIED., *Gen. Ins.*, *fas* 97, p. 23, (1909). MANI, *Rec. Ind. Mus.*, XXXVII, p. 249, (1935).

indica MANI

Pseudochalcis indica, MANI, *Rec. Ind. Mus.*, XXXVII, p. 249, (1935)[1].

[1]Barkuda Island: Chilka Lake.

39. Oxycoryphe KRIECHBAUMER

Oxycoryphe, KRIECHB., *Berlin. Ent. Zeit.*, XXXIX, p. 67, (1894), [Genotype *O. subaenea* KRIECHB.]. ASHM., *Mem. Carnegie Mus.*, I (4), p. 249, (1904). SCHMIED., *Gen. Ins.*, *fas.* 97, p. 23, (1909).

Pilosellus CAMERON

Oxycoryphe pilosellus, CAM., *Entomologist*, p. 110, (1904). RAMAKR., *Spol. Zeyl.*, XIII, p. 240, (1925).

40. Stypiura KIRBY.

Chalcis, Perty, (*partim*), *Del. Anim. Art.*, p. 134, (1834).

Stypiura, KIRB., *Journ. Linn. Soc. London*, (*Zool.*), XVII, p. 59, pl. III, figs. 28-30, (1883), [Genotype *S. conigaSTRA* (Perty)]. ASHM., *Mem. Carnegie Mus.*, I, (4), p. 249, (1904). SCHMIED.; *Gen. Ins.*, *fas.* 97, p. 23, (1909). MANI, *Rec. Ind. Mus.*, XXXVII, p. 250, (1935).

variabilis MANI

Stypiura variabilis, MANI, *Rec. Ind. Mus.*, XXXVII, p. 250, (1935)[1].

[1]Sadiya, Margherita: Assam; Maddathoray: Travancore; Calcutta; Karachi & Madras.

41. Megalocous KIRBY

Halticella, WALK., (*partim*), *Trans. Ent. Soc. London*, I. p. 357, (1862).

Megalocolus, KIRB., *Journ. Linn. Soc. London*, (*Zool*)., XVII, pp. 54, 61, (1883), [Genotype *M. ducator* (WALK.)]. ASHM., *Mem. Carnegie Mus.*, I, (4), p. 249, (1904).

SCHMIED., *Gen. Ins., fas.* 97, p. 24, (1909). MASI, *Ann. Mus. civ. star. Nat.*, X, pp. 182-186, (1924).

Pruinosus CAMERON

Megacolus pruinosus, CAM.,* *Journn. Bombay Nat. Hist. Soc.*, XVII, p. 93, (1906)[1]. RAMAKR., *Spol. Zeyl.*, XIII, p. 238, (1925). [1]Quetta.

41a. Temnata CAMERON

Temnata, CAM., *Mem. Proc. Manchester Lit. Philos. Soc.*, XLI, (4), p. 42, (1897), [Genotype *T. maculipennis* CAM.].

maculipennis CAMERON

Temnata maculipennis, CAM., *Mem. Proc. Manchester Lit. Philos. Soc.*, XLI, (4), p. 42, (1897)[1]. RAMAKR., *Spol. Zeyl.*, XIII, p. 239, (1925). [1]Agra.

42. Brachymeria (Westwood) Stephens.#

Brachymeria, WESTW., *Philos. Mag.*, I, p. 127, (1832), [Genotype *B. minuta* (Linn.)]. Steph., *Cat. Brit. Ins. London*, p. 393, (1829).

Chalcis, Linn., *Syst. Nat.*, (ed. 12), I, ii, p. 952, (1767). FABR., *Mant. Ins.*, I, p. 272, (1787). ASHM., *Mem. Carnegie Mus.*, I, (4), p. 249, (1904). SCHMIED., *Gen. Ins., fas.* 97, p. 25, (1909). CRAW., *Bull. U. S. Dept. Agric. Bur.*, No. 19, (2), *ent.* 30, IV, (1913). WATERST., *Ind. For. Rec.*, Ix, p. 53, (1922). RUSCHKA, *Konowia*, I, pp. 221233, (1922). GAHAN and Fagan, *Bull. U.S.N. Mus.*, CXXIV, p. 24, (1923).

alphius WALKER.

Chalcis alphius, WALK., *Ann. Mag. Nat. Hist.*, XVII, p. 108, (1846)[1]. SCHMIED., *Gen. Ins., fas.* 97, p. 26, (1909). RAMAKR., *Spol. Zeyl.*, XIII, p. 238, (1925). [1]India.

---

* Cameron's *Megacolus* is an error for *Megalocolus*.

\# According to Gahan and Fagan (*Bull. U. S. N. Mus.*, No. 124, pp. 24, 31, 1923) the name *Brachymeria* should be used in the sense of *Chalcis* of authors (not of Fabricius).

amphissa WALKER

Chalcis amphissa, WALK., *List Hymen. Brit. Mus.*, I, p. 83, (1846)[1]. SCHMIED., *Gen. Ins.*, *fas.* 97, p. 26, (1909). RAMAKR., *Spol. Zeyl.*, XIII, p. 238, (1925). — [1]India.

argentifrons Ashmead.

Chalcis argentifrons, ASHM., *Journ. N.Y. Ent. soc.*, XII, p. 136, (1904), GAH., *Proc. U.S.N. Mus.*, LVI, p. 518, (1919)[1]. RAMAKR., *Rep. Proc. Fourth Ent. Meet. Pusa*, p. 364, (1922)[2]; *Spol. Zeyl.*, XIII, p. 239, (1925); *Bull. Ent. Res.*, XVIII, p. 77, (1927)[3]. [Host : *Lasiocampid*]. — [1,2,3]Coimbatore.

aSTRidens WATERSTON

Chalcis aSTRidens, WATERST., *Ind., For. Rec.*, IX, p. 56, (1922)[1]. RAMAKR., *Spol. Zeyl.* XIII, p. 239, (1925). — [1]Dehra Dun.

bengalensis CAMERON

Chalcis bengalensis, CAM., *mem. Proc. Manchester Lit. Philos. Soc.*, XLI, (4), p. 39, (1897)[1]. RAMAKR., *Spol. Zeyl.*, XIII, p. 238, (1925). — [1]Barrackpore: Calcutta.

bilobata CAMERON

Chalcis bilobata, CAM., *Zeit. Hymen. Dipt.*, V, p. 283, (1905). WATERST., *Ind., For. Rec.*, IX, pp. 55-56, fig. 4, (1922)[1]. RAMAKR., *Spol. Zeyl.*, XIII, p. 239, (1925). — [1]Bhowali, Peora, Bhim Tal: Kumaon Himalayas.

criculae KHOL

Chalcis criculae, KHOL., *Ind. Mus., Notes*, I, p. 80, (1889)[1]. SCHMIED., *Gen. Ins.*, *fas.* 97, p. 26, (1909). RAMAKR., *Spol. Zeyl.*, XIII, p. 238, (1925). [Host : wild silkworm *Cricula*]. — [1]Ranchi.

dividens WALKER

Chalcis dividens, WALK., *Ann. Mag. Nat. Hist.*, VI, p. 357, (1860)[1]. SCHMIED. *Gen. Ins.*, *fas.* 97, p. 26, (1909). RAMAKR., *Spol. Zeyl.*, XIII, p. 238, (1925). — [1]Ceylon.

eccentris CAMERON

Chalcis eccentris, CAM.., *Mem. Proc. Manchesier Lit. Philos. Soc.*, XLI, (4), p. 39, (1897)[1]. RAMAKR., *Spol. Zeyl.*, XIII. p. 239, (1925). — [1]Bombay.

elongatula MOTSCHULSKY

Chalcis elongatula, MOTSCH., *Bull. Soc. Imp. Nat. Moscou,* XXXVI, (3), p. 40, (1863)[1]. SCHMIED., *Gen. Ins., fa.* 97, p. 26, (1909). RAMAKR., *Spol. Zeyl.,* XIII, p. 238, (1925). — [1]Mt. Patannas: Ceylon.

euplœæ (Hope) Westwood

Brachymeria euplœæ, WESTW., *Trans. Ent. Soc. London,* II, p. 6, (1837)[1]. — [1]India.

Chalcis euplœæ, COTES,* *Ind. Mus. Notes,* I, p. 319, pl. iii, fig. 1g, (1888)[2]. SCHMIED., *Gen. Ins., fas.* 97, p. 26, (1909). RAMAKR., *Spol. Zeyl.,* XIII, p. 238, (1925). [Host : Moths]. — [2]E. Dooars, Goalpara. (Also Puri: Orissa & Krishnagar: Bengal.)

fulvitarsis CAMERON

Chalcis fulvitarsis, CAM., *Journ. Bombay Nat. Hist. Soc.,* XVII, p. 94, (1906)[1]. WATERST., *Ind. For. Rec.,* IX, p. 56, (1922). RAMAKR. *Spol. Zeyl.,* XIII, p. 239, (1925). — [1]Quetta.

hearseyi KIRBY

Chalcis hearseyi, KIRB., *Journ. Linn., Soc. London, (Zool.),* XVII, p. 76, (1883)[1]. SCHMIED., *Gen. Ins., fas.* 97, p. 27, (1909). RAMAKR., *Spol. Zeyl.,* XIII, p. 238, (1925). — [1]India. (Also Dehra Dun.)

hearseyi var. **Xanthoterus** WATERSTON

Chalcis hearseyi var. *xanthoterus,* WATERST., *Ind., For. Rec.,* IX, p. 58, (19252)[1]. RAMAKR., *Spol. Zeyl.,* XIII, p. 238, (1925). [Hosts : Pupae of *Hypsipyla & Euploea*]. — [1]Dehra Dun.

Koebelei CRAWFORD.

Chalcis koebelei, CRAW., *U.S. Dept. Agric. Ent. Tech. Ser.,* XIX, (2), p. 207, (1910). (Several specimens det. by MASI in the Indian Museum, Calcutta)[1]. — [1]Nilgiris, Bangalore, Puri, Margherita.

lasus WALKER

Chalcis lasus, WALK., *Entomologist,* p. 219, (1841)[1]. SCHMIED., *Gen. Ins., fas.* 97, p. 27, (1929). RAMAKR., *Spol. Zeyl.,* XIII, p. 238, (1925). — [1]India.

---

* On the authority of Cameron, Cotes gives the following synonymy of this species:–

C. lasus, Walk., *Entomologist,* p. 219, (1841).

C. inclinator, Walk., *Trans. Ent. Soc. London,* (3), I, P. 355, (1862).

C. obscurata, Walk., *Trans. Ent. Soc. London,* p. 399, (1871).

Marginata (CAMERON).*

Oncochalcis marginata, CAM., *Entomologist*, p. 162, (1904)[1]; *Zeit. Hymen. Dipt.*, V, p. 284, (1905). SCHMIED., *Gen. Ins., fas.* 97, p. 21, (1909). — [1]India.

Chalcis marginata, Susainathan, *Rep. Proc. Fifth Ent. Meet. Pusa*, p. 32, (1924)[1]. WATERST., *Ind. For. Rec.*, IX, p. 52, (1922)[2]. RAMAKR., *Spol. Zeyl.*, XIII, p. 238, (1925), [Host : Pupa of a Lymantriid]. — [1]Malaya. [2]Dehra Dun.

megaspila CAMERON

Chalcis megaspila, CAM., *Journ. Bombay Nat. Hist. soc.*, XVII, p. 581, (1906)[1]. RAMAKR., *Spol. Zeyl.*, XIII, p. 239, (1925). — [1]Mt. Abu.

nephantidis GAHAN

Brachymeria nephantidis, GAH., *Proc. U.S.N. Mus.*, No, 2831, art. 77, p. 5, (1930)[1]. RAMAKR. and MARGAB., *Journ. Bombay Nat. Hist. Soc.*, XXXVII, p. 194, (1934). [Host : *Nephantis serinopa*]. — [1]Coimbatore, Ayodyapatnam, Mangalore, Cochin and Samalkot.

ornatipes CAMERON

Chalcis ornatipes, CAM., *Journ. Bombay Nat. Hist. Soc.*, XVII, p 290, (1906)[1]. RAMAKR., *Spol. Zeyl.*, XIII, p. 239, (1925). — [1]Kashmere.

pandens WALKER

Chalcis pandens, WALK., *Ann. Mag. Nat. Hist.*, VI, p. 357, (1860)[1]. SCHMIED., *Gen. Ins., fas.* 97, p. 27, (1909). RAMAKR., *Spol. Zeyl.*, XIII, p. 238, (1925). — [1]Ceylon.

paraplesia CRAWFORD

Chalcis paraplesis CRAW., *U.S. Dept. Agric. Ent. Tech. Ser.*, XIX, (2), pp. 23-23, (1910). (Several examples det. by MASI in the collections of the Indian Museum, Calcutta)[1]. — [1]Calcutta. Kasauli, Kurseong and Mungphu.

providens MOTSCHULSKY

Chalcis providens, MOTSCH., *Bull. Soc. Imp. Nat. Moscou*, XXXVI, (3), p. 39, (1863)[1]. SCHMIED., *Gen. Ins., fas* 97, p. 28, (1909). RAMAKR., *Spol. Zeyl.*, XIII, p. 238, (1925). — [1]Mt. Nura Ellia: Ceylon.

---

* Provisional vide *Oneochalcis.*

pulchripes HOLMGREN

Brachymeria pulchripes, Holm., *Eugen. Resa., Ins.,* p. 436, (1868).

Chalcis pulchripes, SCHMIED., *Gen. Ins., fas.* 97, p. 28, (1909). (Several examples det. byb MASI in the collections of the Indian Museum, Calcutta)[1]. — [1]Chilka Lake, Calcutta, Mandalay, Karachi and Bangalore.

responsator WALKER

Chalcis responsator, WALK., *Trans. Ent. Soc. London,* p. 355, (1862). CAM., *Journ. Bombay Nat. Hist. Soc.,* XVII, p. 94, (1906)[1]. SCHMIED., *Gen. Ins., fas.* 97, p. 28, (1909). Husain, *Rep. Proc. Fifth Ent. Meet. Pusa,* p. 35, (1924)[2]. RAMAKR., *Spol. Zeyl.,* XIII, p. 238, (1925). [Host : Cotton *Earias*]. — [1]Quetta. [2]Punjab.

tachardiæ CAMERON

Chalcis tachardiæ, CAM., *Ind. For. Rec.,* IV, p. 93, (1913)[1]. WATERST., *Ind. for. Rec.,* IX, p. 58, (1922)[2]. Husain, *Rep. Proc. Fifth Ent. Meet. Pusa,* p. 35, (1924)[3]. RAMAKR., *Spol. Zeyl.,* XIII, p. 239, (1925). [From lac and from pupa of *Hypsipyla*]. — [1, 2]Dehra Dun [3]Punjab.

tarsalis MOTSCHULSKY

Brachymeria tarsalis, MOTSCH., *Bull. Soc. Imp. Nat. Moscou,* XXXVI, (3), p. 38, (1863)[1]. — Mt. Nura Ellia: Ceylon.

Chalcis tarsalis, SCHMIED., *Gen., Ins., fas.* 97, p. 28, (1909). RAMAKR., *Spol. Zeyl.,* XIII, p. 238, (1925).

TRIBE 2. *CHALCIDARIAE* (= SMICRARIAE)

ASHM., *Mem. Carnegie Mus.,* I, (4), p. 250, (1904). SCHMIED, *Gen. Ins., fas.* 97, p. 29, (1909).

43. Chalcis Fabricius.*

Chalcis, FABR., *Mant. Ins.,* I, p. 272, (1787), [Genotype *C. sispes* (FABR.)=*C. myrifex* (Sulz.)].

Smicra, Spin., *Ann. Mus. Hinst. Nat.,* XVII, p. 147, (1811).

---

* According to Gahan and Fagan (*Bull. U. S. N. Mus.,* No. 124, p. 31, 1923) the names *Chaleis* of Fabricius (not of authors) should be used in the sense of *Simiera* of Spl. nola. Also *vide* foot-note under *Brachymeria.*

KIRB., *Journ. Linn. Soc. London,* XVII, p. 54, (1883). ASHM., *Mem. Carnegie Mus.,* I, (4), p. 250, (1904). SCHMIED., *Gen. Ins., fas.* 97, p. 34, (1909).

nigrorufa (WALKER)

Smicra nigrorufa, WALK., *Ann. Mag. Nat. Hist.,* X, p. 47, (1852)[1]. SCHMIED., *Gen. Inst.,* 97, p. 36, (1909). RAMAKR., *Spol. Zeyl.,* XXII, p. 239, (1925). [1]India.

44. Spilochalcis Thomson†

Chalcis, DALM., *(partim), Svensk. Vet.-Akad. Handl.,* XLI, p. 141, (1820).

Spilochalcis, THOMS., *Hymen. Skandin.,* IV, p. 15, (1875), [Genotype *S. xanthostigma* (DALM.)]. ASHM., *Mem. Carnegie Mus.,* I, (4), pp. 251, 253, 413-447, (1904). schmied., *Gen. Ins., fas.* 97, p. 37, (1909). MANI, *Rec. Ind., Mus.,* XXXVII, p. 251, (1935).

indica MANI

Spilochalcis indica, MANI, *Rec. Ind. Mus.,* XXXVII, p. 252, fig 6, (1935)[1]. [1]Shillong.

Simlaensis CAMERON

Spilochalcis simlaensis, CAM., *Journ. Bombay nat. Hist. Soc.,* XIV, p. 438, (1903)[1]. RAMAKR., *Spol. Zeyl.,* XIII, p. 239, (1925). [1]Simla.

TRIBE 3. *HALTICELLARIÆ*

ASHM., *Mem. Carnegie Mus.,* I, (4), p. 255, (1904). SCHMIED., *Gen. Ins., fas* 97, p. 49, (1909).

45. Hockeria WALKER

Hockeria, WALK., *Ent. Month. Mag.,* II, pp. 21, 34, (1834), [Genotype *H. dexius* WALK.]. ASHM., *Mem. Carnegie Mus.,* I (4), p. 255, (1904). SCHMIED., *Gen. Ins., fas.* 97, p. 53 (1909). MASI, *Boll. Soc. Ent. Ital.,* LIII, pp. 11-16, (1929); *Boll. Lab. Ent. Bologna,* II, p. 180, (1929).

---

† For further records *vide* addendum.

atra MASI

Hockeria atra, MASI, *Boll. lab. Ent. Bologna,* II, p. 180, (1829)[1]. RAMAKR. and MARGAB., *Jaurn. Bombay Nat. Hist. Soc.,* XXXVII, p. 194, (1934). [1]Puri: Orissa.

46. Neochalcis KIRBY

Neochalcis, KIRB., *Journ. Linn. soc. London.* (*Zool.*), XVII, pp. 54, 63, (1883), [Genotype *N. osmicida* Saund.]. ASHM., *Mem. Carnegie Mus.,* I, (4), p. 255, (1904). SCHMIED., *Gen. Ins., fas.* 97, p. 54, (1909).

forticaudis CAMERON

Neochalcis forticaudis, CAM., *Journ. Bombay Nat. Hist. Soc.,* XVII, p. 96, (1606)[1]. RAMAKR., *Spol. Zeyl.,* XIII, p. 240, (1925). [1]Quetta.

47. Euchalcis Dufour

Euchalcis, Duf., *Ann. soc. Ent. France,* I, p. 8, (1861), [Genotype *E. miegii* Duf.]. KIEFF., *Ann. Soc. Ent. France,* pp. 372, 378, (1899), [species tabulated]. ASHM., *Mem. Carnegie Mus.,* I, (4), p. 255, (1904). SCHMIED., *Gen. Ins., fas.* 97, p. 54, (1909).

trichiocephala CAMERON

Euchalcis trichiocephala, CAM., *Journ. bombay Nat. Hist. Soc.,* XVII, p. 579, (1907)[1]. SCHMIED., *Gen. Ins., fas.* 97, p. 55, (1909). RAMAKR., *Spol. Zeyl.,* XIII, p. 239, (1925). [1]Deesa : Bombay.

48. Euchalcidia MASI

Euchalcidia, MASI, *Mem. Soc. Ent. Ital.,* VI, pp. 209, 221, (1927), [Genotype not known]. *Ann. Mus. civ. stor. Nat. Genova,* LIII, p. 201, (1929); *Boll. Lab, Ent. bologna,* II, p. 173, (1929).

crassicornis MASI

Euchalcidia crassicornis, MASI, *Boll. Lab. Ent. Bologna,* II, p. 173, (1929)[1]. RAMAKR. and MARGAB., *Journ. Bombay Nat Hist. Soc.,* XXXVII, p. 193, (1934). [1] Asansol, Katihar Purnea.

49. EugaSTRochalcis MASI

EugaSTRochalcis, MASI, *Boll. Lab. Ent. Bologna*, II, p. 181, (1929), [Genotype *E. breviceps* MASI].

breviceps MASI

EugaSTRochalcis breviceps, MASI, *Boll. Lab. Ent. Bologna*, II, p. 182, (1929)[1]. [1]Bhagalpore.

50. Halticella Spinola

Chalcis, FABR., (*partim*), *Mant. Ins.*, I, p. 272, (1787).

Halticella, Spin., *Ann. Mus. Hist. Nat.*, XVII, p. 148, (1811) [Genotype *H. pusilla* (FABR.)]. KIRB., *Journ. Linn. Soc. London*, (*Zool.*), XVII, p. 64, (1883). ASHM., *Mem. Carnegie Mus.*, I, (4), p. 255, (1904). SCHMIED., *Gen. Ins.*, *fas.* 97, p. 56, (1909). MASI, *Boll. Soc. Ent. Ital.*, LXI, pp. 11-16, (1929).

erythropus CAMERON

halticella erythropus, CAM., *Mem. Proc. Manchester Lit. Philos. Soc.*, XLI, (4), p. 41, (1879)[1]. RAMAKR., *Spol. Zeyl.*, XIII, p. 239, (1925). [1]Agra.

inficiens WALKER

Halticella inficiens, WALK., *Ann. Mag. Nat. Hist.*, VI, p. 357, (1860)[1]. SCHMIED., *Gen. Ins.*, *fas.* 97, p. 57, (1909). RAMAKR., *Spol. Zeyl.*, XIII, p. 239, (1925). [1]Ceylon.

Macrocerus WATERSTON

Halticella macrocerus, WATERST., *Ind. For. Rec.*, IX, p. 73, (1922)[1]. RAMAKR., *Spol. Zeyl.*, XIII, p. 239, (1925). [1]Bhim Tal: Kumaon.

ornatipennis CAMERON

Halticella ornatipennis, CAM., *Journ. Bombay Nat. Hist. Soc.*, XIV, p. 439, (1903)[1]. RAMAKR., *Spol. Zeyl.*, XIII, p. 239, (1925). [1]Deesa : Bombay.0

rufimanus WALKER

Halticella rufimanus, WALK., *Ann. Mag. Nat. Hist.*, VI, p. 357, (1860)[1]. SCHMIED., *Gen. Ins.*, *fas.* 97, p. 57, (1909). RAMAKR., *Spol. Zeyl.*, XIII, p. 239, (1925). [1]Ceylon.

Sepyra WALKER

Halticella sepyra, WALK., *Ann. Mag. Nat. Hist.,* XVII, p. 110 (1846) SCHMIED., *Gen. Ins., fas.* 97, p. 57, (1909) RAMAKR., *Spol. Zeyl.,* XIII, p. 239, (1925). [1]India.

tarsalis WALKER

Halticella tarsalis, WALK., Notes Chalc., p. 44, (1871)[1]. SCHMIED., *Gen. Ins., fas.* 97, p. 57, (1909). RAMAKR., *Spol. Zeyl.,* XIII, p. 239, (1925). [1]India.

51. Antrocephalus KIRBY

Halticella, WALK., (*partim*), *Notes Chalc.,* p. 43, (1871).

Antrocephalus, KIRB., *Journ. Linn. Soc. London* (*Zool.*), XVII, pp. 54, 63, (1883), [Genotype *A. fascicornis* (WALK.)]. ASHM., *Mem. Carnegie Mus.,* I, (4), p. 255. (1904). SCHMIED., *Gen. Ins., fas.* 97, p. 58, (1909).

brevicornutus STRand.

Antrocephalus brevicornutus, STR., *Faun. Exot.,* I, p. 8, (1911)[1]. RAMAKR., *Spol. Zeyl.,* XIII, p. 240, (1925). [1]Ceylon.

deSTRuctor WATERSTON

Antrocephalus deSTRuctor, WATERST., *Ind. for. Rec.,* IX, p. 64, (1922)[1]. RAMAKR., *Spol. Zeyl.,* XIII, p. 240, (1925). [Host : *Hypsipyla robusta* of Toon tree]. [1]Dehra Dun.

divisicornis (WALKER)

Halticella divisicornis, WALK., *Notes Chalc.,* p. 44, (1871)[1]. [1]India.

Antrocephalus divisicornis, KIRB., *Journ. Linn. Soc. London,* (*Zool.*), XVII, p. 63, fig. 24-26, (1883). SCHMIED., *Gen. Ins., fas.* 97, p. 58, (1909). RAMAKR., *Spol. Zeyl.,* XIII, p. 240, (1925).

fascicornis WALKER

Halticella fascicornis, WALK., *Notes Chalc.,* p. 43, (1871)[1]. [1]India.

Antrocephalus fascicornis, KIRB., *Journ. Linn. Soc. London,* (*Zool.),* XVII, p. 63, fig. 24-26, (1883). SCHMIED., *Gen. Ins., fas.* 97, p. 58, (1809) RAMAKR., *Spol. Zeyl.,* XIII, p. 240, (1925).

humilis WATERSTON

Antrocephalus humilis, WATERST., *Ind. For. Rec.*, IX. p. 72, (1922)[1]. RAMAKR., *Spol. Zeyl.*, XIII, p. 240, (1925). [1]Dehra Dun.

longicornutus STRand

Antrocephalus longicornutus, STR., *Faun. Exot.*, I, p. 8, (1911)[1]. RAMAKR., *Spol. Zeyl.*, XIII, p. 240, (1925). [1]Ceylon.

phaeospilus WATERSTON

Antrocephalus phoeospilus, WATERST., *Ind. for : Rec.*, IX, p. 72, (1922)[1]. RAMAKR., *Spol. Zeyl.*, XIII, p. 240, (1925). [1]Bhim Tal: Kumaon.

Renalis WATERSTON

Antrocephalus renalis, WATERST., *Ind. for. Rec.*, IX. p. 69, (1922)[1]. RAMAKR., *Spol. Zeyl.*, XIII, p. 240, (1922). [Host : *Hypsipyla robusta*]. [1]Dehra Dun.

Tristis STRAND

Antrocephalus tristis, STR., *Faun. Exot.*, I, p. 8, (1911)[1]. RAMAKR., *Spol. Zeyl.*, XIII, p. 240, (1925). [1]Ceylon.

varipilosus CAMERON

Antrocephalus varipilosus, CAM., *Journ. Bombay Nat. Hist. Soc.*, XVII, p. 508, (1907)[1]. SCHMIED., *Gen. Ins. fas.* 97, p. 58, (1909). RAMAKR., *Spol. Zeyl.*, XIII, p. 240, (1925). [1]Deesa: Bombay.

52. Stomatoceras KIRBY

Halticella, WALK., (*partim*), *Trans. Ent. Soc. London*, (3), i, p. 361, (1862).

Stomatoceras, KIRB., *Journ. Linn. Soc. London*, (*Zool.*), XVII, p. 61, (1883), [Genotype *S. liberator* (WALK.)]. ASHM., *Mem. Carnegic Mus.*, I, (4), p. 255, (1904). SCHMIED., *Gen. Ins.*, *fas.* 97, p. 58, (1909).

ayyari GAHAN

Stomatoceras ayyari, GAH., *Proc. U.S.N. Mus.*, LVI, p. 518, (1919)[1]. RAMAKR., *Rep. Proc. Fourth Ent. Meet. Pusa*, p. 30, pl. ii, fig. 1; p. 364, (1921)[2], [Recorded on *Parasa lepida*, a pest of mango, castor, etc.]; *Spol, Zeyl.*, XIII, p. 240, (1925); *Bull. Ent. Res.*, XVIII, p. 77, (1927)[3]. [Host : *Parasa lepida.*] [1,2,3]Coimbatore.

sulcatiscutellum Giurault.*

Stomatoceras sulcatiscutellum, RAMAKR., *Rep. Proc. Third Ent. Meet. Pusa,* III, p. 934, (1920)[1]; *Spol. Zeyl.,* XIII, p. 240, (1925); *Bull. Ent. Res.,* XVIII, p. 77, (1927)[2]. [Host : *Nephantis serinopa.*]. [1,2]Coimbatore.

53. Centrochalcis CAMERON

Centrochalcis, CAM., *Zeit. Hymen. Dipt.,* V. p. 230, (1905), [Genotype *C. octodentata* CAM.]. SCHMIED., *Gen. Ins., fas.* 97, p. 60 (1909).

Sp. WATERSTON

Centrochalcis sp., WATERST., in Husain's *Rep. Proc. Fifth Ent. Meet. Pusa,* p. 35, (1924)[1]. [Host : *Earias insulana*]. [1]Punjab.

54. Megachalcis CAMERON

Megachalcis, CAM., *Roy. Asiat. Soc.,* No. 39, p. 96, (1903),. [Genotype *M. fumipennis* CAM.]. SCHMIED., *Gen. Ins., fas.* 97 p. 62, (1909).

fumipennis CAMERON

Megachalcis fumipennis, CAM., *Roy. Asiat. Soc.,* No. 39, p. 97, (1903)[1]. RAMAKR., *Spol. Zeyl.,* XIII, p. 240, (1925). [1]India.

55. Coelochalcis CAMERON

Coelochalcis, CAM., *Entomologist,* XXVII, p. 110, (1904), [Genotype *C. carinifrons* CAM.]. SCHMIED., *Gen. Ins., fas.* 97, p. 65, (1909).

Carinifrons CAMERON

Cœlochalcis carinifrons, CAM., *Entomologist,* XXVII, p. 110, (1904)[1]. SCHMIED., *Gen. Ins., fas.* 97, p. 65, (1909). RAMAKR., *Spol. Zeyl.,* XIII, p. 240, (1925). [1]Sikkim.

Carinigena CAMERON

Coelochalcis carnigena, CAM., *Journ. Bombay Nat. Hist. Soc.,* XVII, p. 579, (1907)[1]. RAMAKR., *Spol. Zeyl.,* XIII, p. 240, (1925). [1]Deesa: Bombay.

---

* The description of this species could not be traced in the literature.

TRIBE 4. *DIRHINARIÆ*

ASHM., *Mem. Carnegie Mus.*, I, (4) p. 257, (1904). SCHMIED., *Gen. Ins.*, *fas.* 97, p. 65, (1909). MASI, *Eos*, III, pp. 29-48 (1927), [Oriental forms].

56. **Dirhinus** DALMAN

Dirhinus, DALM., *Svensk. Vet.-Akad. Handl.*, XXXIX, p. 75, (1818), [Genotype *D. excavatus* DALM.]. KIRB., *Journ. Linn. Soc. London, (Zool.)*. XVII, p. 57, (1883). ASHM., *Mem. Carnegie Mus.*, I, (4), p. 257, (1904). SCHMIED., *Gen. Ins.*, *fas.* 97, p. 66, (1909). MASI, *Ann. Mus. civ. stor. Genova*, XLIII, p. 23, (1919). ROHW., *Philip. Journn. Sc.*, XXII, p. 346, (1923). MASI, *Eos*, III, p. 42, (1927).

auratus Ashmead

Dirhinus auratus, ASHM., *Proc. U.S.N. Mus.*, XXIX, p. 402, (1905). (Two examples det. by MASI in the Indian Museum, Calcutta).[1]

[1]Peradeniya: Ceylon, Puri: India.

Crythroceras CAMERON

Dirhinus crythroceras, CAM., *Journ. Bombay Nat. Hist. Soc.*, XVII, p. 92, (1906)[1]. RAMAKR., *Spol. Zeyl.*, XIII, p. 240, (1925).

[1]Quetta.

himalayanus Westwood

Dirhinus himalayanus, WESTW., in *Royle's IlluStr. Himal. Mt.*, VII, p. 40, (1839)[1]. SCHMIED., *Gen. Ins.*, *fas.* 97, p. 66, (1909).

[1]Himalayas.

Pachycerus MASI

Dirhinus pachycerus, MASI, *Eos*, III, p. 42, (1927)[1]. RAMAKR. and MARGAB., *Journ. Bombay Nat. Hist. Soc.*, XXXVII, p. 193, (1934).*

[1]Calcutta.

57. **Pareniaca** CRAWFORD

Pareniaca, CRAW., *Proc. U.S.N. Mus.*, XLV, p. 312, (1913), [Genotype *P. schwarzi* CRAW.]. MASI, *Eos*, III, p. 31, (1927).

---

* *pachymerum* is an error for *pachycerus* in Ramkar. & Margab.

trichophthalma MASI

Pareniaca trichophthalma, MASI, *Eos,* III, p. 39, (1927)[1]. [1]Gopkuda: Chilka Lake.

FAMILY IV. EURYTOMIDAE

WALK., *Ent. Month. Mag.,* I, p. 12, (1833). WESTW., *Intr. Mod. Class. Ins.,* II, pp. 66, 166, (1840). WALK., *List Chalc. Brit. Mus.,* I, p. 8, (1846). FÖRST., *Hymen. Stud.,* II, pp. 19, 23, 44, (1856). WALSH, *Amer. Ent. bot.,* II, pp. 297, 329-335, 367-370, figs. 10, (1870). WALK., *Notes Chalc.,* I, pp. 1-18, (1871). THOMS., *Hymen. Skandin.,* IV, pp. 11, 25, (1875). GUSTAV MAYR, *Verhy. Zool-Bot. Ges. Wien,* XXVIII, pp. 297-333, (1878). WESTW., *Trans. Ent. Soc. London,* pp. 307, 328, (1882). How, *Ent. Amer.,* I, p. 19, (1886). ASHM., *Proc, Ent. soc. Washington,* pp. 41-43, (1888); *Mem. Carnegie Mus.,* I (4), p. 257, (1904). SCHMIED., *Gen. Ins., fas.* 97, p. 123, (1909). GIR., *Mem. Queensland Mus.,* IV, p. 238, (1915). PHILLIPS., *Tech. bull. U.S. Dept. Agric.,* No. 518, pp. 1-25, pls. 10 figs. 4, (1936), [A second revision of *Isosoma* of America].

TRIBE 1. AXIMARIÆ

CAM., *Biol. Centr. Amer. Hymen.,* I, p. 111, (1884). ASHM., *Mem. Carnegie Mus.,* I, (4), p. 258, (1904). SCHMIED., *Gen. Ins., fas.* 97, p. 124, (1909).

58. Aximopsis ASHMEAD

Aximopsis, ASHM., *Mem. Carnegie Mus.,* I, (4), p. 259, (1904), [Genotype *A. morio* ASHM.]. SCHMIED., *Gen. Ins., fas.* 97, p. 125, (1909).

Sp. RAMAKRISHNA AYYAR.

Aximopsis sp., RAMAKR., *Spol. Zeyl.,* XIII, p. 244, (1925)[1]. [Bred from grass-hopper eggs; probably a hyper-parasite of *Tumidiscapus oophagus.*]. [1]Coimbatore.

TRIBE 2. *ISOSOMARIÆ*

ASHM., *mem. Carnegie Mus.,* I, p. 259, (1904). SCHMIED., *Gen. Ins., fas.* 97, p. 126, (1909). Hedicke, *Arch. Naturg. Berlin,* LXXXVI, H. 11, pp. 1-165, figs. 36, (1921), [Monograph of palaearctic forms].

59. Isosoma WALKER

Ichneumon, FABR., (*partim*), *Suppl. Ent. Syst.,* p. 232, (1798).

Isosoma, WALK., *Ent. Month. Mag.*, I. p. 13, (1832). [Genotype *I. verticellata* (FABR.)]. ASHM., *Mem. Carnegie Mus.*, I. p. 259, (1904). SCHMIED., *Gen. Ins.*, *fas.* 97, p. 128, (1909).

longicornis (MOTSCHULSKY)

Harmolita longicornis, MOTSCH., *Bull. Soc. Imp. Nat. Moscou*, XXXVI, (3), p. 58, (1863)[1]. [1]Mt. Patannas: Ceylon.

Isosoma (?) longicornis, RAMAKR., *Spol. Zeyl.*, XIII, p. 243, (1925).

TRIBE 3. *DECATOMARIÆ*

ASHM., *Mem. Carnegie Mus.*, I, (4), p. 265, (1904). SCHMIED., *Gen. Ins.*, *fas.* 97, p. 132, (1909).

60. Decatoma SPINOLA

Decatoma, Spin., *Ann. Mus. Hist. nat.*, XVII, p. 151, (1811), [Genotype *D. adonidum* Spin.]. ASHM., *Mem. Carnegie Mus.*, I, (4), p. 265, (1904). SCHMIED., *Gen. Ins.*, *fas.* 97, p. 132, (1909).

pigra MOTSCHULSKY

Decatoma pigra, MOTSCH., *Bull. Soc. Imp. Nat. Moscou*, XXXVI, (3), p. 43, (1863)[1]. SCHMIED., *Gen. Ins.*, *fas.* 97, p. 133, (1909). RAMAKR., *Spol. Zeyl.*, XIII, p. 243, (1925). [1]Mt. Patannas: Ceylon.

61. Decatomella MASI

indica MASI

Decatomella indica MASI. (In the Collections of the Zoological Survey of India, Indian Museum, Calcutta, there are specimens bearing this name in Masi's writing. Descriptions could not be traced in the literature available including *Zool.* Rec. up to 1935)[1]. [1]Calcutta.

TRIBE 4. *EURYTOMARIÆ*

ASHM., *Mem. Carnegie Mus.*, I, (4), p. 160, (1904). SCHMIED., *Gen. Ins.*, *fas.* 97, p. 134, (1909).

62. Eurytoma Illiger

Eurytoma, Illig., in Rossi's, *Faun. Etrusca,* (ed. 2), II, p. 128, (1807), [Genotype *E. planata* Illig.]. How., *Proc. Ent. Soc. Washington,* II. p. 66, (1891), [Parasitic habits). Ashm., *Mem. Carnegie Mus.,* I, (4), p. 262, (1904). Schmied., *Gen. Ins., fas.* 97, p. 139, (1909). Crosby, *Bull. N.Y. Expt. Sta.,* 265, pp. 67-98, (1909). Gir., *Mem. Queensland Mus.,* IV, p. 275, (1915).

Tetrasema, Walk., *List Chalc. Brit. Mus.,* II, p. 54, (1848).

Enneatoma, (Spin.), Dahlb., *Oœfvers. Svensk. Vet.-Akad. Forh.,* XIV, p. 292, (1857).

Angustula Motschulsky

Eurytoma angustula, Motsch., *Bull. Soc. Imp. Nat. Moscou,* XXXVI, (3), p. 43, (1863)[1]. Schmied., *Gen., Ins., fas.* 97, p. 140, (1909). Ramakr., *Spol. Zeyl.,* XIII, p. 244, (1925).

[1]Mt. Patannas: Ceylon.

Chrysothrix Waterston

Eurytoma chrysothrix, Waterst., *Bull. Ent. Res.,* XV, p. 69, (1924)[1]. Ramakr., *Spol. Zeyl.,* XIII, p. 244, (1925); *Bull. Ent. Res.,* XVIII, p. 77, (1927).

[1]Walayar: W. Ghats.

Contraria Walker

Eurytoma contraria, Walk., *Ann. Mag. Nat. Hist.,* VI, p. 358, (1860)[1]. Schmied., *Gen. Ins., fas.* 97, p. 141, (1909). Ramakr., *Spol. Zeyl.,* XIII, p. 244, (1925).

[1]Ceylon.

denticoxa Gahan

Eurytoma denticoxa, Gah., *Proc. U.S.N. Mus.,* LVI, p. 515, (1919)[1], Ramakr., *Rep. Proc. Fourth Ent. Meet. Pusa,* p. 364, (1921); *Spol. Zeyl.,* XIII, p. 244, (1925); *Bull. Ent. Res.,* XVIII, p. 77, (1927).[2]

[1,2]Coimbatore.

dentipectus Gahan.

Eurytoma dentipectus, Gah., *Proc. U.S.N. Mus.,* LVI, p. 516, (1929)[1]. Ramakr., *Rep. Proc. Fourth Ent. Meet. Pusa,* p. 364, (1921)[2]; *Spol. Zeyl.,* XIII, p. 244, (1925); *Bull. Ent. Res,* XVIII, p. 77, (1927)[3]. [Host : *Asphondylia sesami* Felt].

[1,2,3,]Coimbatore.

gracilis MOTSCHULSKY

Eurytoma (?) *gracilis*, MOTSCH., *Bull. Soc. Imp. Nat. Moscou*, XXXVI, (3), p. 42, (1863)[1]. SCHMIED., *Gen. Ins.*, *fas.* 97, p. 142, (1909). RAMAKR., *Spol. Zeyl.*, XIII, p. 244, (1925). — [1]Mt. Patannas: Ceylon.

hindupurensis GAHAN

Eurytoma hindupurensis, GAH., *Proc. U.S.N. Mus.*, LVI, p. 518, (1919)[1]. RAMAKR., *Rep. Proc. Fourth Ent. Meet. Pusa*, p. 364, (1921)[2]; *Spol. Zeyl.*, XIII, p. 244, (1925). — [1,2]Hindupur: South India.

indefensa WALKER

Eurytoma indefensa, WALK., *Ann. Mag. Nat. Hist.*, VI, p. 358, (1860)[1]. SCHMIED., *Gen. Ins.*, *fas.* 97, p. 1425, (1909). RAMAKR., *Spol. Zeyl.*, XIII, p. 244, (1925). — [1]Ceylon.

indi Girault

*Eurytoma indi**, GIR. (?) RAMAKR., *Rep. Proc. Third Ent. Meet. Pusa*, I, p. 315, (1920)[1]. — [1]Coimbatore.

maculipes MOTSCHULSKY

Eurytoma maculipes, MOTSCH., *Bull. Soc. Imp. Nat. Moscou*, XXXVI, (3), p. 41, (1863)[1]. SCHMIED., *Gen. Ins.*, *fas.* 97, p. 142, (1909). RAMAKR., *Spol. Zeyl.*, XIII, p. 244, (1925). — [1]Ceylon.

mitsakura Ashmead

Eurytoma mitsakura, ASHM., (?) var. nov. (innominate), GIR., in RAMAKR., *Spol. Zeyl.*, XIII, p. 244, (1925)[1]. — [1]Coimbatore.

pallidiscapus CAMERON

Eurytoma pallidiscapus, CAM., *Ind. For. Rec.*, IV, p. 98, (1912)[1]. RAMAKR., *Spol. Zeyl.*, XIII, p. 244, (1925). — [1]Dehra Dun.

Parasae GAHAN

Eurytoma parasoe, GAH., *Proc. U.S.N. Mus.*, LVI, p. 514, (1919)[1]. RAMAKR., *Rep. Proc. Fourth Ent. Meet. Pusa*, p. 364, (1921)[2]; *Spol. Zeyl.*, XIII, p. 244, (1925); *Bull. Ent. Res.*, XVIII, p. 77, (1927)[3]. [Host : *Parasa lepida*]. — [1,2,3]Coimbatore.

---

* The original description of this species could not be traced in the literature.

pigra Girault

Eurytoma (?) *pigra*, GIR.,* in RAMAKR., *Rep. Proc. Third Ent. Meet. Pusa*, I, p. 321, (1920)[1]. — [1]Coimbatore.

setitibia GAHAN

Eurytoma setitibia, GAH., *Proc. U.S.N. Mus.*, LVI, p. 517, (1919)[1]. RAMAKR., *Rep. Proc. Fourth Ent. Meet. Pusa*, p. 364, (1921)[2]; *Spol. Zeyl.*, XIII, p. 244, (1925); *Bull. Ent. Res.*, XVIII, p. 77, (1927)[3]. — [1,2,3]Coimbatore.

sp. RAMAKRISHNA AYYAR.

Eurytoma sp., RAMAKR., *Spol. Zeyl.*, XIII, p. 244, (1925)[1], [Host : *Alcides bubo*]. — [1]Coimbatore.

sulcata KIEFFER

Eurytoma sulcata, KIEFF., *Ann. soc. Sc. Bruxelles*, XXIX, p. 192, (1905)[1]. RAMAKR., *Spol. Zeyl.*, XIII, p. 244, (1925). [Host : *Cynip of Quercus pachyphylla* gall.] — [1]Bengal (?)

taprobanica Westwood

Eurytoma taprobanica, WESTW., *Trans. Ent. soc. London*, p. 327, (1882)[1]. SCHMIED., *Gen. Ins., fas.* 97, p. 144, (1909). RAMAKR., *Spol. Zeyl.*, XIII, p. 244, (1925). (In galls of *Ficus teila*]. — [1]Kandy: Ceylon.

63. Bruchophagus Ashmead

Bruchophagus, ASHM., *Ent. Amer.*, IV, p. 42, (1888), [Genotype *B. borealis* ASHM.]; *Mem. Carnegie Mus.*, I, (4), p. 262, (1904). SCHMIED., *Gen. Ins., fas.* 97, p. 146, (1909).

Mellipes GAHAN

Bruchophagus mellipes, GAH., *proc. U.S.N. Mus.*, LVI, p. 513, (1919)[1]. RAMAKR., *Rep. Roc. Fourth Ent. Meet. Pusa*, p. 364, (1921)[2]; *Spol. Zeyl.*, XIII, p. 244, (1925); *Bull. Ent. Res.*, XVIII, p. 77, (1927)[3]. [Bred from pods of *daincha* and *Redgram.*] — [1,2,3]Coimbatore.

---

* The description of this species could not be traced in the literature.

FAMILY V. PERILAMPIDÆ

WALK., *List Chalc, Brit. Mus.*, I, p. 103, (1846). FÖRST., *Hymen. Stud.*, II, pp. 19, 22, 46, (1856). THOMS., *Hymen, Skandin.*, IV, pp. 11, 22, (1875). How., *Ent. Amer.*, I, p. 198, (1886). ASHM., *Mem. Carnegie Mus.*, I, (4), p. 265, (1904). schmied., *Gen., Ins., fas.* 97, p. 81, (1909). CRAW., *Proc. Ent. Soc., Washington*, XVI, pp. 69-76, (1914), [Synopsis of N. American forms]. RUSCHKA, *Deutsch. Ent. Zeit.*, 1924, pp. 82-96, (1924), [european forms].

64. Philomides HALIDAY

Philomides, Halid., *Ann. Soc. Ent. France*, (4), II, p. 115, (1862), [Genotype *P. paphius* Halid.]. ASHM., *Mem. Carnegie Mus.*, I, (4), p. 266, (1904). SCHMIED., *Gen. Ins., fas.* 97, p. 83, (1909).

Sp. RAMAKRISHNA AYYAR.

Philomides sp., RAMAKR., *Spol. Zeyl.*, XIII, p. 241, (1925)[1]. [1]Bellary.

65. Monacon WATERSTON

Monacon, WATERST., *Ind. For. Rec.*, IX, p. 74, (1922), [Genotype *M. spinifrons* (CAM.) (?)].

abruptum WATERSTON

Monacon abruptum, WATERST., *Ind., For. Rec.*, IX, p. 70, (1922)[1]. RAMAKR., *Spol. Zeyl.*, XIII, p. 241, (1925). [Host : *Platypus uncinatus* Balndf.]

[1]Tambulbani, Sunderbans: Bengal.

productum WATERSTON

Monacon productum, WATERST., *Ind, For. Rec.*, Ix, p. 75, (1922)[1]. RAMAKR., *spol. Zeyl.*, XIII, p. 241, (1925). pHost : *Diapus furtivus* Samps.].

[1]Apalchand Range, Jalpaiguri, Dooars, Rajabhatkhawa: Bengal.

66. Perilampus LATREILLE

Perilampus, Latr., *Gen. Crust. Ins.*, IV, p. 30, (1890), [Genotype *P. violacea* (FABR.) according to LATREILLE; *P. ruficornis* (FABR.) desig. DALLA torre. Also *vide* GAH. and Fag. 1923]. ASHM., *Mem. Carnegie Mus.*, I, (4), p. 266, (1904). schmied., *Gen. Ins., fas.* 97, p. 83, (1909). smith, *Tech. Ser., U.S. Dept. Agric. Bur. Ent.*, 19, pp. 33-69, fig. 8. (1912).

Smulyan, *Proc. U.S.N. Mus.*, LXXXIII, No. 2990, pp. 369-412, (1936), [Revision].

Cynipsillum, Lamk., *Hist. Anim. S. Vert.*, IV, p. 156, (1817).

hedychroides WALKER

Perilampus hedychroides, WALK., *Notes Chalc.*, IV, p. 67, (1871)[1]. SCHMIED., *Gen. Ins., fas.* 97, p. 83, (1909). RAMAKR., *Spol. Zeyl.*, XIII, p. 241, (1925). — [1]Ceylon.

microgastris FERRIERE

Perilampus microgastris, FERR., *Bull. Ent. Res.*, XXI, pp. 353-354, (1930)[1]. RAMAKR. and MARGAB., *Journ. Bombay Nat. Hist. Soc.*, XXXVII, p. 195, (1934). [Hosts : *Microgaster indicus, Apanteles machaeralis* and an unknown Braconid of *Nephantis serinopa.*] — [1]Rhatgaon, Hosangabad: C. P., Dehra Dun: U. P. & Coimbatore.

Sp. CAMERON

Perilampus sp., CAM., In COTES, *Ind. Mus. Notes,* I, p. 32, pl. ii, fig. lh, (1898)[1]. [Host : *Dasychira thwaitesi* of Tea and *Sal.]* — [1]E. Dooars: Himalayas.

FAMILY VI. EUCHARIDAE

WALK., *List. Chalc. Brit. Mus.*, I, p. 21, (1846), FÖRST., *Hymen., Stud.*, II, pp. 18, 22, 72, (1856). How., *Ent. Amer.*, I, p. 198, (1886). KIRB., *Journ. Linn. Soc. London.* (*Zool.*). XX, pp. 28-37, (1887). ASHM., *Proc. Ent. Soc. Washington,* IV, pp. 235, 245, (1897). DALLA TORRE, *Cat. Hymen.*, V, p. 359, (1898). ASHM., *Mem. Carnegie Mus.*, I, (4), p. 266, (1904). SCHMIED., *Gen. Ins., fas.* 97, p. 67, (1909). GIR., *Mem. Queensland Mus.*, Iv, p. 255, (1915). RUSCHKA, *Deutsch. Ent. Zeit.*, pp. 82-96, figs. 12, (1924).

67. Eucharis LATREILLE

Cynips, FABR., (*partim*), *Mant. Ins.*, I, p. 21, (1787).

Eucharis, latr., *Hist. nat. Crust. Ins.*, III, p. 210, (1802), [Genotype *E. adscendens* (FABR.)]. ASHM., *Mem. Carnegie Mus.*, I, (4), p. 267, (1904), SCHMIED., *Gen. Ins., fas.* 97, p. 71, (1909).

convergens WALKER

Eucharis convergens, WALK., *Ann. Mag. Nat. Hist.*, VI, p. 358, (1860)[1]. MOTSCH., *Bull. Soc. Imp. nat. Moscou,* XXXVI, (3), p. 44, (1863)[2]. SCHMIED., *Gen. Ins., fas.* 97, p. 72, (1909). RAMAKR., *Spol. Zeyl.*, XIII, p. 240, (1925). — [1,2]Ceylon.

Myrmyciæ FOREL

Eucharis myrmyciæ, For., *Ann. Soc. Ent. Belgie*, XX, p. 8, (1890)[1]. SCHMIED., *Gen. Ins., fas.* 97, p. 72, (1909). RAMAKR., *Spo., Zeyl.* XIII, p. 240, (1925). [1]India.

testacea MOTSCHULSKY

Eucharis (?) *testacea*, MOTSCH., *Bull. Soc., Imp. Nat. Moscou*, XXXVI, (3), p. 44, (1863)[1]. SCHMIED., *Gen. Ins., fas.* 97, p 72, (1909). RAMAKR., *Spol Zeyl.*, XIII, p. 240, (1925). [1]Mt. Nura Ellia: Ceylon.

68. Chalcura KIRBY

Eucharis, WALK., (*partim*), *Ann. Mag. Nat. Hist.*, VI, p. 359, (1860).

Chalcura, KIRB., *Journ. Linn. Soc. London*, (*Zool*), XX, p. 30, (1886), [Genotype *C. deprivata* (WALK.)]. ASHM., *Mem. Carnegie Mus.*, I, (4), p. 268, (1904). SCHMIED., *Gen. Ins., fas.* 97, p. 75, (1909).

deprivata (WALKER)

Eucharis deprivata, WALK., *Ann. mag. Nat. Hist.*, VI, p. 359, (1860)[1]. MOTSCH., *Bull. soc. Imp. nat. Moscou*, XXXVI, (3), p. 44, (1863)[2]. [1,2]Ceylon.

Chalcura deprivata, KIRB., *Journ. Linn. soc. London*, (*Zool.*), XX, p. 30, (1886). SCHMIED., *Gen. Ins., fas.* 97, p. 75, (1909). RAMAKR., *Spol. Zeyl.*, XIII, p. 241, (1925).

69. Schizaspidia Westwood

Schizaspidia, WESTW., *Proc. Zool. Soc. London*, III, p. 69, (1835), [Genotype *S. furcifera* WESTW.]. ASHM., *Mem. Carnegie Mus.*, I, (4), pp. 268, 269, (1904). SCHMIED., *Gen. Ins., fas.* 97, p. 76, (1909). MANI, *Rec., Ind. Mus.*, XXXVII, p. 253, (1935).

furcifera Westwood

Schizaspidia furcifera, WESTW., *Proc. Zool. Soc. London*, III, p. 69, (1835)[1]. SCHMIED., *Gen. Ins., fas.* 97, p. 76, (1909). RAMAKR., *Spol. Zeyl.*, XIII, p. 241, (1925). [1]Bengal.

indica MANI

Schizaspidia indica, MANI, *Rec., Ind. Mus.*, XXXVII, p. 254, (1935)[1]. [1]Assam.

Manipurensis Calusen

Schizaspidia manipurensis, Claus., *Proc. Ent. Soc. Washington,* XXX, pp. 83, 85, (1928).[1] RAMAKR. and MARGAB., *Journ., Bombay Nat. Hist. Soc.,* XXXVII, p. 194, (1934).

[1]Manipur: Assam.

scutellaris MANI

Schizaspidia scutellaris, MASI, *Konowia,* V, p. 370, (1927). (There is one specimen in the Indian Museum det. by MASI).[1]

[1]Calcutta.

FAMILY VII. MISCOGASTERIDÆ

WALK., *Ent. Month. Mag.,* I, p. 370, (1833). FÖRST., *Hymen. Stud.,* II, pp. 19, 24, 51, (1856). THOMS., *Hymen. Skandin.,* Iv, pp. 12, 216, (1875). How., *Ent. Amer.,* I, p. 198, (1886). ASHM., *Proc. Ent. Soc. Washington,* Iv, pp. 235-246, (1897); *Proc. U.S.N. Mus.,* XXIII, p. 202, (1900); *Mem. Carnegie Mus.,* I, (4), p. 270, (1904). SCHMIED., *Gen. Ins., fas* 97, p. 269, (1909). GIR., *Mem. Queensland Mus.,* IV, p. 185, (1915). FERR., *Mitt. schweiz. ent. Ges.,* XVI, p. 85, (1935), [key to genera of *Pireninæ.*]

SUBFAMILY I. TRIDYMINAE

WALK., *Ent. Month. Mag.,* II, p. 167, (1835). THOMS., *Hymen. Skandin.,* IV, pp. 12, 192, (1875). How., *Ent. Amer.,* I, p. 198, (1886). ASHM., *Proc. Ent. Soc. Washington.* IV, p. 247, (1899); *Mem Carnegie Mus.,* I, (4), p. 272, (1904). SCHMIED., *Gen. Ins., fas.* 97, p. 273, (1909).

70 Arthrolysis Forster

Arthrolysis, FÖRST., *Hymen, Skandin.,* II, p. 52, (1856), [Genotype *A. scabriculus* (Nees)]. ASHM., *mem. Carnegie Mus.,* I. (4), p. 276, (1904). SCHMIED., *Gen. Ins., fas,* 97, p. 287, (1909).

Picroscytus, THOMS., *Hymen. Skandin.,* V, p. 58, (1878). MASI, *Ann. Mus. civ. stor. nat. Genova,* LII, pp. 263-268, (1927).

birmanus (MASI)

Picroscytus birmanus, MASI, *Ann. Mus. civ. stor. Nat. Genova,* LII, p. 268, (1927)[1].

[1]Burma.

indorum (MASI)

Picroscytus indorum, MASI, *Ann. Mus. civ. stor. Nat. Genova,* LII. p. 267, (1927).[1] — [1]Chilka Lake.

SUBFAMILY II. LELAPINAE

ASHM., *Proc. Ent. Soc. Washington,* IV, p. 247, (1899); *Faun. Hawaii.,* I, p. 311, (1901); *Mem. Carnegie Mus.,* I, (4), p. 279, (1904). SCHMIED., *Gen. Ins., fas* 97, p. 300, (1909). MASI, *Ann. Mus. civ. stor. Nat. Genova,* LI, pp. 149-160, (1924). MERCET., *Eos,* III, pp. 49-63, (1927).

71. Sphærakis MASI

Bruchobius, ASHM., *Mem. Carnegie Mus.,* I, (4), pp. 314, 315, (1904). SCHMIED., *Gen. Ins., fas.* 97, p. 314, (1909).

Sphcerakis, MASI, *Ann. Mus. civ. stor. Nat. Genova,* LI, p. 157, (1924), [Genotype *S. mayari* MASI].

Colemani (CRAWFORD)

Bruchobius colemani, CRAW., *Proc. U.S.N. Mus.,* XLV, p. 250, (1913)[1]. GAH., *Proc. U.S.N. Mus.,* LVI, p. 523, (1919)[2]. RAMAKR., *Rep. Proc. Fourth Ent. Meet. Pusa,* p. 363, (1922)[3]; *Spol. Zeyl.,* XIII, p. 248, (1925); *Bull. Ent. Res.,* XVIII, p. 77, (1927)[4]. [Host: *Bruchus* spp.]. — [1,3,] Bangalore, Mysore. [2,4]Coimbatore.

Sphoerakis colemani, MASI, *Ann. Mus. civ. stor. Nat. Genova,* LI, p. 157, (1924).

maculata MASI

Sphoerakis maculata, MASI, *Ann. Mus. civ. sot. Nat. Genova,* LI, p. 157, (1924)[1]. — [1]Burma, Calcutta.

72. Oedaule WATERSTON

Oedaule, WATERST., *Ind. For. Rec.,* IX, p. 81, (1922), [Genotype *O.* STR*ingifrons* WATERST.]. MASI, *Ann. Mus. civ. stor. Nat. Genova,* LI, pp. 149-160, (1924).

STR**ingifrons** WATERSTON

Oedaule STR*ingifrons,* WATERST., *Ind. For. Rec.,* IX, p. 82, (1922)[1]. [Host : *Caryoborus gonagra* FABR. (?)]. — [1]Dehra Dua.

73. Anysis Howard.*

Anysis, How., *Canad. Ent.,* XXVIII, p. 167, (1896), [Genotype *A. auStRalensis* How.]. Ashm., *mem. Carnegie Mus.,* I. (4), p. 326, (1904). Schmied., *Gen. Ins., fas.* 97, p. 307, (1909). Gir., *Mem. Queensland Mus.,* IV, p. 227, (1915). Smith and Compere, *Univ. Calif. Pub. Ent.,* IV, pp. 309-312, (1928). Masi, *Eos,* VII, p. 439, (1931).

Eurycephalus, Ashm., *Ind. Mus. Notes,* V, p. 61, (1903). [Preocc.].

Eurycranium, Ashm., (non. nov. for *Eurycephalus*), *Proc. U.S.N. Mus.,* XXIX, p. 405, (1905). Gir., *Mem. Queensland Mus.* Iv, p. 227, (1915).

alcocki (Ashmead)

Eurycephalus alcocki, Ashm., *Ind. Mus. Notes,* V, p. 61, (1903)[1]. [1]Calcutta.

Eurycranium alcocki, Schmied., *Gen. Ins., fas.* 97, p. 308, (1909), Ramakr., *Spol. Zeyl.,* XIII, p. 248, (1925).

Anysis alcocki, Masi, *Eos,* VII, pp. 440-442, (1931)[1]. [Host: Coccids]. [1]Waltair.

saisettiae (Ashmead)

Eurycranium saisettioe, Ashm., *proc. U.S.N. Mus.,* XXIX, p. 405, (1905).

Anysis saisettioe, Smith and Comp., *Univ. Calif. Pub. Ent.,* IV, p. 309, (1928). (This apecies appears to be widely diStRibuted in India : there are three specimens in the Indian Museum coll.)[1]. [Host : Different Coccids]. [1]Waltair.

FAMILY VIII. ENCYRTIDAE

Walk., *Ent. Month. Mag.,* IV, p. 439, (1837). West., *Intr. Mod. Class. Ins.,* II, p. 166, (1840). Walk., *List Chalc. Brit. Mus.,* I, pp. 52-53, (1846); II, p. 104, (1848). Först., *Hymen. Stud.,* II, pp, 18, 21, 30, 32, (1856). Thoms., *Hymen. Skandin.,* IV,

* This genus was first placed by Ashmead under Eunctinae of the Pteromalidæ, but according to Smith and Compere *(loc. cit.,)*, should be included in the Miscogasteridæ. There are three specimeans of *A. saisettiae* (Ashm.) in the Indian Museum; they are carded in such a manner as to render complete examination impossible. The location of the genus in the Lelapinae in this catalogue is thus only provisional. girault and Masi consider *Eurycranium* to be a synonym of *Anysis.*

pp. 12, 112, (1875). How., *Ent. Amer.*, I, pp. 198, 216, (1886); *Proc. U.S.N. Mus.*, XV, pp. 361-369, (1892). ASHM., *Proc. Ent. Soc. Washington*, IV, p. 236, (1897). How., *Proc. U.S.N. Mus.*, XXI, pp. 231-248, (1898). ASHM., *Proc. U.S.N. Mus.*, XXII, p. 202, (1900); *Mem. Carnegie Mus.*, I, (4), p. 286, (1904). SCHMIED., *Gen. Ins.*, *fas.* 97, p. 190, (1909). GIR., *Proc. U.S.N. Mus.*, XLV, p. 189, (1913), [A systematic monograph of Signiphorinae]; *Mem. Queensland Mus.*, IV, pp. 1-184, (1915). MERCET., *Bull. soc. Espan. Hist. nat., madrid*, XIX, pp. 470-478, (1919). TIMBERLAKE, *Proc. Hawaiian Ent. soc.*, IV, pp. 183-196, 197-231, (1922). RUSCHKA, *Verh. Zool.-bot. Ges. Wein* LXXII, pp. 1-13, (1923). BOILV., *Bull. Soc. R. Ent. Soc. Egypte*, pp. 3945, (1925). Ishii, *Imp. Plant Quarant. Serv. Yokohama Tech. Bull.*, III, pp. 21-30, (1925). COMP., *Univ. Calif. Pub. Ent.*, IV, pp. 33-50, (1926); pp. 209-230, (1928). GAH. and WATERST., *Bull. Ent. Res.*, XVI, pp. 373-375, (1926). RAMAKR., *Rec. Ind. Mus.*, XXXIV, pp. 287-288, (1932). MERCET, *Soc. Ent. France*, Liver Ceten., pp. 681-687, RAMAKAR., Rec. Ind. Mas., XXXVI,(1932). pp. 219-221, (1934).

SUBFAMILY I. EUPELMINAE

WALK., *Ann. Mag. Nat Hist.*, XVII, p. 114, (1846); *Notes Chalc.*, V, pp. 71-88, (1872). How., *Ent. Amer.*, I, p. 198, (1896). ASHM., *Proc. Ent. Soc. Washington*, IV, p. 238, (1897); *Mem. Carnegie Mus.*, I, (4), p. 287, (1904). SCHMIED., *Gen. Ins.*, *fas.* 97, p. 169, (1909).

TRIBE *EUPELMARIÆ*

ASHM. *Mem. Carnegie Mus.*, I, (4), p. 287, (1904). SCHMIED., *Gen.*, *Ins.*, *fas.* 97, p. 170, (1909).

74. **Metapelma** Westwood

Metapelma, WESTW., *Proc. Zool. Soc. London*, III, p. 69, (1835), [Genotype *M. spectabilis* WESTW.]. ASHM., *Mem. Carnegie Mus.*, I, (4), p. 288, (1904). SCHMIED., *Gen. Ins.*, *fas.* 97, p. 175, (1909).

albisquamulata Enderlein

Metapelma albisquamulata, Ender., *Ent. Mitt.*, I, p. 147, (1912)[1]. — [1]Pankulam: Ceylon.

Obscurata Westwood

Metapelma obscurata, WESTW., *Thesaur. Ent. Oxon.*, p. 150, (1874)[1]. SCHMIED., *Gen.*, *Ins.*, *fas.* 97, p. 175, (1909). RAMAKR., *Spol. Zeyl.*, XIII, p. 247, (1925). — [1]India.

sp. RAMAKRISHNA AYYAR.

Metapelma sp., RAMAKR., *Spol. Zeyl.*, XXII, p. 247, (1925)[1]. [1]Coimbatore.

Taprobanæ Westwood

Metapelma taprobanæ, WESTW., *Thesaur. Ent. Oxon.*, p. 150, (1874)[1]. SCHMIED., *Gen. Ins., fas.* 97, p. 175, (1909). RAMAKR., *Spol. Zeyl.*, XIII, p. 247, (1925). [1]Ceylon, (India also?)

75. Neanastatus Girault

Neanastatus, GIR., *Bull. Wisc. Nat. Hist. Soc.*, XI, p. 35, (1913), [Genotype *N. cinetiventris* GIR.]. *Mem. Queensland Mus.*, IV, p. 29, (1915). MANI, *Rec. Ind., Mus.*, XXXVII, p. 254, (1935).

Solindinelleus, GIR., *Jahrg. Ent. soc.*, XXIX, p. 22, (1914); *Mem. Queensland Mus.*, IV, p. 35, (1915).

pulchericorpus (Girault)

Solindinelleus pulchericorpus, GIR., *Jahrg. Ent. Soc.*, XXIX, p. 22, (1913); *Mem. Queensland Mus.*, IV, p. 35, (1915).

Neanastatus pulchericorpus, GAH., *Proc. U.S.N. Mus.*, LVI, p. 520, (1919)[1]. RAMAKR., *Spol. Zeyl.*, XIII, p. 247, (1925). MANI, *Rec. Ind. Mus.*, XXXVII, p. 255, (1935). [1]Coimbatore.

Ramakrishnai MANI.*

Neanastatus ramakrishnai, MANI, *Rec., Ind. Mus.*, XXXVII, p. 255, (1935)[1]. [1]Howrah: Calcutta.

trochantericus GAHAN

Neanastatus trochantericus, GAH., *Proc. U.S.N. Mus.*, LVI, p. 520, (1919)[1]. RAMAKR., *Spol. Zeyl.*, XIII, p. 247, (1925); *Bull. Ent. Res.*, XVIII, p. 77, (1927)[2]. [1,2]Coimbatore.

76. Eupelmus DALMAN

Eupelmus, DALM., *Svensk. Vet-Akad. Handl.*, XLI, p. 377, (1820), [Genotype *E. Memnonius* DALM.]. ASHM.]. ASHM., *Hem. Carnegie Mus.* I, (4), p. 289, (1904). SCHMIED., *Gen. Ins., fas.* 97, p. 179, (1909).

---

* This species may prove to be a true *Eupelmus.*

amorphococci Ashmead

Eupelmus amorphococci, ASHM.,* RAMAKR., *Spol. Zeyl.*, XIII, p. 247, (1925)[1]. [Host : *Amorphococcus mesuoe*]. — [1]Ceylon.

amphitus WALKER

Eupelmus amphitus, WALK., *Ann. Mag. Nat Hist.*, XVII, p. 114, (1846)[1]. SCHMIED., *Gen. Ins., fas.* 97, p. 179, (1909). — [1]India.

Carinatus KIEFFER

Eupelmus carinatus, KIEFF., *Ann. Soc. Sc., Bruxelles*, XXIX, p. 194, (1905)[1]. RAMAKR., *Spol. Zeyl.*, XIII, p. 247, (1925). [Host : Cecidomyiid of *Artemisia* galls]. — [1]Bengal.

leithi WALKER

Eupelmus leithi, WALK., *Notes Chalc.*, V, p. 83, (1872)[1]. SCHMIED., *Gen. Ins., fas.* 97, p. 181, (1909). RAMAKR., *Spol. Zeyl.*, XIII, p. 247, (1925). — [1]India.

tachardiæ (HOWARD)

Anastatus tachardiæ, How., *Proc. U.S.N. Mus.*, XVIII, p. 641, (1896)[1]. SCHMIED., *Gen. Ins., fas.* 97, p. 188, (1909). RAMAKR., *Spol, Zeyl.*, XIII, p. 247, (1925). — [1]Punduloya: Ceylon.

Eupelmus tachardiæ, MAHDIHASSAN, *Curr. Sc.*, III, pp. 562-564, (1935), [Host]. GLOVER and others, *Curr. Sc.*, IV, pp. 37-39, (1935), [Host]. FERR., *Bull. Ent. Res.*, XXVI, pp. 391-406, (1935)[1]. — [1]Namkum, Lyallpur.

tenuicornis KIEFFER

Eupelmus ienuicornis, KIEFF., *Ann. soc. Sc. Bruxelles*, XXIX, p. 192, (1905)[1]. RAMAKR., *Spol, Zeyl.*, XIII, p. 247, (1925). [Host : *Lasioptera textor*]. — [1]Bengal.

77. Brasema CAMERON

Brasema, CAM., *Biol. Centr. Amer. Hymen.*, I, p. 124, (1884), [Genotype *B. brevispina* CAM.]. ASHM., *mem. Carnegie Mus.*, I, (4), p. 288, (1904). SCHMIED., *Gen. Ins., fas.* 97, p. 182, (1909).

---

*The original description of this species could not be traced in the literature.

annulica udis CAMERON.*

Brasema annulicaudis, CAM., *Ind. For. Rec.,* Iv, p. 98, (1912)[1]. RAMAKR., *spol. Zeyl.,* XIII, p. 247, (1925). [1]Dehra Dun.

78. Bruchocida CRAWFORD

Bruchocida, CRAW., *Proc. U.S.N. Mus.,* XLV, p. 245, (1913), [Genotype *B. vuilleti* CRAW.].

Orientalis CRAWFORD

Bruchocida orientalis, CRAW., *Proc. U.S.N. Mus.,* XLV, p. 247, (1913)[1]. RAMAKR., *Rep. Proc. Third Ent. Meet. Pusa,* III, p. 932, (1920)[2]; *Rep. Proc. Fourth Ent. Meet. Pusa,* p. 363, (1921)[3]; *Spol. Zeyl.,* XIII, p. 247, (1925); *Bull. Ent. Res.,* XVIII, p. 77, (1927).[4] [Host: *Bruchus*]. [1,3,4,]Bangalore. [2] Coimbatore (?).

79. Cacotropia Mostschulsky

Cacotropia, MOTSCH., *Bull. Soc. Imp. Nat. Moscou,* XXXVI, (2), p. 56, (1863), [Genotype *C. echidna* MOTSCH.]. ASHM., *Mem. Carnegie Mus.,* I, (4), p. 289, (1904). SCHMIED., *Gen. Ins., fas* 97, p. 185, (1909).

echidna MOTSCHULSKY

Cacotropia echidna, MOTSCH., *Bull., Soc. Imp. Nat. Moscou,* XXXVI, (2), p. 57, (1863)[1]. SCHMIED., *Gen. Ins., fas.* 97, p. 186. (1909). RAMAKR., *Spol. Zeyl.,* XIII, p. 247, (1925). [1]Mt. Patannas: Ceylon.

80. Anastatus MOTSCHULSKY

Anastatus, MOTSCH., *Etud. Ent.,* VIII, p. 116, (1859), [Genotype *A. mantoidoe* MOTSCH.].* ASHM., *Mem. Carnegie Mus.,* I, (4), p. 290, (1904). SCHMIED., *Gen Ins., fas.* 97, p. 188, (1909). FRANCIES BERNARD, *Bull. soc. Ent. France,* T. XLI, (5), pp. 69-75, (1936), [Forms parasitic on Mantid oothecæ].

blattidarum FERRIERE.†

Anastatus blattidarum, FERR., *Bull. Ent. Res.,* XXI, p. 33, fig. 1, (1930)[1]; *Stylops,* IV, p. 150, (1935), [Male]. [1]Sudan. (Also Mysore.)

---

* Ferriere (*Bull. Ent. Res.,* XXVI, p. 394, (1935) treats this as a synonym of *Eupelmus tachariæ.*

* *Vide* foot-note on *Podagrion mantoidæ* (Motsch.) p. 42.

† I am indebted to Ferriere for synonymy of this species.

Solindenia blattiphagus, MANI, *REc., Ind. Mus.*, XXXVIII, pp. 131-132, fig. 1, (1936)[1]. [Host : cockroach]. [1]Calcutta.

coimbatorensis Girault. (?)

Anastatus coimbatorensis, GIR.* RAMAKR., *Rep. Proc. Third Ent. Meet. Pusa*, III, p. 933, (1920)[1]. Ramachandra Rao, *Rep. Proc. Fourth Ent. Meet. Pusa*, p. 42, (1921)[2]. [1,2]Coimbatore.

colemani CRAWFORD

Anastatus colemani, CRAW., *Proc. U.S.N. Mus.*, XLII, p. 6, (1912).[1] RAMAKR., *Rep. Proc. Fourth Ent. Meet. Pusa*, p. 363, (1921)[2]; *Spol. Zeyl.*, XIII, p. 247, (1925); *Bull. ent. Res.*, XVIII, p. 77, (1927)[3]. Krishna Ayyar, *Bull. Ent. Res.*, XX, pp. 173-178, (1929), [probable parasite of *Tetroda histeroides*]. [Host : *Degonetus serratus*]. [1,2,3]Bangalore.

Sp. RAMAKRISHNA AYYAR.

Anastatus sp., RAMAKR., *Spol. Zeyl.*, XIII, p. 247, (1925)[1]; *Bull. Ent. Res.*, XVIII, p. 77, (1927)[2]. [Host : *Oxya velox*]. [1,2]Coimbatore.

SUBFAMILY II. ENCYRTINAE

How., *Ent. Amer.*, I, pp. 198, 216, (1886). ASHM., *Proc. Ent. Soc. Washington*, IV, p. 236, (1897); *Mem. Carnegie Mus.*, 1, (4) p. 292, (1904). SCHMIED., *Gen. Ins., fas.* 97, p. 191, (1909).

TRIBE 1. *ENCYRTARIAE*

How., *Ent. Amer.*, I, p. 198, (1886). ASHM., *Proc. Ent. Soc. Washington*, IV, p. 248, (1897); *Mem. Carnegie Mus.*, I, (4), p. 297, (1904). SCHMIED., *Gen. Ins., fas.* 97, p. 191, (1909).

81. Encomys Forster

Encyrtus, DALM., *(partim)*, *Svensk. Vet.-Akad. Handl.*, XVI, p. 218, (1795). ASHM., *Mem. Carnegie Mus.*, I, (4), p. 297, (1904).

Eucomys, FÖRST., *Hymen. Stud.*, II, p. 32, (1856), [Genotype *E. swederi* (DALM.)]. SCHMIED., *Gen. Ins., fas.* 97, p.193, (1909).

---

* The original description of this species could not be traced in the literature.

Comys, Först., *Hymen. Stud.*, II, p. 144, (1856).

lecaniorum (Mayr).

Comys lecaniorum, Mayr, *Verh. Zool.-Bot Ges. Wien*, XXV, pp. 740-741, (1875).

Eucomys lecaniorum, Schmied., *Gen. Ins.*, *fas.* 97, p. 193, (1909). Gah., *Proc. U.S.N. Mus.*, LVI, p. 522, (1919)[1]. Ramakr., *Rep. Proc. Fourth Ent. Meet. Pusa*, p. 364, (1921)[2*]; *Spol. Zeyl.*, XIII, p. 244, (1925). — [1,2]Coimbatore.

Encyrtus lecaniorum, Ramakr., *Bull. Ent. Res.*, XVIII, p. 77, (1927)[1]. [Host : *Lecanium nigrum*]. — [1]Coimbatore.

rufescens (Motschulsky)

Cheiloneurus rufescens, Motsch., *Bull. Soc. Imp. Nat. Moscou*, XXXVI, (3) p. 53, (1863)[1]. Schmied., *Gen. Ins.*, *fas.* 97, p. 253, (1909). Ramakr., *Spol. Zeyl.*, XIII, p. 245, (1925). — [1]Mt. Patannas: Ceylon.

Comys rufescens, how., *Proc. U.S.N. Mus.*, XVIII, p. 640, (1896)[1]. [Host : *Lecanium coffeoe*]. — [1]Ceylon.

TRIBE 2. *ECTROMARIAE*

Ashm., *Mem. Carnegie Mus.*, I, (4), p. 292, (1904). SCHMIED. *Gen. Ins.*, *fas.* 97, p. 195, (1909).

82. **Anagyrus** Howard

Anagyrus, How., *Proc. U.S.N. Mus.*, XVIII, p. 638, (1896), [Genotype *A. greeni* How.]. Ashm., *Mem. Carnegie Mus.*, I, (4), p. 294, (1904). Schmied., *Gen. Ins.*, *fas.* 97, p. 201, (1909).

greeni Howard

Anagyrus greeni, How., *Proc. U.S.N. Mus.*, XVIII, p. 639, (1896)[1]. Schmied., *Gen. Ins.*, *fas.* 97, p. 202, (1909). Ramakr., *Spol. Zeyl.*, XIII, p. 244, (1925). [Host : *Maskellia jonata*]. — [1]Pundyloya:Ceylon.

83. **Comperiella** Howard

Comperiella, How., *Ent. News*, pp. 121-122, (1906), [Genotype *C. bifasciata* How.]. Gir., *Insc. Inscit.*

* Ramarkishna Ayyar's *Encomys* is an error for *Eucomys*.

Menstr., V. pp. 29-37, (1917). TIMBERL. *Proc. Ent. Soc. Washington*, XXV, p. 59, (1923). COMP., *Univ. Calif. Pub. Ent.*, IV, pp. 44, 48, (1926). RAMAKR., *Rec. Ind. Mus.*, XXXVI, p. 219, (1934).

Pseudanusia, GIR., *Mem. Queensland Mus.*, IV, p. 155, (1915)

Habrolepistia, MERCET, *Faun. Iberica, Hymen.*, pp. 668-671, (1921).

bifasciata HOWARD.

Comperiella bifasciata, How., *Ent. News*, pp. 121-122, (1906); p. 237, (1907). COMP., *Univ. Calif. Pub. Ent.*, IV, pp. 44-45, 63-71, (1926). RAMAKR., *Rec. Ind. Mus.*, XXXVI, p. 220, (1934). GLOVER, *Journ. Bombay Nat. Hist. Soc.*, XXXVIII, p. 153, (1935)[1]. [Host : *Chrysomphalus*]. [1]Ranchi.

indica RAMAKRISHNA AYYAR.

Comperiella indica, RAMAKR., *Rec. Ind. Mus.*, XXXVI, p. 219, (1934)[1]. [Host : *Aspidiotus tamarindi*]. [1] Coimbatore.

84. Krishnieriella MANI

Krishnieriella, MANI, *Rec. Ind. Mus.*, XXXVII, p. 421, (1935); [Genotype *K. ceroplastodis* MANI].

ceroplastodis MANI

Krishnieriella ceroplastodis, MANI, *Rec. Ind. Mus.*, XXXVII, p. 422, (1935)[1]. [Host : *Ceroplastes cajani*]. [1] Coimbatore.

85. Tetracnemus Westwood

Tetracnemus, WESTW., *Ann. Mag. Nat. Hist.*, I, p. 258, (1837), [Genotype *T. diversicornis* WESTW.]. ASHM., *Mem. Carnegie Mus.*, I, (4), p. 295, (1904). SCHMIED., *Gen. Ins.*, *fas.* 97, p. 206, (1909). RAMAKR., *Rec. Ind. Mus.*, XXXIV, p. 287, (1932).

indicus RAMAKRISHNA AYYAR

Tetracnemus indicus, RAMAKR., *Rec., Ind. Mus.*, XXXIV, p. 287, (1932)[1]. RAMAKR., and MARGAB., *Journ. Bombay Nat. Hist. Soc.*, XXXVII, p. 194, (1934)[2]. [Host : *Pseudococcus citri*]. [1,2]Coimbatore.

TRIBE 3. *MIRINARIÆ*

ASHM., *Mem. Carnegie Mus.*, I, (4), p. 298, (1904). SCHMIED., *Gen. Ins., fas.* 97, p. 208, (1909).

86. **Copidosoma** Ratzeburg

Copidosoma, Ratz., *Ichneum. Forstings.*, I, p. 157, (1844), [Genotype *C. boucheanum* Ratz.]. ASHM., *Mem. Carnegie Mus.*, I, (4), p. 299, (1904). SCHMIED., *Gen. Ins., fas.* 97, p. 223, (1909).

Litomastix, THOMS., *Hymen. Skandin.*, IV, p. 171, (1875).

Clavicornis CAMERON.*

Copidosoma (?) *clavicornis*, CAM., *Ind. For. Rec.*, IV, p. 101, (1912)[1]; *Mem. Ind. For.*, III, (1915). RAMAKR., *Spol, Zeyl.*, XIII, p. 245, (1925). [On lac]. [1]Dehra Dun.

87. **Paracopidosomopsis** Girault.

Berencyrtus, ASHM., (*partim*), *Proc. U.S.N. Mus.*, XXII, p. 365, (1900).

Paracopidosomopsis, GIR., *Psyche*, XXIII, p. 49, (1916), [Genotype *P. floridanus* (ASHM.)].

Javae Girault

Paracopidosomopsis javoe, GIR., *Psyche*, XXIII, p. 49, (1916). GAH., *Proc. U.S.N. Mus.*, LVI, p. 522, (1919)[1], RAMAKR., *Rep. Proc. Fourth Ent. Meet. Pusa*, p. 364, (1921)[2]; *Spol. Zeyl.*, XIII, p. 245, (1925); *Bull. Ent. Res.*, XVIII, p. 77, (1927)[3]. [Host : *Plusia signata*]. [1,2,3]Coimbatore.

88. **Homalotylus** MAYR.

Encyrtus, DALM., (*partim*), *Svensk. Vet.-Alad. Handl.*, XLI, p. 340, (1820).

Homalotylus, MAYR., *Verh. Zool.-bot. Ges. Wien*, XXV, p. 752, (1876), [Genotype *H. flaminius* (Dalm.)]. ASHM., *Mem. Carnegie Mus.*, I, (4), p. 301, (1904). SCHMIED., *Gen, Ins., fas.* 97, p. 235, (1909). TIMBERL., *Proc. U.S.N. Mus.*, LVI, pp. 133-170, (1919).

---

* For recent synonymy of this species *vide* addendum under *Parechthrodryinus*.

Nobrimus, THOMS., *Hymen. Skandin.*, IV, p. 137, (1875) (?).

Flaminius (DALMAN)

Encyrtus flaminius, DALM., *Svensk. Vet. Akad. Handl.*, XLI, p. 340, (1820).

Homalotylus flaminius, SCHMIED., *Gen. Ins.*, *fas.* 97, p. 235, (1909). RAMAKR. and MARGAB., *Journ. Bombay nat. Hist. Soc.*, XXXVII, p. 194, (1934)[1*]. [Host : *Coccinellid*]. [1]Coimbatore.

89. Ooencyrtus Ashmead

Encyrtus, ASHM., (*partim*), *Bull. Ohio Expt. Sta.*, p. 163, (1893).

Ooencyrtus, ASHM., *Proc. U.S.N. Mus.*, XXII, pp. 338, 381, (1900), [Genotype *O. clisiocampoe* (ASHM.)]; *Mem. Carnegie Mus.*, I, (4), p. 302, (1904). SCHMIED., *Gen. Ins.*, *fas.* 97, p. 238, (1909).

pyrillæ CRAWFORD

Ooencyrtus pyrillæ, CRAW., *Insec. Inscit.Menstr.*, IV, p. 102, (1916)[1]. RAMAKR., *Spol. Zeyl.*, XIII, p. 245, (1925). [Host : *Pyrilla aberrans*]. [1]Pusa.

90 Aphycus MAYR.

Encyrtus, DALM., (*partim*), *Svensk. Vet.-Akad. Handl.*, XLI, p. 153, (1820).

Aphycus, MAYR, *Verh. zool.-Bot. Ges. Wien*, XXV, p. 695, (1876), [Genotype *A. apicalis* (DALM.)]. ASHM., *Mem. Carnegie Mus.*, I, (4), p. 302, (1904). SCHMIED., *Gen. Ins.*, *fas.* 97, p. 239, (1909).

fusidorsum GAHAN

Aphycus fusidorsum, GAH., *Proc. U.S.N. Mus.*, LVI, p. 521, (1919)[1]. RAMAKR., *Rep. Proc. Fourth Ent. Meet. Pusa*, p. 364, (1921)[2]; *Spol. Zeyl.*, XIII, p. 245, (1925); *Bull Ent. Res.*, XVIII, p. 77, (1927)[3]. [Host : *Ceroplastes cajani*]. [1,2,3]Coimbatore.

litchensiæ HOWARD

Aphycus litchensiæ, How., *Proc. U.S.N. Mus.*, XVIII, p. 640, (1896)[1]; SCHMIED., *Gen. Ins.*, *fas.* 97, p. 240, [1]Punduloya: Ceylon.

* *fclminus* of Ramakrishna Ayyar and Margabandu is an error for *flaminius*.

(1909), RAMAKR., *Spol. Zeyl.*, XIII, p. 245, (1925). [Host : *Litchensia koebeh*].

91. Encyrtus (LATREILLE) DALMAN

Encyrtus, Latr., *Gen. Crust. Ins.*, IV, p. 31, (1809), [Genotype *E. infidus* (Rossi)]. DALM., *Svensk. Vet.-Akad. Handl.*, XLI, p. 147, (1820). ASHM., *Mem. Carnegie Mus.*, I, (4), p. 297, (1904). SCHMIED., *Gen. Ins., fas.* 97, p. 241, (1909). SILVESTRI, *Boll. Lab. Zool. Portici,* XIII, p. 151, (1919).

Microterys, THOMS., *Hymen. Skandin.*, IV, p. 155, (1820). ASHM., *Proc. U.S.N. Mus.*, XXII, pp. 339, 346, 390, (1900).

adustipennis MOTSCHULSKY

Encyrtus (?) *adustipennis,* MOTSCH., *Bull. Soc. Imp. Nat. Moscou,* XXXVI, (3), p. 55, (1863)[1]. SCHMIED., *Gen. Ins., fas.* 97, p. 242, (1909). RAMAKR., *Spol. Zeyl.*, XIII, p. 245, (1925). [Host : *Lecanium coffeoe*].

[1]Colombo: Ceylon.

barbatus TIMBERLAKE

Encyrtus barbatus, TIMBERL., *Proc. Howaii. Ent. Soc.*, IV, p. 209, (1919). RAMAKR. and MARGAB., *Journ. Bombay Nat. Hist. Soc.*, XXXVII, p. 194, (1934)[1]. [Host : *Lecanium nigrum*].

[1]Coimbatore.

Corvinus MOTSCHULSKY

Encyrtus corvinus, MOTSCH., *Bull. Soc. Imp. Nat. Moscou,* XXXVI, (3), p. 55, (1863)[1]. SCHMIED., *Gen. Ins., fas.* 97, p. 243, (1909). RAMAKR., *Spol. Zeyl.*, XIII, p. 245, (1925).

[1]Mt. Patannas: Ceylon.

flavus HOWARD

Encyrtus flavus, How., *Annual Rep. U.S. Dept. Agric.*, p. 367, (1880); *Proc. U.S.N. Mus.*, XVIII, p. 636, (1896)[1]. RAMAKR., *Spol. Zeyl.*, XIII, p. 245, (1925); *Bull. Ent. Res.*, XVIII, p. 77, (1927)[2*]. [Hosts: *Lecanium and Pulvinaria* spp.]

[1]Ceylon.
[2]Coimbatore.

---

* Ramakrishna Ayyar's *Aphycus flavus.* How. is apprently *Encyrtus* flavus How, in recent years this species has been recorded from america under the name *Euaphycus (Aophycus) flavus* How, as parastic on *Pulvinagria iceryi.*

Kotinskyi (Fullaway)

Apentelicus kotinskyi, FULL., *Annual Rep. Hawaii. Agric. Expt. Sta.*, p. 26, (1913).

Microterys kotinskyi, TIMBERL., *Proc. Hawaii. Ent. Soc.*, IV, p. 184, (1919). RAMAKR. and MARGAB., *Journ. Bombay Nat. Hist. Soc.*, XXXVII, p. 194, (1934)[1]. [Host : *Lecanium nigrum*]. [1]Coimbatore.

litchensiæ HOWARD

Encyrtus litchensiæ, How., *Proc. U.S.N. Mus.*, XVIII, p. 636, (1896)[1]. SCHMIED., *Gen. Ins., fas.* 97, p. 245, (1909). RAMAKR., *Spol. Zeyl.*, XIII, p. 245, (1925). [Host : *Litchensia koebeli*]. [1]Punduloya: Ceylon.

neitneri MOTSCHULSKY

Encyrtus neitneri, MOTSCH., *Bull. Soc. Imp. Nat. Moscou*, XXXVI, (3), p. 54, (1863)[1]. SCHMIED., *Gen. Ins., fas.* 97, p. 245, (1909). RAMAKR., *Spol. Zeyl.*, XIII, p. 245, (1925). [1]Mt. Patannas: Ceylon.

obstructus WALKER

Encyrtus obstructus, WALK., *Ann. Mag. Nat. Hist.*, VI, p. 359, (1860)[1]. MOTSCH., *Bull. Soc. Imp. Nat. Moscou*, XXXVI, (3), p. 55, (1863)[2]. SCHMIED., *Gen. Ins., fas.* 97, p. 245, (1909). RAMAKR., *Spol. Zeyl.*, XIII, 245, (1925). [1]Ceylon. [2]Mt. Patannas: Ceylon.

92. Tachardiaephagus ASHMEAD.

Encyrtus, How., (*partim*), *Proc. U.S.N. Mus.*, XVIII, p. 637, (1896).

Tachardioephagus, ASHM., *Mem. Carnegie Mus.*, I, (4), p. 303, (1904). [Genotype *T. tachardioe* (How.)]. SCHMIED., *Gen. Ins., fas.* 97, p. 249, (1909). GIR., *Jahrg. Soc. Ent. Stuttgart*, XXXI, p. 43, (1916).*

Lissencyrtus, CAM., *Ind. For. Rec.*, IV, p. 100, (1913).

---

* Girault has described *Tachariæphagus* as a new genus; this is invalid under the International Rules of Zoological Nomenclature. Ashmead has very clearly defined the genus and has also indicated the type species and AShmead's *Tachardiæphagus* is thus valid under the rules. Girault only redescribed the type in the U.S.N. Museum; scratching out Ashmead's name on the label and substituting his own for it.

tachardiæ (HOWARD)

Encyrtus tachardiæ, How., *Proc. U.S.N. Mus.,* XVIII, p. 637, (1896)[1]. RAMAKR., *Spol, Zeyl.,* XIII, p. 245, (1925). [1]Punduloya: Ceylon.

Tachardioephagus tachardioe, FERR., *Bull. Ent. Res.,* XIX, p. 171, (1928)[1]. GLOVER., *Bull. Ent. Res.,* XXI, pp. 261-266, pl. 1, (1930), [parasitic on lac insects]. Miller, *Sc. Ser., Dept. Agric. S.S. and F.M.S.,* No. 11, pp. 24, pls. 3, (1933), [parasite of *Laccifer javanus* in Malaya]. RAMAKR. and MARGAB., *Journ. Bombay Nat. Hist. Soc.,* XXXVII, p. 194, (1934). FERR., *Bull. Ent. Res.,* XXVI, pp. 391-406, (1935)[2]. [1,2]Namkum.

tachardioephagus thoracicus, ASHM., *Mem. Carnegie Mus.,* I, (4), p. 303, (1904). SCHMIED., *Gen. Ins., fas.* 97, p. 249, (1909). GIR., *Jahrg. Soc. Ent. Stuttgart,* XXXI, p. 43, (1916)[1]. RAMAKR., *Spol. Zeyl.,* XIII, p. 245. (1925). [1]Ceylon.

Lissencyrtus troupi, CAM., *Ind. For Rec.,* IV, p. 100, (1913)[1]. RAMAKR., *Spol. Zeyl.,* XIII, p. 246, (1925). [1]Dehra Dun.

tachardiae var. **Somervilli** (MAHDIHASSAN)

Lissencyrtus somervilli, Mahd., *Journ. Sc. Assoc. Maharaja's Coll. Vizianagram,* I, p. 1, (1923)[1]. [1]Mysore.

Tachardioephagus somervilli, FERR., *Bull. Ent. Res.,* XIX, p. 171, (1928)[1]. RAMAKR. and MARGAB., *Journ. Bombay Nat. Hist. Soc.,* XXXVII, p. 194, (1934). [1]Namkum.

Tachardioephagus tachardioe var. *somervilli,* FERR., *Bull. Ent. Res.,* XXVI, pp. 391-406, (1935).

93. Erencyrtus MAHDIHASSAN*

Erencyrtus, Mahd., *Journ. Sc. Assoc. Maharaja's Coll. Vizianagram,* I, p. 71, (1923), [Genotype *E. dewitzii* Mahd.]

dewitzii MAHDIHASSAN

Erencyrtus dewitzii, MAHD., *Journ. Sc. Assoc. Maharaja's Coll. Vizianagram,* I, p. 71, (1923)[1]. RAMAKR. and MARGAB., *Journ. Bombay Nat. Hist. Soc.,* XXXVII, p. 194, (1934). FERR., *Bull. Ent. Res.,* XXVI, pp. 391-406, (1935). [1]Mysore.

---

* Location of the genus is provisional

94. **Cheiloneurus** Westwood

Cheiloneurus, Westw., *Philos. Mag.,* (3), III, p. 343, (1833). [Genotype, *C. elegans* (Dalm.)]. Ashm., *Mem. Carnegie Mus.,* I, (4), p. 304, (1904).

Chiloneurus; Först., (*emendation*) *Hymen. Stud.,* II, p. 50, (1856). Schmied., *Gen. Ins., fas.* 97, p. 252, (1909).

Sterrhocoma, Först., *Hymen. Stud.,* II, p. 33, (1856).

Paradiscus Motschulsky

Cheiloneurus (?) *paradiscus,* Motsch., *Bull. Soc. Imp. Nat. Moscou,* XXXVI, (3), p. 52, (1863)[1]. Schmied., *Gen. Ins., fas.* 97, p. 253, (1909). Ramakr., *Spol. Zeyl.,* XIII, p. 245, (1925). [Host : *Lecanium coffeoe*]. Compare 1931, Univ. Calif. Rubod enf. 5, p. 231-34.

[1]Mt. Nura Ellia : Ceylon.

95. **Adelencyrtus** Ashmead

Encyrtus, How., (*partim*), *Proc., U.S.N. Mus.,* XVIII, p. 637, (1896).

Adelencyrtus, Ashm., *Proc. U.S.N. Mus.,* XX, pp. 341, 401, (1900), [Genotype *A. chionaspidis* (How.)]; *Mem. Carnegie Mus.,* I, (4), p. 304, (1904). Schmied., *Gen. Ins., fas.* 97, p. 253, (1909).

chionaspidis (Howard)

Encyrtus chionaspidis, How., *Proc., U.S.N. Mus.,* XVIII, p. 637, (1896)[1].

[1]Punduloya: Ceylon.

Adelencyrtus chionaspidis, Schmied., *Gen. Ins., fas.* 97, p. 253, (1909). Ramakr., *Spol. Zeyl.,* XIII, p. 246, (1925). [Host : *Chionaspis graminis*].

Planchoniæ (Howard)

Encyrtus planchoniæ, How., *Proc. U.s.N. Mus.,* XVIII, p. 637, (1896)[1].

[1]Punduloya: Ceylon.

Adelencyrtus planchonioe, Schmied., *Gen. Ins., fas.* 97, p. 253, (1909). Ramakr., *Spol. Zeyl.,* XIII, p. 246, (1925). [Host : *Planchonia delicata*].

solidus (Howard)

Encyrtus solidus, How., *Proc. U.S.N. Mus.,* XVIII, p. 638, (1896)[1].

[1]Punduloya: Ceylon.

Adelencyrtus solidus, SCHMIED., *Gen. Ins., fas.* 97, p. 253, (1909). RAMAKR., *Spol. Zeyl.*, XIII, p. 246, (1925). [Host : *Eriococcus rhodomyrti*].

96. Callipteroma MOTSCHULSKY

Callipteroma, MOTSCH., *Bull. Soc. Imp. Nat. Moscou*, XXXVI, (3), p. 36, (1863), [Genotype *C. quinqueguttata* MOTSCH.]. ASHM., *Mem. Carnegie Mus.*, I, (4), p. 304, (1904). SCHMIED., *Gen. Ins., fas.* 97, p. 254, (1909).

quinqueguttata MOTSCHULSKY

Callipteroma quinquequiiata, MOTSCH., *Bull. Soc. Imp. Nat. Moscou*, XXXVI, (3), p. 36, (1863)[1]. SCHMIED., *Gen. Ins., fas.* 97, p. 254, (1909). RAMAKR., *Spol. Zeyl.*, XIII, p. 246, (1925).

[1]Mt. Patannas: Ceylon.

sexguttata MOTSCHULSKY

Callipteroma sexguttata, MOTSCH., *Bull. Soc. Imp. Nat. Moscou*, XXXVI, (3), p. 37, (1863)[1]. SCHMIED., *Gen. Ins., fas.* 97, p. 254, (1909). RAMAKR., *Spol. Zeyl.*, XIII, p. 246, (1925).

[1]Mt. Patannas: Ceylon.

testacea MOTSCHULSKY

Callipteroma testacea, MOTSCH., *Bull. Soc. Imp. Nat. Moscou*, XXXVI, (3), p. 37, (1863)[1]. SCHMIED., *Gen. Ins., fas.* 97, p. 254, (1909). RAMAKR., *Spol. Zeyl.*, XIII, p. 246, (1925).

[1]Mt. Patannas: Ceylon.

97. Atroptes HOWARD

Atroptes, How., *Proc. U.S.N. Mus.*, XXI, p. 236, (1898), [Genotype *A. collinsi* How.]. ASHM., *Mem. Carnegie Mus.*, I, (4), p. 305, (1904). SCHMIED., *Gen. Ins., fas.* 97, p. 256, (1909).

hautfeuilli (MAHDIHASSAN)

Microterys hautfeuilli, Mahd., *Journ. Sc. Assoc. Maharaja's Coll. Vizianagram*, II, p. 65, (1923)[1].

[1]Assam.

Atroptes (Micropterys) hautfeuilli, FERR., *Bull. Ent. Res.*, XXVI, pp. 391-406, (1935).

98. Anicetus HOWARD

Anicetus, How., *Proc., U.S.N. Mus.*, XVIII, p. 639, (1896), [Genotype *A. ceylonensis* How.]. ASHM., *Mem.*

Carnegie Mus., I, (4), p. 305, (1904). SCHMIED., *Gen. Ins.*, *fas.* 97, p. 257, (1909).

ceylonensis HOWARD

Anicetus ceylonensis, How., *Proc. U.S.N. Mus.*, XVIII, p. 639, (1896)[1]. SCHMIED., *Gen. Ins.*, *fas.* 97, p. 257, (1909). GAH., *Proc. U.S.N. Mus.*, LVI, p. 522, (1919)[2]. RAMAKR., *Rep. Proc. Fourth Ent. Meet. Pusa*, p. 364, (1921)[3]; *Spol. Zeyl.*, XIII, p. 246, (1925); *Bull. Ent. Res.*, XVIII, p. 77, (1927)[4] [Hosts : *Ceroplastodes and Pulvinaria*].

[1]Punduloya: Ceylon.
[2,3,4]Coimbatore.

Caravterocerus ceylonensis, How. (?), RAMAKR., *Spol. Zeyl.*, XIII, p. 246, (1925).

dodonia FERRIERE

Anicetus dodonia, FERR., *Bull. Ent. Res.*, XXVI, pp. 391-406, (1935)[1].

[1]India.

99. Psylledontus CRAWFORD

Psylledontus, CRAW., *Proc. U.S.N. Mus.*, XXXVIII, p. 88, (1910), [Genotype *P. insidiosus* CRAW.].

Secundus Girault

Psylledontus secundus, GIR., *Ann. Ent. Soc. Amer.*, XIII, p. 28, (1915)[1]. RAMAKR., *Spol. Zeyl.*, XIII, p. 246, (1925). [Host : unknown gall-forming Psyllid].

[1]Ceylon.

100. Homalopoda HOWARD

Homalopoda, How., *Journ. Linn. Soc. London*, (*Zool.*), XXV, p. 90, (1894), [Genotype *H. cristata* How.]. ASHM., *Mem. Carnegie Mus.*, I, (4), p. 305, (1904). SCHMIED., *Gen. Ins.*, *fas.* 97, p. 259, (1909).

cristata HOWARD

Homalopoda cristata, How., *Journ. Linn. Soc. London*, (*Zool.*), XXV, p. 91, (1894); *Proc. U.S.N. Mus.*, XVIII, p. 640, (1896)[1]. SCHMIED., *Gen. Ins.*, *fas.* 97, p. 259, (1909). RAMAKR., *Spol. Zeyl.*, XIII, p. 246, (1925). [Host : *Aspidiotus secretus*].

[1]Ceylon.

TRIBE 4. *ARRHENOPHAGARIAE*

ASHM., *Proc. U.S.N. Mus.;* XXII, p. 408, (1900); *Mem. Carnegie Mus.*, I, (4), p. 311, (1904). SCHMIED., *Gen. Ins., fas.* 97, p. 260, (1909).

101. **Arrhenophagus** Aurivillius

Arrhenophagus, Auriv., *Ent. Tidskr.*, IX, p. 144, (1888), [Genotype *A. chionaspidis* Auriv.]. ASHM., *Mem. Carnegie Mus.*, I, (4), p. 311, (1904). SCHMIED., *Gen. Ins., fas.* 97, p. 260, (1909).

albitibiae Girault

Arrhenophagus albitibioe, GIR., *Trans. N.Y. Ent. soc.*, XXXII, p. 240, (1915)[1]. RAMAKR., *Spol. Zeyl.*, XIII, p. 246, (1925). [1]Ceylon.

chionaspidis Aurivillius

Arrhenophagus chionaspidis, Auriv., *Ent. Tidskr.*, IX, p. 146, (1888). RAMAKR., *Spol. Zeyl.*, XIII, p. 146, (1925)[1]. [Host : *Fiorinia saprosoma*]. [1]Ceylon.

FAMILY IX. PTEROMALIDAE

WALK., *Ent. Month. Mag.*, II, p. 286, (1835). WESTW., *Intr. Mod. Class. Ins.*, II, p. 166, (1840). WALK., *List Chalc. Brit. Mus.*, I, p. 23. (1846). FÖRST. *Hymen. Stud.*, II, pp. 19, 25, 59, 63, (1856). THOMS. *Hymen. Skandin.*, IV, pp. 12, 217, (1875). How., *Ent. Amer.*, I, p. 108, (1886). ASHM., *Proc. Ent. Soc. Washington,* IV, p. 246, (1897). Rubsaam., *Ergebn. Magalhaens. Sammelreise,* VI, p. 4, pl. i, (1902). ASHM., *Mem. Carnegie Mus.*, I, (4), p. 311, (1904). SCHMIED., *Gen. Ins., fas.* 97, p. 303, (1909). GIR., *Mem. Queensland Mus.*, IV, pp. 313-346, (1915).

SUBFAMILY I. EUNOTINAE

MOTSCH., *Bull. soc. Imp. Nat. Moscou,* XXXVI, (2), p. 69, (1863). WALK., *Notes Chalc.*, IV, p. 100, (1872). DALLA TORRE, *Cat. Hymen.*, V, p. 87, (1898). ASHM., *Mem. Carnegie Mus.*, I, (4), p. 325, (1904). SCHMIED., *Gen. Ins., fas.* 97, p. 304, (1909). MASI, *Eos,* VII, pp. 411-459, (1931).

102. **Scutellista** MOTSCHULSKY

Scutellista, MOTSCH., *Etud., Ent.*, VIII, p. 172, (1859), [Genotype *S. cyanea* MOTSCH.]. ASHM., *mem. Carnegie Mus.*, I, (4), p. 325, (1904). SCHMIED., *Gen. Ins., fas.* 97, p. 305, (1909). MASI, *Eos,* VII, p. 456, (1931).

Aspidocoris, A. COSTA, *Boll. Acad. Sc. Napoli*, p. 24, (1863).

Cyanea MOTSCHULSKY

Scutellista cyanea, MOTSCH., *Etud., Ent.*, VIII, pp. 171-173, (1859)[1]. How., *Rev. Patol. Veg.*, V, pp. 81-87, (1896); *Bull. U.S. Dept. Agric. Div. Ent.*, 17, pp. 13-16, (1898); *ibid.*, 26, pp. 16-17, (1900). MARLATT, *U.S. Dept. Agric., Farmers' Bull.*, 172, pp. 29-30, (1903); *Bull. U.S. Dept. Agric. Div. Ent.*, 44,pp. 50-56, (1904). Issacs, *First Bienn. Rep. Calif. Comm. Hort.*, pp. 93-94, (1903-4). Jeffrey, *First Bienn. Rep. Calif. Comm. Hort.*, pp. 162-164, (1903-4). MASI, *Boll. Lab. Zool. Portici*, I, pp. 266-269, (1907). SILVESTRI and MARTELLI, *Boll. Lab. Zool. Portici*, II, 326-341, (1908). Sehmied., *Gen. Ins., fas.* 97, p. 305, (1909). QUAYLE, *Journ, Econ. Ent.*, III, pp. 446-451, (1910). MERCET, *Rev. Accad. Sc. Madrid*, IX, pp. 185-190, (1910). HOW. and FISKE, *Bull. U.s. Dept. Agric. Bur. Ent.*, 91, pp. 31-33, (1911). QUAYLE, *Bull. Calif. Agric. Expt. Sta.*, 214, pp. 456-459, (1911); *ibid.*, 223, pp. 179-187, (1911). ESSIG, *Journ. Ent. Zool.*, V, p. 55, (1913). MASI, *Boll. lab. Zool. Portici*, IX, pp. 307-308, (1915). RAMAKR., *Rep. Proc. Third Ent. Meet. Pusa*, III, p. 935, (1920)[2]; *Rep. Proc. Fifth Ent. Meet. Pusa*, p. 264, pl. xxii, fig. 1-4, (1924)[3]; *Spol. Zeyl.*, XIII, p. 248, (1925); *Bull. Ent. Res.*, XVIII, p. 77, (1927)[4]. Kuwana and Ishii, *Kontyu*, II, pp. 17-24, (1927) (In Jap.). Smith and COMP., *Univ. Calif. Pub. Ent.*, IV, pp. 322-332, (1928). MASI, *Eos*, VII, pp. 457-459, (1931)[5]. Balachowsky, *Bull. Soc. Hist. Nat. Afric. N.*, XXI, pp. 115-116, (1930), [Coccid parasite]. Compere, *Univ. Calif. Pub. Ent.*, V, (12), pp. 247-255, (1931), [Parasite of the Black Scale]. BOUHELIER and others, *Def. cult. Dir. Gen. Agric. Comm. Colon.* (Morocco), No. 6, pp. 1-60, figs. 10, (1932), [Parasite of scale insects in Africa].

[1]Ceylon.

[2,3,4]Coimbatore.

[5]Waltair.

Aspidocoris cyaneus, A. COSTA, *Boll. Ann. Accad. Aspir. nat. napoli*, pp. 24-26, (1863); *Notizi Osserv. Geofaun. Mem. sesta, Napoli*, pp. 34-35, (1886); *Boll. Soc. Ent. Ital.*, XXII, p. 120, (1890).

Aspidocoris myrti, (HALIDAY) COSTA, *Boll Ann. Accad. Aspir. nat. Napoli*, p. 24, (1863). [Hosts : *Saisettice* spp., *Ceroplastes rusci*, *Phœnococcus artemisœ*, *Coccus* sp., etc.].

103. Muscidea MOTSCHULSKY

Muscidea, MOTSCH., *Bull. Soc. Imp. Nat. Moscou,* XXXVI, (2), p. 70, (1863), [Genotype *M. pubescens* MOTSCH.]. ASHM., *Mem. Carnegie Mus.,* I, (4), p. 326, (1904). SCHMIED., *Gen. Ins., fas.* 97, p. 306, (1909).

pubescens MOTSCHULSKY

Muscidea pubescens, MOTSCH., *Bull. Soc. Imp. Nat. Moscou,* XXXVI, (2), p. 71, (1863)[1]. schmied., *Gen. Ins., fas.* 97, p. 306, (1909). RAMAKR., *Spol. Zeyl.,* XIII, p. 248, (1925).

[1]Mt. Patannas: Ceylon.

104. Cardiogaster MOTSCHULSKY

Cardiogaster, MOTSCH., *Bull. Soc. Imp. Nat. Moscou,* XXXVI, (2), p. 72, (1863), [Genotype *C. fusciveniris* MOTSCH.]. ASHM., *Mem. Carnegie Mus.,* I, (4), p. 326, (1904). SCHMIED., *Gen. Ins., fas.* 97, p. 307, (1909).

Fusciventris MOTSCHULSKY

Cardiogaster fusciventris, MOTSCH., *Bull. Soc. Imp. nat. Moscou,* XXXVI, (2), p. 72, (1863)[1]. SCHMIED., *Gen. Ins., fas.* 97, p. 307, (1909). RAMAKR., *Spol. Zeyl.,* XIII, p. 248, (1925). [Host : *Lecanium coffeoe].*

[1]Colombo.

105. Solenoderus MOTSCHULSKY

Solenoderus, MOTSCH., *Bull. soc., Imp. Nat. Moscou,* XXXVI, (3), p. 71, (1863), [Genotype *S. cyaniventris* MOTSCH.]. ASHM., *Mem. Carnegie Mus.,* I, (4), p. 326, (1904). SCHMIED., *Gen. Ins., fas.* 97, p. 307, (1909).

Cyaniventris MOTSCHULSKY.

Solenoderus cyaniventris, MOTSCH., *Bull. Soc. Imp. Nat. Moscou,* XXXVI, (3), p. 71, (1963)[1]. SCHMIED., *Gen. Ins., fas.* 97, p. 307, (1909). RAMAKR., *Spol. Zeyl.,* XIII, p. 248, (1925).

[1]Mt. Patannas: Ceylon.

106. Mnoonema MOTSCHULSKY

Mnoonema, MOTSCH., *Bull. Soc. Imp. Nat. Moscou,* XXXVI, (3), p. 59, (1863), [Genotype *M. timida* MOTSCH.]. ASHM., *Mem. Carnegie Mus.,* I, (4), p. 326, (1904). SCHMIED., *Gen. Ins., fas.* 97, p. 307, (1909).

timida MOTSCHULSKY

Mnoonema timida, MOTSCH., *Bull. Soc. Imp. Nat. Moscou*, XXXVI, (3), p. 60, (1863)[1]. SCHMIED., *Gen. Ins., fas.* 97, p. 307, (1909). RAMAKR., *Spol. Zeyl.*, XIII, p. 248, (1925).

[1]Mt. Patannas: Ceylon.

107. Cephaleta MOTSCHULSKY

Cephaleta, MOTSCH., *Etud. Ent.*, VIII, p. 173, (1859), [Genotype *C. purpureiventris* MOTSCH.]. ASHM., *Mem. Carnegie Mus.*, I, (4), p. 326, (1904). SCHMIED., *Gen. Ins., fas.* 97, p. 308, (1909).

brunniventris MOTSCHULSKY

Cephaleta brunniventris, MOTSCH., *Etud. Ent.*, VIII, p. 174, (1859); *Bull. Soc. Imp. Nat. Moscou*, XXXVI, (3), p. 73, (1863)[1]. RAMAKR., *Spol. Zeyl.*, XIII, p. 248, (1925).

[1]Ceylon.

purpureiventris MOTSCHULSKY

Cephaleta purpureiventris, MOTSCH., *Etud. Ent.*, VIII, p. 174, (1859); *Bull. soc. Imp. Nat. Moscou*, XXXVI, (3), p. 73, (1863)[1]. SCHMIED., *Gen. Ins., fas.* 97, p. 308, (1909). RAMAKR., *Spol. Zeyl.*, XIII, p. 248, (1905).

[1]Ceylon.

SUBFAMILY II. PTEROMALINAE

ASHM., *Proc. Ent. Soc. Washington*, IV, p. 236, (1897); *Proc. U.S.N. Mus.*, XXIII, p. 248, (1900); *Mem. Carnegie Mus.*, I, (4), p. 312, (1904). SCHMIED., *Gen. Ins., fas.* 97, p. 308, (1909).

TRIBE 1. *METOPONARIAE*

ASHM., *Mem. Carnegie Mus.*, I, (4), p. 313, (1904). SCHMIED.,*Gen. Ins., fas.* 97, p. 309, (1909).

108. Etroxys Westwood

Etroxys, WESTW., *Ann. Mag. Nat. Hist.*, VI, p. 121, (1833), [Genotype *E. scenicus* WESTW.]. ASHM., *mem. Carnegie Mus.*, I, (4), p. 313, (1904). SCHMIED., *Gen. Ins., fas.* 97, p. 311, (1909).

GaSTRacanthus, WESTW., *Ann. Mag. Nat. Hist.*, VI, p. 121, (1833).

Hetroxys, FÖRST., *Hymen. Stud.*, II, p. 66, (1856).

Aetroxys, WALK., *Notes Chalc.*, VII, p. 109, (1872).

Hetroxys, DALLA TORRE, *Cat. Hymen.*, V, p. 101, (1898).

marginicollis CAMERON

Etroxys marginicollis, CAM., *Journ. Bombay Nat. Hist. Soc.*, XVII, p. 97, (1906)[1]. RAMAKR., *Spol. Zeyl.*, XIII, p. 248, (1925). [1]Quetta.

Xanthopus CAMERON

Etroxys xanthopus, CAM., *Journ. Bombay Nat. Hist. Soc.*, XVII, p. 97, (1906)[1]. RAMAKR., *Spol. Zeyl.*, XIII, p. 248, (1925). [1]Quetta.

TRIBE 2. *PTEROMALARIAE*

ASHM., *Mem. Carnegie Mus.*, I, (4), p. 318, (1904). SCHMIED., *Gen. Ins., fas.* 97, p. 327, (1909).

109. Pteromalus Swederus

Ichneumon, Linne, (*partim*), *Syst. Nat.* (ed. 10), p. 567, (1758).

Pteromalus, Swed., *Svensk. Vet.-Akad. handl.*, XVI, p. 200, (1795), [Genotype *P. puparum* (Linn.)]. ASHM., *Mem. Carnegie Mus.*, I, (4), p. 321, (1904). schmied., *Gen. Ins., fas.* 97, p. 335, (1909).

Iuzonensis GAHAN

Pteromalus luzonensis, GAH., *Philip. Journ. Sc.*, XXVII, p. 99, (1925). (There are six specimens det. by MASI in the Ind. Museum collection)[1]. [Host : *Papilio* sp.]. [1]Calcutta.

magniceps WALKER

Pteromalus magniceps, WALK., *Ann. Mag. Nat. Hist.*, VI, p. 359, (1860)[1]. SCHMIED., *Gen. Ins., fas.* 97, p. 347, (1909). RAMAKR., *Spol. Zeyl.*, XIII, p. 248, (1925). [1]Ceylon.

Oryzae CAMERON.*

Pteromalus oryzoe, CAM., *Mem. Proc. Lit. Philos. Soc. Manchester*, IV, p. 184, (1891)[1]. SCHMIED., *Gen. Ins., fas.* 97, p. 349, (1909). RAMAKR., *Spol. Zeyl.*, XIII, p. [1]India. [2]Coimbatore.

* This species may prove to be identical with *Aplastomorpha calandræ* (How.)

248, (1925); *Bull. Ent. Res.*, XVIII, p. 77, (1927)[2]. [Host : *Calandra granaria*].

110. Aplastomorpha CRAWFORD

Pteromalus, HOW., (*partim*), in Comstock, *Rep. Ent. U.S. Dept. Agric. Washington*, 1880, p. 273, (1883).

Meraporus, Tucker, (*partim*), *Canad. Ent.*, XLII, p. 343, (1910).

Aplastomorpha, CRAW., *Proc. U.S.N. Mus.*, XLV, p. 252, (1913), [Genotype *A. calandraoe* (HOW.)]. GIR., *Insec. Inscit. Menstr.*, V, p. 152, (1917). GAH., *Proc. Ent. Soc. Washington*, XXV, p. 188, (1923).

Calandrae (HOWARD)

Pteromalus calandroe, HOW., in Comstock *Rep. Ent. U.s. Dept. Agric. Washington*, 1880, p. 273, (1883). SCHMIED., *Gen. Ins., fas.* 97, p. 338, (1909). [Host : *Calandra oryzoe*].

Aplastomorpha calandroe, GAH., *Proc. Ent. Soc. Washington*, XXV, p. 188, (1923). RAMAKR., *Spol. Zeyl.*, XIII, p. 249, (1925)[1]; *Bull. Ent. Res.*, XVIII, p. 77, (1927)[2]. — [1,2]Madras Coimbatore.

Aplastomorpha pratti, CRAW., *loc. cit.*

Meraporus vandinei, Tuck., *Caned. Ent.*, XLII, p. 343, (1910). GIR., *Insec. Inscit. Menstr.* V, p. 152, 91917)[1]. GAH., *Proc. U.S.N. Mus.*, LVI, p. 523, (1919)[2]. RAMAKR., *Rep. Proc. Fourth Ent. Meet. Pusa*, p. 364, (1921)[1]. [Host: *Sitodrepa*]. — [1,2]Coimbatore, Madras. [1]Madras.

111. Eupteromalus Kurdjnmov.*

Pteromalus, THOMS., (*partim*), *Hymen. Skandin.*, V, p. 155, (1878).

Eupteromalus, KURDJ., *Rev. russe Ent.*, XIII, p. 12, (1913), [Genotype *E. nidulans* (THOMS.), not *nidulans* (FÖRST.) of authors].

Parnarae GAHAN

Eupteromalus parnaroe, GAH., *Proc. U.S.N. Mus.*, LVI, p. 522, (1919)[1]. RAMAKR., *Spol. Zeyl.*, XIII, p. 249, (1925); *Bull. Ent. Res.*, XVIII, p. 77, (1927). [Host: *Parnara mathias*]. — [1]Karvetnagar: S. India.

---

* Location provisional.

TRIBE 3. *ROPTROCERARIÆ*

ASHM., *Mem. Carnegie Mus.*, I, (4), p. 322, (1904). SCHMIED., *Gen. Ins., fas.* 97, p. 359, (1909).

112. **Roptrocerus** Ratzeburg

Pachyceras, Ratz., *Ichneum. Forstins.*, I, P. 217, (1844).

Roptrocerus, Ratz., *Ichneum. Forstins.*, II, p. 209, (1848), [Genotype *R. xylophagorum* (Ratz.)]. ASHM., *Mem. Carnegie Mus.*, I, (4), p. 323, (1904). SCHMIED., *Gen. Ins., fas.* 97, p. 300, (1909). WATERST., *Ind. For. Rec.*, IX, p. 84, (1922).

Sulcatus WATERSTON

Roptrocerus sulcatus, WATERST., *Ind. For. Rec.*, IX, pp. 85-89, (1922)[1]. RAMAKR., *Spol. Zeyl.*, XIII, p. 249, (1925).

[1]Bhowali, Kumaon: Himalayas.

testaciventris MOTSCHULSKY

Roptrocerus testaciventris, MOTSCH., *Bull. Soc. Imp. Nat. Moscou*, XXXVI, (3), p. 49, (1863)[1]. SCHMIED., *Gen. Ins., fas.* 97, p. 360, (1909). RAMAKR., *Spol. Zeyl.*, XIII, p. 249, (1925).

[1]Mt. Nura Ellia: Ceylon.

SUBFAMILY III. SPHEGIGASTERINAE

FÖRST., *Hymen. Stud.*, II, p. 24, (1856). THOMS., *Hymen. Skandin.*, V, p. 16, (1878). ASHM., *proc. Ent. Soc. Washnigton*, IV, p. 248, (1897); *Mem. Carnegie Mus.*, I, (4), p. 327, (1904). SCHMIED., *Gen. Ins., fas.* 97, p. 367, (1909).

TRIBE SPHEGIGASTERARIAE

ASHM., *Mem. Carnegie Mus.*, I, (4), p. 330, (1904). SCHMIED., *Gen. Ins., fas.* 97, p. 374, (1909).

113. **Acroclisoides** Girault

Acroclisoides, GIR., *Mem. Queensland Mus.*, III, p. 334, (1915), [Genotype *A. megacephalus* GIR.].

indicus FERRIERE.

Acrolisoides indicus, FERR., *Bull. Ent. Res.*, XXII, pp. 279-280, (1931)[1]. RAMAKR. and Margag., *Journ. Bombay Nat. Hist. Soc.*, XXXVII, p. 195, (1934), [Notes].

[1]Dehra Dun.

114. **TrigonogaStra** Ashmead.

TrigonogaStra, ASHM., *mem. Carnegie Mus.*, I, (4), p. 330, (1904), [Genotype *T. aurata* ASHM.]. SCHMIED., *Gen. Ins., fas.* 97, p. 378, (1909).

brunneicornis FERRIERE

TrigonogaStra brunneicornis, FERR., *Bull. Ent. Res.*, XXI, pp. 356-357 (1930)[1]. RAMAKR. and MARGAB., *Journ. Bombay Nat. Hist. Soc.*, XXXVII, p. 195, (1934)[2]. [Host : *Agromyza of Hibiscus esculantus*]. [1,2]Peradeniya: Ceylon.

megacephala WATERSTON

TrigonogaStra megacephala, WATERST., *Bull. Ent. Res.*, V, p. 328, (1914)[1]. RAMAKR., *Spol. Zeyl.*, XIII, p. 249, (1925). [1]Ceylon.

rugosa WATERSTON

TrigonogaStra rugosa, WATERST., *Bull. Ent. Res.*, V, p. 326, (1914)[1]. RAMAKR., *Spol. Zeyl.*, XIII, p. 249, (1925). [Host : *Agromyza phaseoli*]. [1]Ceylon.

115. **Polycystus** Westwood

Polycystus, WESTW., *Intr. Mod. Class. Ins.*, II, p. 68, (1840). [Genotype *P. mathewsii* WESTW.]. ASHM., *Mem. Carnegie Mus.*, I, (4), p. 331, (1904). schmied., *Gen. Ins., fas.* 97, p. 380, (1909).

propinquus WATERSTON

Polycystus propinquus, WATERST., *Bull. Ent. Res.*, V, p. 325, (1914)[1]. RAMAKR., *Spol. Zeyl.*, XIII, p. 249, (1925). [Host : *Agromyza* sp.]. [1]Ceylon.

116. **Agiommatus** CRAWFORD

Agiommatus, CRAW., *Proc. U.S.N. Mus.*, XLI, p. 278, (1911), [Genotype *A. sumatraensis* CRAW.].

acherontiæ FERRIERE

Agiommatus acherontiæ, FERR., *Bull. Ent. Res.*, XXII, p. 280, (1931)[1]. RAMAKR., and MARGAB, *Journ. Bombay Nat. Hist. Soc.*, XXXVII, p. 195, (1934). [Host : *Acherontia styx*]. [1]Dehra Dun.

GENERA NOT REFERRED TO THEIR EXACT TRIBES.

117. Leptorhopala MOTSCHULSKY

Leptorhopala, Mostsch., *Bull. Soc. Imp. Nat. Moscou,* XXXVI, (2), p. 60, (1863), [Genotype *L. cuprifrons* MOTSCH.].

cuprifrons MOTSCHULSKY

Leptorhopala cuprifrons, MOTSCH., *Bull. Soc. Imp. Nat. Moscou,* XXXVI, (2), p. 61, (1863)[1]. RAMAKR., *Spol. Zeyl.,* XIII, p. 249, (1925). — [1]Mt. Patannas: Ceylon.

118 Tetrarhopala MOTSCHULSKY

Tetrarhopala, MOTSCH., *Bull. Soc. Imp. Nat. Moscou,* XXXVI, (3), p. 62, (1863), [Genotype *T. nigra* MOTSCH.].

nigra MOTSCHULSKY

Tetrarhopala nigra MOTSCH., *Bull. Soc. Imp. Nat. Moscou,* XXXVI, (3), p. 62, (1863)[1]. RAMAKR., *Spol. Zeyl.,* XIII, p. 249, (1925). — 1Mt. Patannas: Ceylon.

FAMILY X. ELASMIDAE

FÖRST., *Hymen., Stud.,* II, pp. 19, 25, 71, (1856). THOMS., *Skandin.,* V, p. 180, (1878). HOW., *Ent. Amer.,* I, p. 198, (1886); *Proc. Ent. Soc. Washington,* III, p. 13, (1894). ASHM., *Proc. U.S.N. Mus.,* IV, p. 236, (1897); *Mem. Carnegie Mus.,* I, (4), p. 335, (1904). SCHMIED., *Gen. Ins., fas.* 97, p. 388, (1909).

119. Elasmus Westwood

Eulophus, Fonscol., (*partim*), *Ann. Sc. Nat.,* XXVI, p. 298, (1832).

Elasmus, WESTW., *Philos. Mag.,* III, p. 343, (1833), [Genotype *E. flabellatus* (Fonscol.)]. ASHM., *Mem. Carnegie Mus.,* I, (4), p. 336, (1904). SCHMIED., *Gen. Ins., fas.* 97, p. 338, (1909) FERR., *Bull. Ent, Res.,* XX, pp. 411-423, (1929), [Descriptions and Keys to African and Asiatic spp.].

Cyclopleura, CAM., *Ind. For. Rec.,* IV, p. 96, (1913).

albomaculatus GAHAN

Elasmus albomaculatus, GAH., *Philip. Journ.* Sc., XVII, p. 347, (1920). (Specimens det. by FERRIERE in the — [1]Dehra Dun.

collections of the Forest Research Inst., Dehra Dun)[1]. [Host : *Apanteles malevolus* of *Hybloea puera* defoliating the teak].

anticles WALKER

Elasmus anticles, WALK., *Ann. Mag. Nat. Hist.,* XVII, p. 181, (1846)[1]. SCHMIED., *Gen. Ins., fas.* 97, p. 389, (1909). RAMAKR., *Spol. Zeyl.,* XIII, p. 249, (1925). FERR., *Bull. Ent. Res.,* XIX, p. 172, (1928); *ibid.,* XX, p. 416, (1929). [1]Bombay.

brevicornis GAHAN

Elasmus brevicornis, GAH., *Treubia,* III, p. 50, (1922). FERR., *Bull. Ent. Res.,* XX, p. 413, *1929)[1]. RAMAKR. and MARGAB., *Journ. Bombay Nat. Hist. Soc.,* XXXVII, p. 195, (1934)[2], [Notes]. [1]Dehra Dun. [2]Kuala Lumpur.

Claripennis (CAMERON)

Cyclopleura claripennis, CAM., *Ind. for. Rec.,* IV, p. 97, (1913)[1]. [1]Dehra Dun.

Elasmus colemani, Mahd., *Journ. Sc. Assoc. Maharaja's Call. Vizianagram,* I, p. 69, (1923)[1]. [1]Mysore.

Elasmus claripennis, FERR., *Bull. Ent. Res.,* XIX, p. 171-172, (1928)[1]; *ibid.,* XX, p. 418, (1929). Miller, *Sc. Ser., Dept. Agric. S.S. and F.M.S.,* No. 11, pp. 1-24, pls. 3, (1933), [Parasite of *Eublemma amabilis* in Malaya]. RAMAKR. *and* MARGAB., *Journ. Bombay Nat. Hist. Soc.,* XXXVII, p. 195, (1934), [Notes]. [1]Namkum.

Ceylonicus FERRIERE

Elasmus ceylonicus, FERR., *Bull. Ent, Res.,* XX, p. 418, (1929)[1]. RAMAKR., and MARGAB., *Journ. Bombay Nat. Hist. Soc.,* XXXVII, p. 195, (1934). [Host : *Psyche albipes ?*]. [1]Pelamadulla: Ceylon.

Fumipennis (CAMERON)

Cyclopleura fumipennis, CAM., *Ind. For. Rec.,* IV, p. 97, (1913)[1]. FERR., *Bull. Ent. Res.,* XX, P. 416, (1929), *stulops,* II (4-5), pp. 86-96, 97-108, figs. 12, (1933), [Parasite of *Tirathaba* in Java]. [1]Dehra Dun.

Homonae FERRIERE

Elasmus homonoe, FERR., *Bull. Ent. Res.,* XX, p. 415, (1929)[1]. KING, *Bull, Tea Res. Inst. Ceylon,* No. 10, [1]Pelamadulla: Ceylon.

pp. 27-33, (1933), [Parasite of *Homona coffiaria* in Ceylon]. RAMAKR. and MARGAB., *Journ. Bombay Nat. Hist. Soc.*, XXXVII, p. 195, (1934).

hutsoni FERRIERE

Elasmus hutsoni, FERR., *Bull. Ent. Res.*, XX, p. 412, (1929)[1]. RAMAKR. and MARGAB., *Journ. Bombay Nat. Hist. Soc.*, XXXVII, p. 195, (1934). [Host : *Psyche albipes?*]. [1]Pelamadulla: Ceylon.

hyblææ FERRIERE

Elasmus hyblææ, FERR., *Bull. Ent. Res.*, XX, p. 414, (1929)[1]. RAMAKR. and MARGAB., *Journ. Bombay Nat. Hist. Soc.*, XXXVII, p. 195, (1934). [Host : *Hybloea puera*]. [1]Aravallicadu: Nilambur, S. India.

indicus Rohwer

Elasmus indicus, ROHW., *Ann. Mag. Nat. Hist.*, XVII, p. 124, (1921)[1]. RAMAKR., *Rep. Proc. Fourth Ent. Meet. Pusa*, p. 365, (1921)[2]; *Spol. Zeyl.*, XIII, p. 249, (1925); *Bull. Ent. Res.*, XVIII, p. 77, (1927)[3] FERR., *Bull. Ent. Res.*, XX, p. 415, (1929). [Host : *Anomalococcus*]. [1, 2,3]Coimbatore.

Johnstoni FERRIERE

Elasmus johnstoni, FERR., *Bull. Ent. Res.*, XX, p. 258, (1929)[1]. RAMAKR., and MARGAB., *Journ. Bombay Nat. Hist. Soc.*, XXXVII, p. 195, (1934). [Hosts : *Earias* spp. and *Platyedra gossypiella*]. [1]Karnal: Punjab.

Nephantidis Rohwer

Elasmus nephantidis, ROHW., *Ann. Mag. Nat. Hist.*, XVII. p. 125, (1921)[1]. RAMAKR., *Rep. Proc. Fourth Ent. Meet. Pusa*, pp. 23, 365, 91921)[2]; *Spol. Zeyl.*, XIII, p. 249, (1925)[3]; *Bull. Ent. Res.*, XVIII, p. 77, (1927)[4]. Ramachandra Rao and Cherian, *Year book Madras Agric. Dept.*, 1926, pp. 39-49, (1927). FERR., *Bull. Ent. Res.*, XX, p. 415, (1929). [Host : *Nephantis serinopa*]. [1,2,3,4]Coimbatore.

FAMILY XI. EULOPHIDÆ

WESTW., *Intr. Mod. Class. Ins.*, p. 166, (1840). HALID., *Trans. Ent. Soc., London*, III, p. 296, (1843). FORST., *Hymen. Stud.*, II, pp. 18, 19, 21, 26, 30, 72, 74, 78, 83, (1856). THOMS., *Hymen. Skandin.*, V, p. 180, (1878). HOW., *Ent. Amer.*, I, p. 198, (1885); *ibid.*, II, p. 98, (1886). ASHM., *Proc. U.S.N. Mus.*, XXIII, p. 203, (1900); *Mem. Carnegie Mus.*, U, (4), p. 336, (1904). SCHMIED.,

Gen. Ins., fas. 97, p. 390, (1909). GIR., *Mem. Queensland Mus.*, II, pp. 294-295, (1913), [Table of subfamilies].

SUBFAMILY I. ELACHERTINÆ

FÖRST., *Hymen. Stud.*, II, pp. 19, 26, 72, (1856). thoms., *Hymen. Skandin.*, V, p. 180, (1878). HOW., *Ent. Amer.*, I, p. 198, (1886). ASHM., *Proc. Ent. Soc. Washington*, IV, p. 249, (1897); *mem. Carnegie Mus.*, I, (4), p. 350, (1904). SCHMIED., *Gen. Ins., fas.* 97, p. 391, (1909).

TRIBE 1. *ELACHERTARIÆ*

ASHM., *Mem. Carnegie Mus.*, I, (4), p. 354, (1904). SCHMIED., *Gen. Ins., fas.* 97, p. 391, (1909).

120. Cirrospilus Westwood

Cirrospilus, WESTW., *Philos. Mag.*, (3), I, p. 128, (1832), [Genotyupe *C. elegantissimus* WESTW.]. ASHM., *Mem. Carnegie Mus.*, I, (4), p. 354, (1904). SCHMIED., *Gen. Ins., fas.* 97, p. 397, (1909).

coccivorus MOTSCHULSKY

Cirrospilus coccivorus, MOTSCH., *Bull. Soc. Imp. Nat. Moscou*, XXXVI, (3), p. 68, (1863)[1]. SCHMIED., *Gen. Ins., fas.* 97, p. 398, (1909). RAMAKR., *Spol. Zeyl.*, XIII, p. 249, (1925). [Host : *Lecanium caffeoe*].

[1]Colombo: Ceylon.

Viridifrons MOTSCHULSKY

Cirrospilus viridifrons, MOTSCH., *Bull. Soc. Imp. Nat. Moscou*, XXXVI, (3), p. 68, (1863)[1]. SCHMIED., *Gen. Ins., fas.* 97, p. 399, (1909). RAMAKR., *Spol. Zeyl.*, XIII, p. 249, 91925).

[1]Mt. Patannas: Ceylon.

121. Trichospilus FERRIERE

Trichospilus, FERR., *Bull. Ent. Res.*, XXI, p. 358, (1930). [Genotype *T. pupivora* FERR.].

Pupivora FERRIERE

Trichospilus pupivora, FERR., *Bull Ent. Res.*, XXI, p. 358, (1930)[1]. RAMAKR. and MARGAB., *Journ. Bombay Nat. Hist. Soc.*, XXXVII, p. 196, (1934). ANANTHA-

[1]Coimbatore: S. India; Peradeniya, Lunu wila, Passra

NARAYANAN, *Bull. Ent. Res.*, XXV, pp. 55-61, fig. 1, (1934), [Bionomics]. [Hosts : *Nephantis serinopa, Thosea cervina, Spodopatera mauritia,* puparia of a Techinid of *nacobia annubilata, Tirathaba* spp.]. and Kurunegala: Ceylon.

122. **Euryscotolinx** Girault.*

Euryscotolinx, GIR., *Mem. Queensland Mus.*, II, p. 266, (1913), [Genotype *E. guttativertex* GIR.].

coimbatorensis Rohwer

Euryscotolinx coimbatorensis, ROHW., *Ann. Mag. Nat. Hist.*, VII, p. 134, (1921)[1]. RAMAKR., *Rep. Proc. Fourth Ent. Meet. Pusa*, p. 365, (1921)[2]; *Spol.; Zeyl.*, XIII, p. 250, (1925)[3]; *Bull. Ent. Res.*, XVIII, p. 78, (197)[4]. [Host: *Cyphosticha coerulæ*]. [1,2,3,4]Coimbatore.

TRIBE 2. *EUPLECTRAIAE*

ASHM., *Mem. Carnegie Mus.*, I, (4), p. 351, (1904). SCHMIED., *Gen. Ins., fas.* 97, p. 491, (1909).

123. **Euplectrus** Westwood

Euplectrus, WESTW., *Phiols. Mag.*, (3), I, p. 128, (1832), [Genotype *E. maculiventris* WESTW.]. ASHM., *Mem. Carnegie Mus.*, I, (4), p. 351, (1904). SCHMIED., *Gen. Ins., fas.* 97, p. 402, (1909). MANI, *Rec. Ind. Mus.*, XXXVII, p. 256, (1935), [Key to spp.].

Diplectron, Dahlb., *Oef. Vet.-Akad. Forh.*, XIV, p. 292, (1857).

Busyi CRAWFORD

Euplectrus busyi, CRAW., *Proc. U.S.N. Mus.*, XLI, p. 279, (1909)[1]. RAMAKR., *Spol. Zeyl.*, XIII, p. 250, (1925); *Bull. Ent. Res.*, XVIII, p. 78, (1927)[2*]. MANI, *Rec. Ind. Mus.*, XXXVII, p. 257, (1935). [1,2] Ceimbatore.

ceylonesis HOWARD

Euplectrus Ceylonensis, HOW., *Proc. U.S.N. Mus.*, XVIII, p. 641, (1896)[1]. RAMAKR., *spol. Zeyl.*, XIII, p. 250, (1925). MANI, *Rec. Ind. Mus.*, XXXVII, p. 256, (1935). [1]Ceylon.

---

* Location provisional.

* Ramakrishna Ayyar's *busbyi* is an error for *busyi.*

euplexiæ Rohwer

Eupectrus euplexiæ, ROHW., *Ann. mag., nat. Hist.*, VII, p. 135, (1921)[1]. RAMAKR., *Rep. Proc. Fourth Ent. Meet. Pusa*, p. 365, (1921)[2]; *Spol. Zeyl.*, XIII, p. 250, 91925); *Bull. Ent. Res.*, XVIII, p. 78, (1927)[3]. MANI, *Rec. Ind. Mus.*, XXXVII, p. 257, (1935). [Host : *Perigea capensis*]. — [1,2,3]Coimbatore.

himalayaensis MANI

Euplectrus himalayansis, MANI, *Rec., Ind. Mus*, XXXVII, p. 257, (1935)[1]. — [1]Darjeeling: Himalayas.

Leucostomus Rohwer

Euplectrus leucostomus, ROHW., *Ann. Mag. nat. Hist.*, VII, p. 134, (1921)[1]. RAMAKR., *Spol. Zeyl.*, XIII, p. 250, (1925).[2] MANI, *Rec. Ind. Mus.*, XXXVII, p. 256,* (1935). — [1,2]Coimbatore.

nyctemerae CRAWFORD

Euplectrus nyctemeroe, CRAW., *Proc. U.S.N., Mus.*, XLII, p. 9, (1912)[1]. RAMAKR., *Rep. Proc. Fourth Ent. Meet. Pusa*, p. 363, (1921)[2]; *Spol. Zeyl.*, XIII, p. 250, (1925); *Bull. Ent. Res.*, XVIII, p. 78, (1927)[3]. MANI, *Rec. Ind. Mus.*, XXXVII, p. 257, (1935). [Host : *Nyctemera lacticinia*]. — [1,2,3]Bangalore: Mysore.

SUBFAMILY II. EULOPHINÆ

ASHM., *Mem. Carnegie Mus.*, I, (4), p. 355, (1904). SCHMIED.,*Gen. Ins.*, *fas.* 97, p. 408, (1909).

TRIBE *EULOPHARIÆ*

ASHM., *Mem. Carnegie Mus.*, I, (4), p. 356, (1904). SCHMIED., *Gen. Ins.*, *fas.* 97. p. 409, (1909)

124. **Sympiesis** Förster

Eulophus, Nees, (*partim*), *Hymen. Affin. Mon.*, IIm p. 168, (1834).

Sympiesis, FÖRST., *Hymen. Stud.*, II, p. 74, (1856), [Genotype *S. sericeicornis* (Nees)]. ASHM., *Mem, Carnegie Mus.*, I, (4), p. 357, (1904). SCHMIED., *Gen, Ins.*, *fas.* 97, p. 411 (1909).

Sympiezus, THOMS., *Hymen. Skandin.*, V, p. 208, (1878).

Purpureus WATERSTON

Sympiesis purpureus, WATERST., *Bull Ent. Res.,* V, p. 333, (1914)[1]. RAMAKR., *Spol. Zeyl.,* XIII, p. 250, (1925). [1]Ceylon.

125. **Asympiesiella** Girault.*

Asympiesiella, GIR., *Arch. Naturg. Jahrg.,* LXXIX, A, H. 6, p. 78, (1913), [Genotype *A. nelsonensis* GIR.].

india Girault.*

Asympiesiella india, GIR., *Canad. Ent.* XLVIII, p. 341, (1916)[1]. RAMAKR., *Rep. Proc. Third Ent. Meet. Pusa,* III, p. 934, (1920). Dutt, *Rep. Proc. Fourth Ent. Meet. Pusa,* p. 288, pl. XLIX, figs. 1-6, (1921)[2] [biology]. RAMAKR., *Spol. Zeyl.,* XIII, p. 250, (1925); *Bull. Ent. Res.,* XVIII, p. 78, (1927)[3], [Host : *Gracilaria* of red-gram]. [1,2,3]Pusa.

126. **Eulophus** Geoffroy

Ichneumon, Linne,. *Syst. Nat.,* (ed. 10), I, p. 568, (1758). FABR., *(partim), Spec. Ins.,* p. 441, (1781).

Eulophus, Geof., *Hist. Abr. Ins.,* II, p. 312, pl. 15, fig. 3, (1762), [Genotype *E. ramicornis* (FABR.) desig. latr., *E. pectinicornis* (Linn.) desig. WESTW.]. ASHM., *Mem. Carnegie Mus.,* I, (4), p. 357, (1904). SCHMIED., *Gen. Ins., fas.* 97, p. 413, (1909).

brama MOTSCHULSKY

Eulophus brama, MOTSCH., *Bull. Soc. Imp. Nat. Moscou,* XXXVI, (3) p. 64, (1863)[1]. SCHMIED., *Gen. Ins., fas.* 97, p. 414, (1909). RAMAKR., *Spol. Zeyl.,* XIII, p. 250, (1925). [1]Mt. Patannas: Ceylon.

fraternus MOTSCHULSKY

Eulophus fraternus, MOTSCH., *Etud. Ent.,* VIII, p. 118, (1859)[1]. SCHMIED., *Gen. Ins., fas.* 97, p. 415, (1909). RAMAKR., *Spol. Zeyl.,* XIII, p. 250, (1925). [1]Ceylon.

mantiechthrus Mooschulsky

Eulophus mantiechthrus, MOTSCH., *Etud. Ent.,* VIII, p. 117, (1859); *Bull. Soc. Imp. Nat. Moscou,* XXXVI, (3), p. [1]Ceylon.

---

* Location provisional.

* Probably a typographical error for *indica.*

65, (1863)[1]. Schmied., *Gen. Ins., fas.* 97, p. 416, (1909). Ramakr., *Spol. Zeyl.*, XIII, p. 250, (1925).

paria Motschulsky

Eulophus paria, Motsch., *Bull. Soc. Imp. Nat. Moscou*, XXXVI, (3), p. 63, (1863)[1]. Schmied., *Gen. Ins., fas.* 97, p. 416, (1909). Ramakr., *Spol. Zeyl.*, XIII, p. 250, (1925). [1]Mts. Nura Ellia and Patannas: Ceylon.

tardescens Motschulsky

Eulophus tardescens, Motsch., *Bull. Soc. Imp. Nat. Moscou*, XXXVI, (3), p. 65, (1863)[1]. Schmied., *Gen. Ins., fas.* 97, p. 418, (1909). Ramakr., *Spol, Zeyl.*, XIII, p. 250, (1925). [1]Mt. Nura Ellia: Ceylon.

ventricosus Motschulsky

Eulophus ventricosus, Motsch., *Bull. Soc. Imp. nat. Moscou*, XXXVI, (3), p. 63, (1863)[1]. Schmied., *Gen. Ins., fas.* 97, p. 418, (1909). Ramakr., *Spol. Zeyl.*, XIII, p. 250, (1925). [1]Mt. Patannas: Ceylon.

Vischnu Motschulsky

Eulophus vischnu, Motsch., *Bull. Soc. Imp. Nat. Moscou*, XXXVI, (3), p. 65, (1863)[1]. Schmied., *Gen. Ins., fas.* 97, p. 418, (19090. Ramakr., *Spol. Zeyl.*, XIII, p. 250, (1925). [1]Mt. Nura Ellia: Ceylon.

SUBFAMILY III. ENTEDONINÆ

Ashm., *Proc. Ent. Soc. Washington*, IV, p. 236, (1897); *Mem. Carnegie Mus.*, I, (4), p. 337, (1904). Schmied., *Gen. Ins., fas.* 97, p. 431, (1909).

TRIBE 1. *TETRACAMPARIÆ*

Ashm., *Mem. Carnegie Mus.*, I, (4), p. 337, (1904), Schmied., *Gen. Ins., fas.* 97, p. 421, (1909). Kryger, *Notul. Ent.*, XI, pp. 150-152. (1932).

127. Cassidocida Crawford

Cassidocida, Craw., *Proc. U.S.N. Mus.*, XLV, p. 253, (1913), [Genotype *C. aspidomorphoe* Craw.].

aspidomorphae Crawford

Cassidocida aspidomorphoe, Craw., *Proc. U.S.. Mus.*, XLV, p. 253, (1913)[1]. Ramakr., *Rep. Proc. third Ent. Meet. Pusa*, III, p. 934, (1920)[2]; *Spol. Zeyl.*, XIII, p. 250, [1,2,3]Bangalore, Mysore, Coimbatore.

(1925); *Bull. Ent. Res.*, XVIII, p. 78, (1927). [Host : *Aspidomorpha miliaris*.]

TRIBE 2. *OMPHALARIÆ*

ASHM., *Mem. Carnegie Mus.*, I, (4), p. 338, (1904). SCHMIED., *Gen. Ins.*, *fas.* 97, p. 424, (1909).

128. **Closterocerus** Westwood

Closterocerus, WESTW., *Ann. Mag. nat. Hist.*, VI, p. 419, (1833), [Genotype *C. trifasciatus* WESTW.]. ASHM., *Mem. Carnegie Mus.*, I, (4), p. 340, (1904). SCHMIED., *Gen. Ins.*, *fas.* 97, p. 428, (1909).

insignis WATERSTON

Closterocerus insignis, WATERST., *Bull. Ent. Res.*, V, p. 330, (1914)[1]. RAMAKR., *Spol. Zeyl.*, XIII, p. 250, (1925). [on *Oscinis theoe*].

[1]Ceylon.

TRIBE 3. *ENTEDONARIÆ*

ASHM., *Mem. Carnegie Mus.*, I, (4), p. 340, (1904). SCHMIED., *Gen. Ins.*, *fas.* 97, p. 430, (1909).

129. **Pleurotropis** Förster

Pleurotropis, FÖRST., *Hymen. Stud.*,II, pp. 78, 82, (1856); *Progr. Realsch. Aachen*, XXXVII, No. 58, (1861), [Genotype *P. isomera* FÖRST]. ASHM., *Mem. Carnegie Mus.*, I, (4), p. 386, (1904). SCHMIED., *Gen. Ins.*, *fas.* 97, p. 435, (1909).

detrimentosus Galhan

Pleurotropis detrimentosus, GAH., *Proc.*, *U.S.N. Mus.*, No. 2831, art. 77, p. 9, (1930). RAMAKR., and MARGAB., *Journ. Bombay Nat. Hist. Soc.*, XXXVII, p. 196, (1934).[1]

[1]India.

epilachnae Rohwer

Pleurotropis epilachnce, ROHW., *Ann. Mag. nat. Hist.*, VII, p. 126, (1921)[1]. RAMAKR., *Rep. Proc. fourth Ent. Meet. Pusa*, p. 365, (1921)[2]; *Spol. Zeyl.*, XIII, p. 251, (1925)*; *Bull. Ent. Res.*, XVIII, p. 77, (1927)[3]. [Host: *Epilachna* sp.].

[1,2,3]Coimbatore.

* *Pleurotrupis* of Ramakrishna Ayyar is an error for *Pleurotropis*.

foveolatus CRAWFORD

Pleurotropis foveolatus, CRAW., *Proc. U.S.N. Mus.,* XLII, p. 7, (1912)[1]. RAMAKR., *Rep. Proc. Fourth Ent. Meet. Pusa,* p. 363, (1921)z; *Spol. Zeyl.,* XIII, p. 251, (1925); *Bull. Ent. Res.,* XVIII, p. 78, (1927)[3]. [Host: *Epilachna* sp.]. — [1,2,3]Bangalore.

130. Entedon DALMAN

Entedon, DALM., *Svensk. Vet.-Akad. Handl.,* XLI, pp. 136, 181, (1820), [Genotype *E. cyanellus* DALM.]. ASHM., *Mem. Carnegie Mus.,* I, (4), p. 342, (1904). SCHMIED., *Gen. Ins., fas.* 97, p. 432, (1909).

imbreus WALKER

Entedon imbreus, WALK., *Ann. Mag. Nat. Hist.,* XVII, p. 184, (1846)[1]. SCHMIED., *Gen. Ins., fas.* 97, p. 439, (1909). RAMAKR., *Spol, Zeyl.,* XIII, p. 250, (1925). — [1]India.

TRIBE 4. *PEDIOBARIÆ*

ASHM., *Mem. Carnegie Mus.,* I, (4), p. 343, (1904). SCHMIED., *Gen. Ins., fas.* 97, p. 445, (1909).

131. Pediobopsis Girault

Pediobopsis, GIR., *Mem. Queensland Mus.,* II, p. 155, (1913). [Genotype *P. spenceri* GIR.].

Locustivora Rohwer

Pediobopsis locustivora, ROHW., *Ann. Mag. Nat. Hist.,* VII, p. 125, (1921)[1]. RAMAKR., *Rep. Proc. Fourth Ent. Meet. Pusa,* p. 365, 91921)[2]; *Spol. Zeyl.,* XIII, p. 251, (1925); *Bull. Ent. Res.,* XVIII, p. 77, (1927)[3]. [Host : Locust]. — [1,2,3]Coimbatore.

SUBFAMILY IV. APHELININAE

FÖRST., *Hymen. Stud.,* II, pp. 18, 21, 30, (1856). Gernet, *Bull. Soc. Imp., Nat. Moscou,* XXXVI, (3), pp. 154-174, (1863). THOMS., *Hymen, Skandin.,* V, pp. 12, 183, (1876). HOW., *Ent. Amer.,* I, p. 198, (1886); *Bull. U.S. Agric. Ent. Tech. Ser.,* I, pp. 44, (1895). [Revision of N. American forms]. HOW. and ASHM., *Proc. U.S.N. Mus.,* XVIII, pp. 633-648, (1896), [Ceylonese spp.]. HOW., *Proc. Ent. Soc. Washington,* IV, pp. 133-139, (1877), [New parasites of Coccids]. ASHM., *Proc. Ent. Soc. Washington,* IV, p. 236, (1897). HOW., *Canad. Ent.* XXXII, pp. 167-168, (1900). Embileton, *Trabs. Ent. Soc. London,* pp. 129-229, (1902), [Economic importance]. ASHM., *Mem. Carnegie Mus.*, I, (4), p. 344, (1904). SCHMIED., *Gen. Ins., fas.* 97, p. 448, (1909). IMMS, *Quart. Journ. Micr. Sc.*, LXIII, pp. 293-374, figs. 34, (1918). GAH., *Proc. U.S.N. Mus.*, LV, pp. 403-407, (1919). LUNDIE, *Mem. Cornell Univ. Agric. Expt. Sta.*, No, 79, pp. 127, figs. 6, (1924), [Biology of *Aphelinus mali* H.]. COMP., *Bull. Calif. Acad. Sc.*, XXIII, pp. 113-123, figs. 6, (1924), [Parasites of Coccids]; *Univ. Calif. Pub. Ent.*, III, pp. 295-326, pl. vi, (1925). [Parasites and hyperparasites of *Saisetia oleoe* Ber.]; *ibid.*, IV, pp. 1-30, (1926), [Coccid-inhabiting forms]. COMP. and SMITH, *Univ. Calif. Pub. Ent.*, IV, pp. 51-61, (1926). [Life history notes and introduction of parasites in California]; *ibid.*, IV, pp. 63-73, (1927), [Life history notes on Oriental forms]. COMP. *ibid.*, IV, pp. 209-230, (1928), [African and Californian Coccid-inhabiting forms]. MERCET, *Rev. Biol. Limnol. Ser.* B., I, pp. 3-28, figs. 7, (1928), [Characters, keys, biology, etc.]; *Rev. Fitopat.*, IV-VI, pp. 6-9, (1926-1928), [Keys]. Nowicki, *Beitr. Syst. Ins.*, IV, pp. 153-4160, (1929), [Notes]. Compere, *Univ. Calif. Pub. Ent.*, VI, 112, pp. 277-322, figs. 19, (1936), [Notes on classification].

TRIBE 1. *APHELINARIÆ*

ASHM., *Mem. Carnegie Mus.*, I, (4), p. 345, (1904). HOW., *Bull. U.S. Dept. Agric., Div. Ent., Tech. Ser.*, No. 12, p. 71, (1907). SCHMIED., *Gen. Ins., fas.* 97, p. 449, (1909).

132. Aphelinus DALMAN

Entedon, DALM., *(partim)*, *Svensk., Vet.-Akad. handl.*, XLI, p. 181, pl. viii, figs. 55, 56, (1820).

Aphelinus, DALM., *Svenk. Vet.-Akad. Handl.*, XLI, p. 181, (1820), [Genotype *A. abdominalis* (DALM.)]. ASHM., *Mem. Carnegie Mus.*, I, (4), p. 346, (1904). schmied., *Gen. Ins., fas.* 97, p. 451, (1909). MERCET, *Eos*, VIII, pp. 353-365, figs. 9, (1932), [Acepts Timberlake's suggestions for reSTRicting *Aphelinus* to spp. infesting Aphids and using *Aphytis* HOW. for coccidiphagous forms].

Agonioneurus, WESTW., *Ann. Mag. Nat. Hist.*, VI, p. 122, (1833).

Myina, Nees, *Hymen,. Ichneum. Affin. Mon.*, II, p. 189, (1834).

Coccobius, Ratz., *(partim)*, *Ichneum. Forstins.*, III, p. 195, (1852).

Eriophilus, Haldeman, *Pennsylv. Farm. Journ.* p. 130, (1851); *Proc. Boston Soc. Nat. Hist.*, VI, p. 402, (1860).

Misina, Rond., *Arch. Zool.*, II, p. 11, (1870).

diaspidis HOWARD

Aphelinus diaspidis, HOW., in Comstock, *Rep. Ent. U.S. Dept. Agric.,* 1880, pp. 355, (1881). SCHMIED., *Gen. Ins., fas.* 97, p. 452, (1909). QUAYLE, *Journ. Econ. Ent.,* III, pp. 398-401, (19140). RAMAKR., *Spol. Zeyl.,* XIII, p. 251, (1925).[1] OGILVIE, *Rep. Dept. Agric. Bermuda,* 1927, pp. 26-37, (1928), [Introduction into Bermuda against *Aulacaspis pentagona.*]. [Host : Different scale-insects.].

[1]Ceylon.

fuscipennis HOWARD

Aphelinus fuscipennis, HOW., in comstock, *Rep. Ent. U.S. Dept. Agric. Washington,* 1880, p. 356, (1881). SCHMIED., *Gen Ins., fas.* 97, p. 452, (1909). Misra, *Rep. Proc. Fifth Ent. Meet. Pusa,* p. 130, (1924)[1]. [Host: *Aleyrodes* sp.]

[1]Pusa.

mytilaspidis Le Baron

Aphelinus mytilaspidis, LE BARON, *Amer. Ent.,* II, p. 360, fig. 220, (1870). How., *Proc. U.S.N. Mus.,* XVIII, p. 635, (1896)[1]. SCHMIED., *Gen. Ins., fas.* 97, p. 452, (1909). IMMS, *Quart. Journ. Micr. Sc.,* LXI, pp. 215-274, pl. XIX-XX, (1916), [Biology]. RAMAKR., *Spol. Zeyl.,* XIII, p. 251], (1925). Griswold, *Mem. Cornell Univ. Agric. Expt. Sta.,* 93, pp. 56-57, (1925), [Biology]. Widiez, *Agric. Prat. Pays Chauds,* No. 24, pp. 415-445, (1932). [Parasite of *Lepidosaphes ficus* in Algeria]. Suter, *Mitt. Schweiz. Ent. Ges.,* XV, (9), pp. 347-420, figs. 66, graph 1, (1933), [Parasite of *Lepidosaphes ulmi* in Switzerland], Alden, *Bull. Georgia St. Board Ent.* No. 73, pp. 32, figs. 14, (1933), [Parasite of *Aspidiotus perniciosus* in U.S.A.]. [Host: Different scale-insects.].

[1]Ceylon.

theæ CAMERON

Aphelinus theæ, CAM., *Mem. Proc. Manchester Lit. Philos. Soc.,* IV, p. 183, pl. i, fig. 5, (1891)[1]. cotes, *Ind. Mus. Notes,* III, pp. 24, 132, (1896), [figure]. SCHMIED., *Gen. Ins., fas.* 97, p. 452, (1909). RAMAKR., *Spol. Zeyl.,* XIII, p. 251, (1925). [Host : *Aspidiotus theoe.*].

[1]Kangra Valley.

133. Physcus HOWARD

Coccophagus, How., (*partim*), in Comstock, *Rep. Ent. U.S. Dept. Agric.,* p. 360, (1880).

Physcus, HOW., *Bull. U.S. Dept Agric., Div. Ent., Tech. Ser.*, I, p. 43, (1895), [Genotype *A. varicornis* (How.)]. ASHM., *Mem. Carnegie Mus.*, I, (4), p. 346, (1904). SCHMIED., *Gen. Ins., fas.* 97, p. 454, (1909).

Varicornis (HOWARD)

Coccophagus varicornis, HOW., in Comstock, *Rep. ent. U.S. Dept. Agric.*, p. 360, (1980).

Physcus varicornis, HOW., *U.S. Dept. Agric., Div. Ent., Tech. Ser.*, I, p. 43, 91895); *Proc., U.S.N. Mus.*, XVIII, p. 635, (1896)[1]. SCHMIED., *Gen. Ins., fas.* 97, 9. 455, (1909). RAMAKR., *Spol. Zeyl.*, XIII, p. 251, (1925). [Host : *Chionaspis*].

[1]Caylon.

134. Encarsia Forster

Encarsia, FÖRST., *Jahrg. Verh. naturh. Ver., preuss. Rheinl.*, XXXV, p. 65, (1878), [Genotype *E. tricolor* FÖRST.]. ASHM., *Mem. Carnegie Mus.*, I, (4), p. 345, (1904). SCHMIED., *Gen. Ins., fas.* 97, p. 457, (1909).

aonidiae HOWARD

Encarsia aonidice, HOW., *Proc. U.s.N. Mus.*, XVIII, p. 636, (1896)[1]. SCHMIED., *Gen. Ins., fas.* 97, p. 456, (1909). RAMAKR., *Spol. Zeyl.*, XIII, p. 251, (1925). [Host : *Aonidia.*].

[1]Punduloya: Ceylon.

planchoniae HOWARD

Encarsia planchonioe, HOW., *Proc. U.S.N. Mus.*, XVIII, p. 635, (1896)[1]. SCHMIED., *Gen. Ins., fas.* 97, p. 456, (1909). RAMAKR., *Spol. Zeyl.*, XIII, p.)1925). [Host : *Planchonia*].

[1]Punduloya: Ceylon.

135. Aspidiotiphagus HOWARD

Coccophagus, CRAW., (*partim*), *DeStruct. Ins.*, p. 28, (1891).

Aspidiotiphagus, HOW., *Ins. Life*, VI, p. 229, (1894), [Genotype *A. citrinus* (CRAW.)]. ASHM., *Mem. Carnegie Mus.*, I, (4), p. 345, (1904). SCHMIED., *Gen. Ins., fas.* 97, p. 457, (1909). Paoli, *Boll. Soc. Ent. Ital.*, LVIII, pp. 97-105, (1925), [Revision, key, etc.].

Prospaltoides, Brethes, *Nunquam Otissus*, p. 19, (1914).

Citrinus (CRAWFORD)

Coccophagus citrinus, CRAW., *DeSTRuct. Ins.*, p. 28, (1891).

Encarsia citrinus, Riley, in HOW., *Ins. Life,* IV, p. 168, (1891).

Aspidiotiphagus citrinus, HOW., *Ins. LIfe,* VI, p. 231, (1894); *Proc. U.S.N. Mus.*, XVIII, p. 635, (1896)[1]. SCHMIED., *Gen. Ins. fas.* 97, p. 457, (1909). Malenotii, *Berlese Redia,* XII, pp. 15-18, (1917), [Food habits]. RAMAKR., *Spol. Zeyl.*, XIII, p. 251, (1925). SilveSTR., *Nuov. Ann. Agric.*, VI, pp. 97-101, (1926), [Parasite of *Chrysomphalus dictyosperminis* in Italy]. FERR., *Act. Soc. Helvet. Sc. Nat.*, CVII, (2), pp. 223-224, (1926), [Parasite of *Aspidiotus hederoe* in Switzerland]. Vandenberg, *Rep. Guam. Agric. Expt. Sta.*, 1925, pp. 17-20, (1926), [Parasite of *Aspidiotus deSTRuctor*]. Tothill, *Fiji Legis. Coun. paper,* No. 38, 1927, pp. 4. (1927), [Introduction into Fiji for the control of *Aspidiotus deSTRuctor*]. Bouhelier and others, *Def. cult. Dir. Gen. Agric. Commn Colon,* (Moroco), No. 6, pp. 1-60, figs. 10, (1932), [Parasite of *Aspidiotus hederoe* in Algeria]. Alden, *Bull. Georgia St. Board Ent.* No. 73, pp. 32, figs. 14, (1933), [Para site of *Aspidiotus perniciosus* in U.S.A.].

[1]Ceylon.

Prospaltoides howardi, Brethes, *Ann. Mus. Buenos Aires,* XXVII, p. 492, (1916).

136. Prospaltella HOWARD

*Prospalta,** HOW., *Ins. Life,* VII, p. 6, (1874), [Genotype *P. murtfeldtii* (HOW.)]. ASHM., *Mem. Carnegie Mus.*, I, (4), p. 345, (1904). MASI, *Boll. Lab. Zool, Portici,* III, p. 142, (1908). SCHMIED., *Gen. Ins., fas.* 97, p. 457, (1909).

Prospaltella, ASHM., *Proc. Ent. Soc. Washington,* VI, p. 126, (1904).

lahorensis HOWARD

Prospaltella lahorensis, HOW., *Journ. Econ. Ent.*, IV, p. 132, (1911)[1]. Misra, *Rep. Proc. Fifth Ent. Meet. Pusa,* pp. 129-135, (1924), [Descr. and biology]. RAMAKR.,

[1]Lahore.

* According to Girault (*Mem. Queensland, Mus.*, IV, p. 47, 1915), *Prospalta* should be considered a synonym of *Coccophayus* Westw.

Spol. Zeyl., XIII, p. 251, (1925). [Hosts : *Dialeurodes citri* and *Aleyrodes ricini.*]

137. Aneristus HOWARD

Aneristus, HOW., *Cand. Ent.*, XXVII, p. 351, (1895), [Genotype *A. ceroplastoe* HOW.]. ASHM., *Mem. Carnegie Mus.*, I, (4), p. 345, (1904). SCHMIED., *Gen. Ins.*, *fas.* 97, p. 457, (1909). DOZIER, *Journ. Dept. Agric. Peurto Rico*, XVI, pp. 93-102, (1932), [Notes].

Ceroplastae HOWARD

Aneristus ceroplaste, HOW., *Canad. Ent.*, XXVII, p. 351, (1895). SCHMIED., *Gen. Ins.*, *fas.* 97, p. 458, (1909). GAH., *Proc. U.S.N. Mus.*, LXV, p. 12, (1921)[1]. DOZIER, *Journ, Dept. Agric. Peurto Rico*, Ix, p. 365, (1925), [Further description]. Box, *Journ. Dept. Agric. Peurto Rico.* IX, (4), pp. 291-356, pls. 2, figs, 19, (1925), [parasiters of Coccids]. RAMAKR., *Spol. Zeyl.*, XIII, p. 252, (1925)[2]; *Bull. Ent. Res.*, XVIII, p. 78, (1927)[3]. SMITH and COMP., *Univ. Calif. Pub. Ent.*, IV, (9), pp. 231-334, figs, 63, (1928), [Parasite of the Black Scale]. Shiraki, *Journ. Trop. Agric.*, VI, (1-2), pp. 29-36, 187-194, (1934), [Parasite of *Pulvinaria polygonata* in Formosa].

[1]Coimbatore.

[2,3]Coimbatore.

Coccophagus orientalis, HOW., *Proc. U.S.N. Mus.*, XVIII, p. 633, (1896)[1]. SCHMIED., *Gen. Ins.*, *fas.* 97, p. 459, (1909). COMP. and SMITH, *Univ. Claif. Pub. Ent.*, IV, p. 53, (1926). [Host : *Lecanium viridoe.*].

[1]Punduloya: Ceylon.

138. Coccophagus Westwood

Entedon, DALM., (*partim*), *Svensk. Vet.,-Akad. Handl.*, XLVI, p. 365, (1825).

Coccophagus, WESTW., *Philos. Mag.*, III, p. 344, 91833), [Genotype *C. scutellaris* (DALM.) desig. WESTW., *C. pulchellus* WESTW. desig. ASHM.]. HOW., *Rep. Ent. U.S. Dept. Agric.*, 1880, pp. 353-354, (1881); *Bull. Dept. Argic., Bur. Ent.*, V, pp. 24-25, (1885); *Div. Ent., U.S. Dept. Agric., Tech. Ser.*, 1, pp. 10-11, 13-32, (1895). ASHM., *Mem. Carnegie Mus.*, I, (4), p. 345, (1904). MASI, *Boll. Lab. Zool. Portici*, I, pp. 238-246, (1907). SCHMIED., *Gen. Ins.*, *fas.* 97, p. 458, (1909). MERCET, *Trab. Mus. Cienc. Nat.*, No. 10, pp. 219-220, (1912). GIR., *Mem. Queensland Mus.*, IV, p. 47, (1915). COMP.,

Proc. U.S.N. Mus., LXXVIII, art. 7, pp. 1-132, pl. 14, (1931), [Revision].

Flavescens HOWARD

Coccophagus flavescens. HOW., *Proc. U.S.N. Mus.*, XVIII, p. 634, (1896)[1]. SCHMIED., *Gen. Ins., fas.* 97, p. 458, (1909). MERCET, *Trub. Mus. Cienc. Nat.*, No. 10, pp. 227-228, (1912). RAMAKR., *Spol. Zeyl.*, XIII, p. 252, (1925). COMP. *Proc. U.S.N. Mas.*, LXXVIII, art. 7, p. 31, pl. 7, fig. 22, (1931), [Redeseription, notes, etc.]. [Host : *Lecanium hemisphericum* of coffee.].

[1]Punduloya: Ceylon.

Longifasciatus HOWARD

Coccophagus longifasciatus, HOW., *Bull. U.S. Dept. Agric., Div. Ent., Tech. Ser.*, No. 12, p. 80, fig. 17, (1907)[1]. SCHMIED., *Gen. Ins., fas.* 97, p. 458, (1909). GIR., *Psyche*, XVI, p. 80, (1909), [Redescription]. MERCET, *Trab. Mus. Cienc. Nat.*, No. 10, p. 238, (1912). RAMAKR., *Spol. Zeyl.*, XIII, p. 252, (1925). COMP. and SMITH, *Univ. Calif. Pub. Ent.*, IV, p. 254, (1928). COMP., *Proc. U.S.N. Mus.*, LXXVIII, art, 7, p. 74, pl. 3, fig. 50, (1931), [Notes]. [Host : *Lecanium nigrum.]*.

[1]Mannar: Ceylon.

tschirchi MAHDIHASSAN

Coccophagus tschirchi, Mahd., *Journ. Sc. Assoc. Maharaja's Coll. Vizianagram*, I, p. 82, (1923)[1]. FERR., *Bull. Ent. Res.*, XIX, p. 173, fig. 2, (1928)[2], [Redescription, notes, etc.]. COMP., *Proc. U.S.N. Mus.*, LXXVIII, art. 7, p. 41, pl. 7, figs. 29, (1931),. [Notes]. RAMAKR. and MARGAB., *Journ. Bombay Nat. Hist. Soc.*, XXXVII, p. 196, (1934].[3] [Host : lac insect of *Butea*.].

[1]Khonde Bhavi; Kolar Dist., Mysore. [2]Kundari: Benga; [3]Coimbatore.

Zebratus HOWARD

Coccophagus zebratus, HOW., *Bull. U.S. Dept. Agric., Div. Ent., Tech. Ser.*, No. 12, p. 81, (1907)[1]. SCHMIED., *Gen. Ins., fas.* 97, p. 459, (1909). RAMAKR., *Spol. Zeyl.*, XIII, p. 252, (1925)[1].

[1]Ceylon.

TRIBE 2. *PTEROPTRICARIAE*

ASHM., *Mem. Carnegic Mus.*, I, (4), p. 347, (1904). HOW., *Bull. U.S. Dept. Agric., Div. Ent., Tech. Ser.*, No. 12, p. 72, (1907). SCHMIED., *Gen. Ins., fas,* 97, p. 459, (1909).

139. Merietta Motschulsky

Marietta, Motsch., *Bull. soc. Imp. nat. Moscou*, XXXVI, (3), p. 51, (1863), [Genotype *M. leopardina* Motsch.]. Ashm., *Mem. Carnegic Mus.*, I, (4), p. 346, (1904). Schmied., *Gen. Ins.*, *fas.* 97, p. 461, (1909).

leopardina Motschulsky

Marietta leopardina, Motsch., *Bull. Soc. Imp. nat. Moscou*, XXXVI, (3), p. 52, (1863)[1]. Schmied., *Gen. Ins.*, *fas.* 97, p. 461, (1909). Ramakr., *Spol. Zeyl.*, XIII, p. 252, (1925). [1]Ceylon.

SUBFAMILY V. TETRASTICHINAE

Först., *Hymen, Stud.*, II, pp. 26, 83, (1856). Thoms., *Hymen. Skandin.*, V, p. 180, (1878). How., *Ent. Amer.*, I, p. 199, (1886). Ashm., *Proc. Ent. Soc. Washington*, IV, p. 249, (1897); *Mem. Carnegi Mus.*, I, (4), p. 347, (1904). Schmied., *Gen. Ins.*, *fas.* 97, p. 463, (1909). Kurdin., *Rev. russ. ent.*, XIII, pp. 243-256, (1913), translation by Novicky in *Eos*, III, pp. 513-517, (1927), [Notes on validity of some genera].

TRIBE 1. *CERATONEURARIAE*

Ashm., *Mem. Carnegie Mus.*, I, (4), p. 347, (1904). Schmied., *Gen. Ins.*, *fas.* 97, p. 463, (1909).

140. Ceratoneura Ashmead

Ceratoneura, Ashm., *Journ. Linn. Soc. London*, (*Zool*), XXV, p. 178, (1894), [Genotype C. *petiolata* Ashm.]; *Mem. Carnegie Mus.*, I, (4), p. 347, (1904). Schmied., *Gen. Ins.*, *fas.* 97, p. 463, (1909).

indi Girault

Ceratoneura indi. Gir., *Descr. Hymen. Chalc. Observ.* (private publication), V, p. 10, (1917)[1]. Ramakr., *Spol. Zeyl.*, XIII, p. 252, (1925)[2]; *Bull. Ent. Res.*, XVIII, p. 78, (1927)[3], [bred from *Sesbania* shoots]. [1,2,3]Coimbatore.

indica Rohwer

Ceratoneura indica, Rohw., *Ann. Mag. Nat. Hist.*, VII, p. 127, (1921)[1]. Ramakr., *Spol. Zeyl.*, XIII, p. 252, (1925)[2]; *Bull, Ent. Res.*, XVIII, p. 78, (1927)[3]. [1,2,3]Coimbatore.

TRIBE 2. *TETRASTICHARIÆ*

ASHM., *Mem. Carnegie Mus.*, I, (4), p. 347, (1904). SCHMIED., *Gen. Ins., fas.* 97, p. 464, (1909). KURDIN., *Rev. russ. ent.*, XIII, pp. 243-256, figs. 1-4, (1913), [synops. and synonym of genera].

141. **Hyperteles** Förster

Eulophus, FÖRST., *Beitr. Mon. Pteromal.*, *p. 41, (1841).*

Hyperteles, FÖRST., *Hymen. Stud.*, II, p. 84, (1856), [Genotype *H. elongatus* (FÖRST.)] ASHM., *Mem. Carnegie Mus.*, I, (4), p. 348, (1904). SCHMIED., *Gen. Ins., fas.* 97, p. 467, (1909).

Oxymorpha, FÖRST., *Hgmen. Stud.*, II, p. 145, (1856).

Longicauda KIEFFER

Hyperteles longicauda, KIEFF., *Ann. Soc. Sc. Bruxelles*, XXIX, p. 195, 91905)[1]. RAMAKR., *Spol. Zeyl.*, XIII, p. 252 (1925). [Host: *Cecidomyiid* of *Artemisia* galls]. [1]Bengal.

142. **Syntomosphyrum** Forster

Syntomosphyrum, FÖRST., *Jahrg. Verh. Naturh. ver. preuss. Rheinl.*, XXXV, p. 60, (1878), [Genotype *S. fulvipes*, FÖRST.]. ASHM. *Mem. Carnegie Mus.*, I, (4), p. 349, (1904). SCHMIED., *Gen. Ins., fas.* 97, p. 469, (1909).

indicum SILVESTRI

Syntomosphyrum indicum, Silv., *Boll, Lab. Zool. Portici*, IV, p. 228, (1910)[1]. FULL-away, *Rep. Proc. Third Ent. Meet. Pusa*, II, p. 628, (1920). RAMAKR., *Spol. Zeyl.*, XIII, p. 252, (1925); *Bull. Ent. Res.*, XVIII, p. 78, (1927)[2]. [Host: Unknown fruit-fly.]. [1,2]Bangalore.

taprobanes WATERSTON

Syntomosphyrum taprobanes, Waterest., *Bull. Ent. Res.*, V, p. 337, (1914)[1]. RAMAKR., *Spol. Zeyl.*, XIII, p. 252, (1925). [Host: *Scymnus*.]. [1]Ceylon.

143. **Ceranisus** WALKER

Ceranisus, WALK., *Entomologist*, I, p. 6, pl. N, fig. 2, (1840), [Genotype *C. pacuvius* WALK., transferred to *Cirrospilus* by DALLA TORRE]. ASHM., *Mem. Carnegic Mus.*, I, (4), p. 349, (1904). SCHMIED., *Gen. Ins., fas.* 97, p. 469, (1909).

nigricornis MOTSCHULSKY

Ceranisus (?) *nigricornis*, MOTSCH., *Bull. Soc. Imp. Nat. Moscou*, XXXVI, (3), p. 67, (1863)[1]. SCHMIED., *Gen. Ins., fas.* 97, p. 470, (1909). RAMAKR., *Spol. Zeyl.*, XIII, p. 252, (1925).

[1]Mt. Patannas: Ceylon.

Semicetaceus MOTSCHULSKY

Ceranisus (?) *semicetaccus*, MOTSCH., *Bull. Soc. Imp. Nat. Moscou*, XXXVI, (3), p. 67, (1863)[1]. SCHMIED., *Gen. Ins., fas.* 97, p. 470, (1909). RAMAKR., *Spol. Zeyl.*, XIII, p. 252, (1925).

[1]Mt. Nura Ellia: Ceylon.

subconicus MOTSCHULSKY

Ceranisus (?) *subconicus*, MOTSCH., *Bull. soc. Imp. Nat. Moscou*, XXXVI, (3), p. 66, (1863)[1]. SCHMIED., *Gen. Ins., fas.* 97. p. 470, (1909). RAMAKR., *Spol. Zeyl.*, XIII, p. 252, (1925).

[1]Mt. Nura Ellia: Ceylon.

144. Tetrastichus HALIDAY

Eulophus, Nees, *(partim)*, *Hymen. Ichneum. Affin. Mon.*, II, p. 175, (1834).

Tetrastichus, Halid., *Trans. Ent. Soc. London*, III, p. 297, (1843), [Genotype *T. miser* (Nees)]. ASHM., *Mem. Carnegie Mus.*, I, (4), p. 349, (1904). SCHMIED., *Gen. Ins., fas.* 97, p. 471, (1909). GAH., *Proc. U.S.N. Mus.*, XLVIII, p. 168, (1914).

Trichoceras, Ratz., *Ichneum. Forstins.*, I, p. 171, (1848).

Geniocerus, Ratz., *Ichneum. Forstins.*, I, p. 175, (1848).

Oomyzus, Rond., *Bull. Comm. Agrar. Parma*, III, p. 140, (1870).

Tetrastichodes, ASHM., *Trans. Amer. Ent. Soc.*, XIV, p. 203, (1887); *Mem. Carnegie Mus.*, I, (4), p. 348, (1904). SCHMIED. *Gen. Ins., fas.* 97, p. 467, (1909). GAH., *Proc. U.S.N. Mus.*, XLVIII, p. 168, (1914).

Balttotetrastichus, GIR., *Ent. News*, XXVIII, p. 257, (1917).

asthenogmus (WATERSTON)

Tetrastichodes asthenogmus, WATERST., *Bull. Ent. Res.,* V, p. 340, (1914)[1]. RAMAKR., *Spol. Zeyl.,* XIII, p. 253, (1925). [Host: Cockroach.]. — [1]Ceylon.

ayyari Rohwer

Tetrastichus ayyari, ROHW., *Ann. Mag. nat. Hist.,* VII, p. 128, (1921)[1]. Ballard, *Rep. Proc. Fourth Ent. Meet Pusa,* p. 22, (1921)[2]. RAMAKR., *Rep. Proc. Fourth Ent. Meet. Pusa,* p. 365, (1921)[3]. *Spol. Zeyl.,* XIII, p. 253, (1925); *Bull. Ent. Res.,* XVIII, p. 78, (1927)[4]. [Host: *Chilo simplex.*]. — [1,2]Coimbatore. [3,4]Coimbatore.

coimbatorensis Rohwer

Tetrastichus coimbatorensis, Rowh., *Ann. Mag. Nat. Hist.,* VII, p. 133, (1921)[1]. Ballard, *Rep. Proc. Fourth Ent. Meet. Pusa,* p. 22, (1921)[2]. RAMAKR., *Rep. Proc. Fourth Ent. Meet. Pusa,* p. 365, (1921)[3]; *Spol. Zeyl.,* XIII, p. 253, (1925); *Bull. Ent. Res.,* XVIII, p. 78, (1927)[4]. [Host: Cholam gall-fly.]. — [1,2,3,4]Coimbatore.

colemani CRAWFORD

Tetrastichus colemani, CRAW., *Proc. U.S.N. Mus.,* XLII, p. 8, (1912)[1]. RAMAKR., *Rep. Proc. Fourth Ent. Meet. Pusa,* p. 363, (1921)[2]; *Spol. Zeyl.,* XIII, p. 253, (1925); *Bull. Ent. Res.,* XVIII, p. 78, (1927)[3]. [Host: *Aspidomorpha miliaris.*]. — [1,2,3]Bangalore.

ellia MOTSCHULSKY

Tetrastichus ellia, MOTSCH., *Bull. Soc. Imp. Nat. Moscou,* XXXVI, (3), p. 69, (1863)[1]. SCHMIED., *Gen. Ins., fas.* 97, p. 474, (1909). RAMAKR., *Spol. Zeyl.,* XIII, p. 252, (1925). — [1]Nura Ellia: Ceylon.

gardneri FERRIERE

Tetrastichus gardneri, FERR., *Bull. Ent. Res.,* XXII, p. 291, (1931)[1], [Descr. and notes]. RAMAKR., and MARGAB., *Journ. Bombay Nat. Hist. soc.,* XXXVII, p. 196, (1934). — [1]Dehra Dun.

issaci Rohwer

Tetrastichus issaci, ROHW., *Ann. Mag. Nat. Hist.,* VII, p. 130, (1921)[1]. RAMAKR., *Spol. Zeyl.,* XIII, p. 253, (1925); *Bull. Ent. Res.,* XVIII, p. 78, (1927)[2]. — [1,2]Coimbatores.

nyemitanus Rohwer

Tetrastichus nyemitanus, Rohw., *Ann. Mag. Nat. Hist.,* VII, p. 131, (1921).[1] Ballard, *Rep. Proc. Fourth Ent, Meet. Pusa,* p. 22, 91921)[2]. Ramakr., *Rep. Proc. Fourth Ent. Meet. Pusa,* p. 365, (1921)[3]; *Spol. Zeyl.,* XIII, p. 253, (1925); *Bull. Ent. Res.,* XVIII, p. 78, (1927)[4].* [Host: Cholam shoot *Anthomyid.*]. — [1,2,3,4]Coimbatore.

okawus Rohwer

Tetrastichus okawus, Rohw., *Ann. Mag. Nat. Hist.,* VII, p. 128, (1921)[1]. Ramakr., *Spol. Zeyl.,* XIII, p. 2536 (1925); *Bull. Ent. Res.,* XVIII, p. 78, (1927)[2]. — [1,2]Coimbatore.

Ophiusæ Crawford

Tetrastichus ophiusoe, Craw., *Proc. U.S.N. Mus.,* XLII, p. 8, (1912)[1]. Ramakr., *Rep. Proc. Third Ent, Meet. Pusa,* III, p. 932, (1920)[2]; *Rep, Proc. fourth Ent. Meet. Pusa,* p. 363, (1921)[3]; *Spol. Zeyl.,* XIII, p. 253, (1925); *Bull. Ent. Res.,* XVIII, p. 78,(1927)[4]. [Host: *Achoea janata.*]. — [1,2,3,4]Bangalore.

Patannas Motschulsky

Tetrastichus patannas, Motsch., *Bull. soc. Imp. Nat. Moscou,* XXXVI, (3) p. 69, (1963)[1]. Schmied., *Gen. Ins., fas.* 97, p. 477, (1909). Ramakr., *Spol. Zeyl.,* XIII, p. 253, (1925). — [1]Mt. Patannas: Ceylon.

purpureus (Cameron)

Hadrothrix purpureus, Cam., *Ind. For. Rec.,* IV, p. 103, (1913)[1]. Ramakr., *Spol. Zeyl.,* XIII, p. 251, (1925). Glover, *Bull. Ind. Lac. Res. Inst.,* No. 16, p. 23, figs. 2, pl. 1, (1933), [bred from lac of *Zizyphus*]. — [1]Dehra Dun.

Tetrastichus immsii, Mahd., *Journ. Sc. Assoc. Maharaja's Coll. Vizianagaram,* I, pp. 47-99, (1923).

Tetrastichus (Geniocerus) purpureus, Ferr., *Bull. ent. Res.,* XIX, pp. 174-175, (1928)[1]. Ramakr. and Margab., *Journ. Bombay Nat. Hist. Soc.,* XXXVII, p. 196, (1934). — [1]Namkum.

Pyrillæ Crawford

Tetrastichus pyrilloe, Craw., *Insec. Inscit. Menstr.,* IV, p. 101, (1916)[1]. Ramakr., *Spol. Zeyl.,* XIII, p. 253, (1825). [Host: *Pyrilla aberrans.*]. — [1]Pusa.

---

* Ramakrishna Ayyar's *nyemitauras* is an error for *nyemitanus.*

radiatus WATERSTON

Tetrastichus radiatus, WATERST., *Bull. Ent. Res.*, XIII, p. 55, (1922)[1]. Husain and Mathur, *Rep. Proc. Fifth Ent. Meet. Pusa*, p. 121, (1924)[2]. Husain and Dina Nath, *ibid.*, pp. 122-128, (1924)[3], [Biology]. RAMAKR., *Spol. Zeyl.*, XIII, p. 253, (1925). [Host: *Diaphorina citri.*]. — [1,2,3]Lyallpur, Sargodha, Gujranwala: Punjab.

Schoenobii FERRIERE

Tetrastichus schoenobii, FERR., *Bull, Ent. Res.*, XXII, pp. 290-291[1]. RAMAKR. and MARGAB., *Journ. Bombay Nat. Hist. Soc.*, XXXVII, p. 196, (1934)[2]. [Hosts : *Schoenobius bipunctifer and Spodoptera mauritia.*] — [1]Malay States, Siam, Ceylon. [2]Coimbatore.

Spirabillis WATERSTON

Tetrastichus spirabilis, WATERST., *Ind. For. Rec.*, IX, pp. 89-94, (1922)[1]. RAMAKR., *Spol. Zeyl.*, XIII, p. 253, (1921). [Host: *Hypsipyla robusta.*]. — [1]Dehra Dun.

145. Ootetrastichus Perkins

Ootetrastichus, Perk., *Bull. Hawaii. Sug. Plant. Assoc.*, I, p. 263, (1906), [Genotype *O. beatus* perk.].

Sp. RAMAKRISHNA AYYAR

Ootetrastichus sp., RAMAKR., *Spol. Zeyl.*, XIII, p. 252, 91925)[1]; *Bull. Ent. Res.*, XVIII, p. 78, (1927)[2]. [Host: *Fulgorid* sp.]. — [1,2]Coimbatore.

146. Thriptoctenus CRAWFORD*

Thriptoctenus, CRAW., *Proc. Ent. Soc. Washington*, XIII, p. 233, (1911), [Genotype *T. russelli* CRAW.].

Maculatus WATERSTON

Thriptoctenus maculatus, WATERST., *Ann. Mag. Nat. Hist.*, V. pp. 243-244, (1930)[1]. RAMAKR. and MARGAB., *Journ. Bombay Nat. Hist. Soc.*, XXXVII, p. 196, (1934). [Host: *Rhipiphorothrips cruentatus*]. — [1]Lyallpur: Punjab.

147. Tumidiscapus Girault.*

Tumidiscapus, GIR., *Journ. N.Y. Ent. Soc.*, XIX, p. 182, (1911), [Genotype *T. flavus* Gir.].

---

* Location of these genera provisional.

Oophagus Girault

[1]Coimbatore.

Tumidiscapus oophagus, GIR., *Proc. U.S.N. Mus.,* LIII, p. 445, (1917)[1]. Ramachandra Rao, *Rep. Proc. Fourth Ent. Meet. Pusa,* p. 42, (1921), [Notes]. RAMAKR., *Spol. Zeyl.,* XIII, p. 251, (1925). [Host: *Oxya velox.*].

[1]Ceylon.

FAMILY XII. TRICHOGRAMMIDÆ

WALK., *List Chalc. Brit. Mus.,* I, p. 62, (1846). FÖRST. *Hymen, Stud.,* II, pp. 20, 26, 87, (1856). AURIV., *Ent. Tidskr.,* XVIII, p. 250, (1897). ASHM., *Mem. Carnegie Mus.,* I, (4), p. 358, (1904). SCHMIED., *Gen. Ins., fas.* 97, p. 480, (1909). GIR., *Trans. Amer. Ent. Soc.,* XXXVII, pp. 43-55, (1911). Kurdin., *Rev. russ, ent.,* XII, p. 283, 91912), [Synonymy]. GIR., *Archiv. Naturg.,* LXXIX, A, p. 69, (1913); *Bull. Wisc. Soc.,* XI, pp. 150-179, (1913); *ibid.,* XII, pp. 59-71, (1914); *Mem. Queensland Mus.,* III, pp. 142-153, (1915); *Ent. News,* XXVII, pp. 4-8, (1916). DEN DOOP, *Med. van het Deli Profst.,* XIX, pp. 213, 220, (1918). KRYGER, *Kobenhav. Ent. Medd.,* XII, pp. 257-354, (1919); *ibid,* XIII, pp. 183-188, (1920). GIR., *Insec. Inscit. MenSTR.,* VIII, pp. 199-203, (1920). BLOOD, *Ann. Rep. Bristol Nat. Soc.,* (4), V, pp. 253-258, (1923). Marchal, *Compt. Rend. Acad. Sc. France,* CLXXXV, pp. 489-493, 521-523, (1927). BLOOD and KRYGER, *Kobenhav. Ent. Medd.,* XVI, pp. 203-222, (1928). DOZIER, *Proc. Ent. Soc. Washington,* XXXIV, pp. 29-37, (1932).

SUBFAMILY I. OLIGOSITINAE

ASHM., *Mem. Carnegie Mus.,* I, (4), p. 359, (1904). SCHMIED., *Gen. Ins., fas.* 97, p. 430, (1909).

148. Oligosita WALKER

Oligosita, (Halid. MS.), WALK., *Ann. Mag. Nat. Hist.,* VII, p. 211, (1851), [Genotype *O. collina* WALK.]. ASHM., *Mem. Carnegie Mus.,* I, (4), p. 359, (1904). SCHMIED., *Gen. Ins., fas.* 97, p. 482, (1909).

Nodicornis Westwood

Oligosita nodicornis, WESTW., *Trans. Linn. Soc. London,* (*Zool.*), I, p. 592, (1879)[1]. SCHMIED., *Gen. Ins., fas.* 97, p. 483, (1909). RAMAKR., *Spol. Zeyl.,* XIII, p. 253, (1925).

[1]Ceylon.

Staniforthi Westwood

Oligosita staniforthi, WESTW., *Trans. Linn. Soc. London,* (*Zool.*), I, p. 591, (1879)[1]. SCHMIED., *Gen. Ins., fas.* 97, p. 483, (1909). RAMAKR., *Spol. Zeyl.,* XIII, p. 253, (1925).

[1]Ceylon.

149. **Westwoodella** Ashmead

Oligosita, WESTW., (*partim*), *Truns. Linn. Soc. London*, (*Zool.*), I, p. 591, (1879).

Westwoodella, ASHM., *Mem. Carnegie Mus.*, I, (4), p. 359, (1904), [Genotype *W. subfasciata* (WESTW.)]. SCHMIED., *Gen. Ins.*, *fas.* 97, p. 483, (1909).

Subfasciata (Westwood)

Oligosita subfasciata, WESTW., *Trans. Linn. Soc. London*, (*Zool.*), I, p. 591, (1897)[1]. [1]Ceylon.

Westwoodella subfasciata, SCHMIED., *Gen. Ins.*, *fas.* 97, p. 483, (1909). RAMAKR., *Spol. Zeyl.*, XIII, p. 253, (1925).

SUBFAMILY II. TRICHOGRAMMINÆ

ASHM., *Mem. Carnegie Mus.*, I, (4), p. 360, (1904). *SCHMIED*, *Gen. Ins.*, *fas.* 97, p. 483, (1909).

150. **Trichogramma** Westwood

Microma (?) Curtis, *Guide Brit. Mus. Ins.*, p. 137, (1827).

Trichogramma, WESTW., *Philos. Mag.*, II, p. 444, (1833), [Genotype *T. evanescens* WESTW.]. ASHM., *Mem. Carnegie Mus.*, I, (4), p. 360, (1904). SCHMIED., *Gen. Ins.*, *fas.* 97, p. 485, (1909). Bose, *Rep. Proc. Fifth Ent. Meet. Pusa*, pp. 280-288, pl. XXVI, figs. 3-6, (1924), [parasitie on *Danais limniace* at Pusa]. Marchal, *Compt. Rend. Acad. Sc. France*, 185, pp. 489-521, (1927). WOLFF, *Ent. Jahrb.*, XXXIx, pp. 156-157, (1929), [Notes]. FLANDERS, *Hilgardia*, IV, (16), pp. 466-501, (1930), [Mass production]. FERR., *Bull. Ent. Res.*, XXI, p. 37, (1930), [Synonymy]. PETERSON, *Journ. N.Y. Ent. Soc.*, XXXVIII, p. 584, (1930). MARCHAL, *Compt. Rend. Acad. Sc. France*, 191, p. 379, (1930). OLDROYD and RIBBANDS, *Proc. roy. Ent. Soc. London*, V, (B), vii, pp. 148-152, (1936), [Validity of trichiation as a systematic character in *Trichogramma*].

Calleptiles, HALID., *Ent. Month. Mag.*, I, p. 340, (1833).

Ophineurus, RATZ., (*partim*), *Ichneum. Forstins.*, III, p. 196, (1852).

Trichogrammatoidea, GIR., *Trans. Amer. Ent. soc.*, XXXVII, p. 13, (1911). FERR., *Bull. Ent. Res.*, XXI, p. 37, (1930).

Evanescens Westwood

Trichogramma evanescens, WESTW., *Philos. Mag.*, II, p. 444, (1833). Enoch, *Entomologist,* XXVIII. p. 283, (1895). SCHMIED., *Gen. fas.* 97, p. 485, (1909). Gatenby, *Quart. Journ. Micr. Sc.,* LXII, pp. 149-187, pl. X-XII, 91917); *ibid.,* LXIII, pp. 161-174, pl. XIII, (1918), [Embryonic development as egg-parasite of *Donacia simplex* FABR.]. Husain, *Rep. Proc. Fifth Ent. Meet. Pusa,* p. 35, (1924)[1], [Eggpparasite of the cotton boll-worm *Earias Minsulana,* life-history notes, etc.]. Husain and Mathur, *Rep. Proc. Fifth Ent. Meet. Pusa,* p. 121, (1924)[2], [Recorded on cotton boll-worms]. RAMAKR., *Spol. Zeyl.,* XIII, p. 253, (1925). Hase, *Arb. Biol. Reichasanstl. Land. Forstw. Berlin,* XIV, H. 2, pp. 171-224, fig. 9, (1925), [Life-history]. Hintzolmann, *ibid.,* pp. 225-230, (1925), [Morphology]. Schulze, *Zeit. Morph. Oecol. tiere,* VI, pp. 553-587, (1926), [Fertility]. Marchal, *Compt. Rend. Acad. Sc. Paris,* 185, pp. 489-493, (1927), [biology compared]. Zorin, *Defens des plantes,* IV, pp. 316-319, (1927), [Reared under artificial conditions]. Zulucta, *Zeit. Indukt. Astammn. u. Vererb Lehre Suppl.,* II, p. 1606, (1928), [Polymorphism]. Hase, *Internat. Corn-borer Invest. Sc. Rep.,* II, pp. 85-89, (1929), [Notes]. Hase, *Z. angew. Ent.,* (2), pp. 345-385, figs. 16, (1930), [Control of the Maize-borer]. Hey and Thomas, *Ann. Appl. Biol.,* XX, (3), pp. 439-462, figs. 13, pls. ii, (1933), [Bionomics as parasite of *Tortrix cratoegana* in Britain]. Marcus, *Z. angew. Ent.,* XX, (2), pp. 169-203, fig. 5, (1933), [Parasite of *Cephaleia arvensis* in Denmark]. King, *Bull. Tea Res. Inst. Ceylon,* No. 10, pp. 27-33, (1933)[3], [Introduction and breeding trials in Ceylon]. Salt, *Proc. Roy. Sac.,* 114, (1934); 177, (1935). Salt and Laing, *Nature,* 135, p. 792, 91935), [Capacity to distinguish previously parasitized eggs]. Flanders, *Pan-Pac. Ent.,* XI, (4), pp. 175-177, (1936), [Proficiency and size influenced by host.].

[1,2]Lyallpur: Punjab.

[3]Ceylon.

minutum Riley.*

Trichogramma minutum, Riley, *Third Ann. Rep. Ins. Mussouri,* p. 158, (1870). HOLLOWAY, *Ent. News,* XXIII, p. 329, (1912). GIR., *Rev. Russ, Ent.,* XIII, p. 295, (1913); *Ent. News,* XXVII, p. 5, (1916), [Hosts]. HARLAND, *West Ind. Bull.,* XV, pp. 168, 175, (1916), [Notes]. BAIRD, 56th *Ann. Rep. Ent. soc. Ontario,* 1925, p. 78, (1926), [Parasite of *Pyrausta nubilalis*]. HOLLOWAY and HALEY, *U.S. Bur. Ent., Facts about Sugar,* XXII, (2), pp. 42-43, (1927), [Parasite of the Sugarcane Moth Borer]. *Box., Dept. Sc. and Agric.,* pp. 5, (1927), [Parasite of the Sugarcane Moth Borer in the Barbados]. GARMAN, *Bull. Connect. Agric., Expt. Sta.,* No. 285, pp. 234-239, pl. i (1927), [Control of the Oriental Peach-moth]. CHOPRA, R.L., *Rep. Agric. Punjab,* 1925-6, (2), I, pp. 67-125, (1928)[1], [Hosts in the Punjab]. MARTIN, *Ent. Amer.,* VIII, (3), pp. 105-156, pls. 3, (1928), [Bionomics in U.S.A.]. HINDS and SPENCER, *Journ. Econ. Ent.,* XXI, (2), pp. 273-278, pl. i, (1928), [Methods of rearing]. FLANDERS, *Journ. Econ. Ent.,* XXI, (3), p. 512, (1928), [Methods of rearing]. CLEARE JR., *Bull. Ent. Res.* XIX, (1), pp. 31-33, pls. 2, figs. 3, (1928), [Methods of rearing]. LIGHT, *Bull. Tea Res. Inst. Ceylon,* No. 3, pp. 37-48, (1929), [Suggestions for importation into Ceylon against the Coffee-scale]. FLANDERS, *Journ. Econ.,* XXII, (1), pp. 245-248, (1929), [Production and diSTRibution.]. TUCKER, *Trop. Agric.,* VI (8), pp. 224-226, (1929), [Control of Sugarcane Borers]. HINDS and SPENCER, *Journ. Econ. Ent.,* XXII, (4), pp. 633-636, (1929), [Experiments in 1928 for the control of the Sugarcane Borers]. PETERSON, *Journ.* N.Y. Ent. Soc., XXXVIII, pp. 1-8, pls. 2, (1930), [Forms]. TUCKERS, *Trop. Agric.,* VII, (11), pp. 292-295, (1930), [Control of *Diatroea saccharalis* in the Barbados by mass production and liberation]; *Rep. Dept. Sc. Agric., Barbados,* 1929-30, pp. 109-114, (1930), [Outline of work]; *ibid.,* 1930-31, pp. 80-97, pls. x, map 1, (1931), [Report on mass rearing in connection with control of *Diatroea saccharalis* in *Barbados].* Steenburgh, *60th Ann. Rep. Ent. soc.* Ontario, pp. 124-130, (1930), [Enemy of the Oriental Fruit-moth in Ontario]. VAN DER GOOT, *Proc.*

[1]Punjab.

* Flanders (*Pan. Pac. Ent.,* XI, p-79, 1935) cosiders *T. minutum* to be a synonym of *T. evanescens.*

Fourth Pac. Sc. Congr., Java, 1929, IV, Agric. papers, pp. 1-7, (1930), [Control of *Chilo simplex* in the Dutch East Indies]. HAEUSSLER, *Journ. Agric. Res.,* XLI, No. 5, pp. 365-377, figs. 3, (1930), [Enemy of the Oriental peach-moth in N. America]. FLANDERS, *Journ. Econ. Ent.,* XXIII, (5), pp. 837-841, pl. 1 (1930), [Recent developments in mass production]; *ibid.,* pp. 886-887, (1930), [Evaluatio of liberations]. PETERSON, *Tech. Bull. U.S. Dept. Agric.,* No. 215, pp. 21, figs. 9, (1930), [Biological study as an egg parasite of the Oriental Peach-moth]. FLANDERS, *Pan-Pac. Ent.,* VII, (1), pp. 20-21, (1930), [Races]. STREENBURGH, 61st *Ann. Rep. Ent. Soc. Ontario,* 1930, pp. 57-65, (1931), [Observation on its effects on the Oriental Peach-moth in Ontario]. GARMAN, *Journ. Econ. Ent.,* XXIV, (1), pp. 315-316, (1931), [Summary of breeding experimnts in Connecticut]. ALDEN and FARLINGER, *Journ. Econ. Ent.,* XXIV, (2), pp. 480-483, pls. ii, (1931), [Artificial rearing and colonization]. GARMAN and SCHREAD, *Ann. Ent. Soc. Amer.,* XXIV, (2) pp. 424-426, (1931), [Sex-ratio importance in breeding]. PETERSON, *Journ. Econ. Ent.,* XXIV, (5), pp. 1070-1074, (1931), [Refrigeration, notes, etc.]. FLANDERS, *Hilgardia,* V. (12), pp. 395-406, figs. 4, (1931), [Temperature relationship as a basis for racial segregation]. JAYNES, *Journ. Econ. Ent.,* XXV, (1), pp. 64-68, (1932), [Parasite of the Sugarcane Borer in S. America]. XWEZEY and PEMBERTON, *Entomology, Rep. comm. Expt. Sta. Hawaiian Sugar Plant. Assoc.,* 1930-31, pp. 19-32, figs. 8 (1932), [In the Hawaiian Islands]. KING, *Bull. Tea Res. Inst. Ceylon,* No. 8, pp. 20-29, (1932), [Introduction in Ceylon]. CLARK, *New Zealand Journ. Sc. Technol.,* XIII, (4), pp. 235-243, figs. 4, (1932), [Parasite on a pest of *Pinus radiata* in New Zealand]. GILLIAT, *Sc. Agric.,* XII, (6), pp. 357-371, pls. iii, (1932), [Control of *Eucosma (spilonota) ocellana* Schiff. in Nova Scotia]. WILLE, *Anz. Schadlingsk.,* VIII, (3) pp. 25-29, figs. 2, (1932), [Control of the Sugarcane borer west of Peru]. ALDEN, *Bull. Georgia St. Board Ent.,* No. 76, pp. 22, Charts 2, (1932), [Experimental results in the laboratory on its parasitism of fruit pests]. HINDS and OSTERBEGER, *Journ. Econ. Ent.,* XXV, (1), pp. 57-64, (1932), [colonization on the Sugarcane Borer in Louisiana in 1931]. List and Davis, *Journ. Econ. Ent.* XXV, (5), pp. 981-985, (1932), [Use in the control of the Codling-moth in Colorado]. TUCKER, *Trop. Agric.*, VIII, (11), pp. 283-288, pls. 9, (132), [Mass breeding for the control of the Sugarcane Borer in the Barbados in 1930]. DOZIER, *Proc. Ent. Soc. Washington,* XXXIV, (3), pp. 29-37, (1932), [From West Indies]. TUCKER, *Agric. Journ. Barbados,* I, (1), pp. 29-36, (1932), [Status as a control of the Sugarcane Borer in Barbados]. BOX, *Bull. Imp. Inst.*, XXX, (2), pp. 185-197, (1932), [Control of the Sugarcane Borer Moth in Antigua]. TUCKER, *Journ. Dept. Sc. Agric. Barbados,* I, (2), pp. 24-25, (1932), [Report]. JAYNES, *Tech. Bull. U.S. Dept. Agric.*, No. 363, pp. 26, figs. 10, 91933), [In America]. TOMASZEWSKI and HASE, *Nachr. Bl. deuts. Pfl. Sch. Dieust,* XIII, (6), pp. 41-43, figs. 5 (1933), [Mass breeding in Germany]. HINDS, OSTERBERGER and DUGAS, *Bull. Louisiana Agric. Expt. Sta.*, No. 235, pp. 36, graphs 2, (1933), [Review of work of controlling Sugarcane Borer Moth by field colonization]. COLLINS, *Journ. Econ. Ent.*, XXVI, (1), pp. 54-57, (1933), [Parasite of the Oriental Moth *Cnidocampa flavescens* WALK. in Massachusetts]. LIST, *Journ. Econ. Ent.*, XXVI, (2), pp. 373-377, (1933), [In relation to Codling-Moth control]. SCHREAD, *Journ. Econ. Ent.*, XXVI, (2), pp. 402-404, pl. 1, (1933), [Breeding methods in Connecticut]. Holloway, *Journn. Econ. Ent.*, XXVI, (5), pp. 974-977, (1933), [Method of avoiding deSTRuction of *Trichogramma* in Sugarcane fields]. HINDS and OSTERBERGER, *Journ. Econ. Ent.*, XXVII, (4), p. 788, (1934). GARMAN and SCHREAD, *Journ. N.Y. Ent. Soc.*, XLII, p. 263, (1934). RAMAKR. and MARGAB., *Journ. Bom. Nat. Hist. Soc.*, XXXVII, p. 196, (1934), [Hosts in India]. BARE, *Journ. Econ. Ent.*, XXVIII, pp. 8036-815, (1935), [some remarks on races, mass production, liberation, etc.].

151. Chætostricha WALKER

Chætostricha, (Halid. MS.), WALK., *Ann. Mag. Nat. Hist.*, VII, p. 210, (1851), [Genotype *C. dimidiata* WALK.]. ASHM., *Mem. Carnegie Mus.*, I, (4), p. 360, (1904). SCHMIED., *Gen. Ins., fas.* 97, [. 486 (1909). MANI, *Rec. Ind. Mus.*, XXXVII, p. 337, (1935).

Chætostricha, FÖRST., *Hymen, Stud.*, II, p. 87, (1856).

mukerjii MANI

Choetostricha mukerjii, MANI, *Rec. Ind. Mus.,* XXXVII, p. 337, fig. 1, (1935)[1]. MUKERJI and BHUYA, *Sc. Cult.,* I, (4), pp. 211-212, figs. 1-2, (1935), [Notes on possible economic importance]; *Proc. Twenty Second Ind. Sc. Cong., Calcutta,* p. 379, (14935), [Notes on life-history]; *Zool. Anz.,* CXV, pp. 209-212, figs. 1-4, (1936), [Detailed account of life-history]. [Host: *Bruchus* spp.].

[1]Calcutta.

152. Aprobosca Westwood

Aprobosca, WESTW., *Trans. Linn. Soc. London,* (*Zool.*), I, p. 592, (1878), [Genotype *A. erosicornis* WESTW.]. ASHM., *Mem. Carnegie Mus.,* I, (4), p. 360, (1904). SCHMIED., *Gen. Ins., fas.* 97, p. 487, (1909).

erosicornis Westwood

Aprobosca erosicornis, WESTW., *Trans. Linn. Soc. London,* (*Zool.*), I, p. 593, p;. 73, fig. 24-25, (1878)[1]. SCHMIED., *Gen. Ins., fas.* 97, p. 487, (1909). RAMAKR., *Spol. Zeyl.,* XIII, p. 253, (1925).

[1]Ceylon.

FAMILY XIII. MYMARIDAE

Halid., *Ent. Month. Mag.,* I, p. 341, (1833). WESTW., *Intr. Mod. Class. Ins., Synop.,* p. 78, (1840). FÖRST., *Hymen. Stud.,* II, pp. 20, 27, 116, (1856). ASHM., *Proc. Ent. Soc. Washington,* IV, p. 236, (1897); *Mem. Carnegie Mus.,* I, (4), p. 361, (1904). SCHMIED., *Gen. Ins., fas.* 97, p. 488, (1909). GIR., *Mem. Queensland Mus.,* III, p. 96, (1915); *Insec. Inscit. Men*STR., VIII, p. 96, (1920). GAH., *Pan-Pacific Ent., San Francisco,* III, p. 180, (1927). DOZIER, *Journ. Dept. Agric. Peurto Rico, San Juan,* XVI, p. 81, (1932). CLOSE WATS., *Micr. Rec.,* XXIX, p. 3, (1933). Blackbourn, *Vict. nat.,* LI, pp. 213-216, figs. 4, (1935), [Morphology and Biology].

SUBFAMILY I. GONATOCERINÆ

ASHM., *Mem. Carnegie Mus.,* I, (4), p. 362, (1904). SCHMIED., *Gen. Ins. fas.* 97, p. 489, (1909).

Tribe GONATOCERINÆ

ASHM., *Mem. Carnegie Mus.,* I, (4), p. 362, (1904). SCHMIED., *Gen. Ins. fas.* 97, p. 491, (1909).

153. **Gonatocerus** Nees

Gonatocerus Nees, Hymen. Ichneum. Affin. Mon., II, p. 192, (1814), [Genotype *G. longicornis* Nees]. ASHM. *mem. Carnegie Mus.*, I, (4), p. 362, (1904), SCHMIED., *Gen., Ins., fas.* 97, p. 491, (1909).

longicrus KIEFFER

Gonatocerus longicrus, KIEFF., *Rec. Ind. Mus.*, IX, p. 201, (1913)[1]. RAMAKR., *Spol. Zeyl.*, XIII, p. 253, (1925). [1]Orissa.

154. **Litus** HALIDAY

Litus, Halid., *Ent. Month. Mag.*, I, p. 345, (1833), [Genotype *L. cynipseus* Halid.]. ASHM., *Mem. Carnegie Mus.*, I, (4), p. 362, (1904). SCHMIED., *Gen. Ins., fas.* 97, p. 492, (1909).

falfaltia (?) Meun., *Ann. Soc. Sc. Bruxelles*, XXV, p. 2, (1901).

Enocki HOWARD.

Litus enocki, HOW., *Proc. U.S.N. Mus.*, XVIII, p. 643, (1896)[1]. SCHMIED., *Gen. Ins., fas.* 97, p. 492, (1909). RAMAKR., *Spol. Zeyl.*, XIII, p. 254, (1925). [1]Punduloya: Ceylon.

155. **Alaptus** Westwood

Alaptus, WESTW., *Intr. Mod. Class. Ins.*, II, p. 79,. (1840), [Genotype *A. minimus* WALK., *A. buscalus* Halid. designated by Ashmead is a typographical error for *A. fusculus* WALK.]. ASHM., *Mem. Carnegie Mus.*, I, (4), p. 362, (1904). SCHMIED., *Gen. Ins., fas.* 97, p. 493, (1909).

magnanimus Annandale

Alaptus magnanimus, Annand., *Rec. Ind. Mus.*, III, pp. 299-300, pl. XXIII, (1909)[1]. [1]Calcutta.

SUBFAMILY II. MYMARINÆ

ASHM., *Mem. Carnegie Mus.*, I, (4), p. 363, (1904). SCHMIED., *Gen. Ins., fas.* 97, p. 493, (1909).

TRIBE 1. *MYMARARIÆ*

ASHM., *Mem. Carnegie Mus.*, I, (4), p. 363, (1904). SCHMIED., *Gen. Ins., fas.* 97, p. 494, (1909).

156. **Mymarilla** Westwood

Mymar, Ward., (*partim*), *Ent. Month. Mag.*, XII, p. 197, (1875).

Mymarilla, WESTW., *Trans. Linn. Soc. London*, (*Zool.*), I, p. 585, (1878), [Genotype *M. taprobanicus* (Ward)]. ASHM., *Mem. Carnegie Mus.*, I, (4), p. 363, 91904). SCHMIED., *Gen. Ins., fas.* 97, p. 495, (1909).

taprobanicus Ward

Mymar taprobanicus, Ward, *Ent. Month. Mag.*, XII, p. 197, (1875)[1].

[1]Ceylon.

Mymar (*Mymarilla*) *taprobanica*, WESTW., *Trans. Linn. Soc. London*, (*Zool.*), I, p. 585, (1878).

Mymarilla taprobanica, SCHMIED., *Gen. Ins., fas.* 97, p. 495, (1909). RAMAKR., *Spol. Zeyl.*, XIII, p. 254, (1925).

TRIBE 2. ANAPHESARIÆ

ASHM., *Mem. Carnegie Mus.*, I, (4), p. 363, (1904). SCHMIED., *Gen. Ins., fas.* 97, p. 498, (1909).

157. **Anthemus** HOWARD

Anthemus, HOW., *Proc. U.S.N. Mus.*, XVIII, p. 643, (1896), [Genotype *A. chionaspidis* HOW.]. ASHM., *Mem. Carnegie Mus.*, I, (4), p. 363, (1904). SCHMIED., *Gen. Ins., fas.* 97, p. 498, (1909).

chionaspidis HOWARD

Anthemus chionaspidis, Who., *Proc. U.S.N. Mus.*, XVIII, p. 643, (1896)[1]. SCHMIED., *Gen. Ins., fas.* 97, p. 498, (1909). RAMAKR., *Spol. Zeyl.*, XIII, p. 254, (1925).

[1]Punduloya: Ceylon.

ADDENDUM

FAMILY TORYMIDAE

SUBFAMILY TORYMINAE

1. Lioterphus Thomson

Torymus, BOHEMAN, (*partim*), *Svensk. Vet. Akad. Handl.*, LIV, p. 363, (1833).

Lioterphus, THOMS., *Hymen. Skandin.*, IV, p. 99, (1875), [Genotype *L. pallidicorn* (Bohem.)]. ASHM., *Mem. Carnegie Mus.*, I, (4), p. 241, (1904). SCHMIED., *Gen. Ins.*, *fas.* 97, p. 101, (1909).

Orissaensis MANI

Lioterphus orissoensis, MANI, *Rec. Ind. Mus.*, XXXVIII, p. 125, (1936)[1].

[1]Puri: Orissa.

SUBFAMILY PODAGRIONINÆ

2. Pachytomus Westwood.*

Pachytomus, WESTW., *Trans. Ent. Soc. London*, IV, p. 260, (1847), [Genotype *P. klugianus* WESTW.]. ASHM., *Mem. Carnegie Mus.*, I, (4), p. 244, (1904). SCHMIED., *Gen. Ins.*, *fas.* 97, p. 117, (1909).

mantisiphagus MANI

Pachytomus mantisiphagus, MANI, *rec. Ind. Mus.*, XXXVIII, p. 126, (1936)[1]. [Host: Mantid].

[1]Barkuda: Chilka Lake.

* Dalla Torre (*Cat. Hymen.*, V, p. 369, 1898) considers this to be a synonym of Podagrion but Ashead (1904) does not agree with him.

FAMILY CHALCIDIDÆ

SUBFAMILY LEUCOSPIDINÆ

3. Polistomorpha Westwood

paivai MANI

Polistomorpha paivai, MANI, *Rec. Ind. Mus.,* XXXVIII, p. 338, (1936)[1].

[1]Katihar: Purneah.

4. Leucospis Fabricius

assamensis MANI

Leucospis assamensis, MANI, *Rec. Ind. Mus.,* XXXVIII, pp. 339-340, (1936)[1].

[1]Sibsagar: Assam.

bengalensis var. **mackenziei** MANI

Leucospis bengalensis mackenziei, MANI, *Rec. Ind. Mus.,* XXXVIII, p. 339, (1936)[1].

[1]Chapra.

ornatifrons WELD

Leucospis ornatifrons, WELD, *Proc. U.S.N. Mus.,* LXI, art. 6, p. 22, (1922). MASI, *Boll. soc. Ent, Italina,* LXVII, p. 41, (1935), MANI, *Rec. Ind. Mus.,* XXXVII, p. 245, (1935)[1].

[1]Salem.

SUBFAMILY CHALCIDINÆ

TRIBE BRACHYMERARIÆ

5. Meyeriella Krausse

Meyeriella, KRAUS., *Archiv Naturgesch.,* LXXXII, Abt. A, Heft 1, pp. 93-95, figs. 1-5, (1916), [Genotype *M. indica* Kraus.].

indica Krausse

Meyeriella indica, KRAUS., *Archiv Naturgesch.,* LXXXIII, Abt. A, Heft 1, pp. 93-95, figs. 1-5, (1916)[1]. RAMAKR., *Spol, Zeyl.,* XIII, p. 240, (1925). [Host: *Delias eucharis*].

[1]Trichinopoly.

6. Brachymeria (Westwood) Stephens

fiskei CRAWFORD

Chalcis fiskei, CRAW., *Proc. Ent. Soc. Washington,* XI, p. 16, figs. 13-14, (1910)[1]. [There --is one example of this species in the Indian Museum, Calcutta].[2] — [1]Japan. [2]Kurseong.

7. Tumidicoxoides Girault

Tumidicoxoides, GIR., *Mem. Queensland Mus.,* Iv, p. 325 (1915), [Genotype *T. kurandaensis* GIR.].

Sp. RAMAKRISHNA AYYAR

Tumidicoxoides sp., RAMAKR., *Spol. Zeyl.,* XIII, p. 245, (1925)[1]. [Host : *Carea subtilis*]. — [1]Coimbatore.

TRIBE *CHALCIDARIÆ*

8. Spilochalcis Thomson

fletcheri MANI

Spilochalcis fletcheri, MANI, *Rec. Ind. Mus.,* XXXVIII, p. 340, (1936)[1]. [Host: Rose Tenthredinid]. — [1]Shillong.

TRIBE *HALTICELLARIÆ*

9. Neochalcis KIRBY

myrmeleonæ MANI

Neochalcis myrmeleonæ, MANI, *Rec. Ind. Mus.,* XXXVIII, p. 471, (1936)[1]. [Host: ant-lion]. — [1]Madras.

10. Euchalcis Dufour

myrmeleonae MANI

Euchalcis myrmeleonæ, MANI, *Rec. Ind. Mus.,* XXXVIII, p. 470, (1936)[1]. [Host: ant-lion]. — [1]Madras.

11. Stomatoceroides Girault

Stomatoceroides, GIR. *Canad. Ent.,* XLV, p. 140, (1913), [Genotype *S. bicolor* GIR.]; *Mem. Queensland Mus.,* IV, p. 314, (1915).

indicus MANI

Stomatoceroides indicus, MANI, *Rec. Ind. Mus.,* XXXVIII, p. 471, (1936)[1].

[1]Waltair.

TRIBE *CHALCITELLARIÆ*

ASHM., *Mem, Carnegie Mus.,* I, (4), p. 254, 91904). SCHMIED., *Gen. Ins., fas.* 97, p. 48, (1909). GIR., *Mem. Queensland Mus.,* IV, pp. 347. 352, (1915).

12. Anacryptus KIRBY

Epitranus, WALK., *(Partim), Trans. Ent. Soc. London,* (3), I, p. 348, (1862).

Anacryptus, KIRB., *Journ. Linn. soc. London. (Zool.),* XVII, p. 56, (1883), [Genotype *A. impulsator* (WALK.)]. ASHM., *Mem. Carnegie Mus.,* I, (4), p. 254, (1904). SCHMIED., *Gen. Ins., fas.* 97, p. 48, (1909). MANI, *Rec. Ind. Mus.,* XXXVIII, p. 126, (1936).

Chilkænsis, MANI

Anacryptus chilkænsis, MANI, *Rec. Ind. Mus.,* XXXVIII, p. 127, (1936)[1]. [Host: Probably *Camponotus compressus* (?)].

[1]Barkuda: Chilka Lake.

13. Paranacryptus Girault

Paranacryptus, GIR., *Mem. Queensland Mus.,* IV, p. 349, (1915), [Genotype *P. sanguineus* GIR.] MANI, *Rec. Ind. Mus.* XXXVIII, p. 127, (1936).

Sanguineus Girault

Paranacryptus sanguineus, GIR., *Mem. Queensland Mus,* IV, p. 349, (1915). MANI, *Rec. Ind. Mus.,* XXXVIII, pp. 127-128, (1936)[1], [Notes and redescription].

[1]Port Blair: Andamans.

14. Pararretoceroides, gen. nov.*

Arretoceroides, MANI, *(partim), Rec. Ind., Mus.,* XXXVIII, p. 128, (1936).

---

* This genus is closely related to *Arretoceroides* Girault *(Mem. Queensland Mus.* IV. p. 351, 1915) from which it is easily distinguished by the hind femur not being finely serrate as in the genus *Stomatoceras* Kirby but having only 10 blunt teeth.

ceylonensis (MANI)

Arretoceroides ceylonensis, MANI, *Rec. Ind. Mus.*, XXXVIII, pp. 128-129, (1936)[1]. [Genotype]. [1]Peradeniya: Ceylon.

FAMILY EUCHARIDÆ

15. Psygmatocera Enderlein

Psygmatocera, ENDERL., *Ent. Mitteil.*, I. p. 146, (1925), [Genotype *P. ceyloni* Enderl.].

ceyloni Enderlein

Psygmatocera ceyloni, ENDERL., *Ent. Mitteil.*, I, p. 146, (1925)[1]. RAMAKR., *Spol. Zeyl.*, XIII, p. 241, (1925).† [1]Bentotta, Kandy: Ceylon.

FAMILY ENCYRTIDÆ

16. Mesocomys CAMERON

Mesocomys, CAM., *Trans. S. Africa Philos. Soc.*, XV, p. 210, (1905), [Genotype *M. pulchripes* CAM.].

orientalis FERRIERE

Mesocomys orientalis, FERR., *Stylops*, IV, p. 150, (1935).[1] [1]Burma.

17. Parechthrodryinus Girault

Parechthrodryinus, GIR., *Proc. U.S.N. Mus.*, LI, p. 480, (1916), [Genotype *P. convexus* GIR.]

clavicornis (CAMERON)

Copidosoma clavicornis, CAM., *Ind. For. Rec.* IV, p. 101, (1912)[1]. Also *vide* p. 88 above. [1]Dehra Dun.

Parechthrodryinus clavicornis, FERR., *Bull. Ent. Res.*, XXVI, pp. 391-406, (1935).

18. Proleuroceus FERRIERE

Proleuroceus, FERR., *Bull. Ent. Res.*, XXVI, p. 403, (1935), [Genotype *P. fulgoridis* FERR.]

---

† *ceylonica* of Ramakr. is an error for *ceyloni.*

fulgoridis FERRIERE

Proleurocerus fulgoridis, FERR., *Bull. Ent. Res.,* XXVI, p. 403, (1935)[1] — [1]India.

19. Paralitomastix MERCET

Encyrtus, Nees, (*partim*), *Hymen. Ichneum. affin. Monogr.,* II, p. 214, (1834).

Paralitomastix, Merc., *Faun. Iber., Hymen.,* XII, (25), pp. 70, 438, (1921), [Genotype *P. varicornis* (Ness)].

varicornis (Ness)

Encyrtus varicornis, Nees, *Hymen. Ichneum. affin. Monogr.,* II, p. 214, (1834)[1]. MAYR, *Verh. Zool.-Bot. Ges. Wien,* XXV, p. 705, (1875). — [1]Europe.

Encyrtus varicornis, DALLA TORRE, *Cat. Hymen.,* V. p. 265, (1898). SCHMIED., *Gen. Ins., fas.* 97, p. 247, (1909). According to RAMAKRISHNA AYYAR, this species occurs in India.[2] [Host: *Dolichomeris evidantis*]. — [2]Chichawati: Punjab.

20. Metaplopoda MASI

Metaplopoda, MASI, *Konowia,* V, p. 276, (1926), [Genotype *M. grallaria* MASI].

grallaria MASI

Metaplopoda grallaria, MASI, *Konowia,* V, p. 276, figs. 21-22, (1926)[1]. [There is one specimen of this species determined by MASI in the Indian Museum, Calcutta[2]]. — [1]Formosa. [2]Puri.

FAMILY PTEROMALIDÆ

SUBFAMILY PTEROMALINÆ

TRIBE *PTEROMALARIÆ*

21. Neocatolaccus Ashmead

Catolaccus, ASHM., (*partim*), *Bull, Ohio Expt. Sta.,* I, p. 161, (1864).

Neocatolaccus, ASHM., *Mem. Carnegie Mus.*, I, (4), p. 320, (1904), [Genotype *N. tylodermoe* (ASHM.)]. MASI, *Boll. Lab. Zool. Portici*, III, p. 139, (1908), [Redefined in a footnote]. SCHMIED., *Gen. Ins., fas.* 97, p. 356, (1909).

sphenopteeræ FERRIERE

Neocatolaccus sphenopteræ, FERR., *Bull. Ent. Res.*, XXII, pp. 130-131, (1931)[1]. [Host: *Sphenoptera gossypii*].[2]

[1]Sudan.
[2]Also Punjab (?)

FAMILY EULOPHIDAE

SUBFAMILY ENTEDONINAE

TRIBE *OMPHALARIAE*

22. Euderus HALIDAY

Entedon, WALK., (*partim*), *Mon. Chalc.*, I, p. 106, (1839).

Euderus, Halid., *Trans. Ent. Soc. London*, III, (4), p. 298, (1843), [Genotype *E. amphis* (WALK.)]. ASHM., *Mem. Carnegie Mus.*, I, (4), p. 339, (1904). SCHMIED., *Gen. Ins., fas.* 97, p. 426, (1909). GIR., *Insec. Inscit. Menstr.*, XII, p. 94, (1924), [Characters].

Gossypii FERRIERE

Euderus gossypii, FERR., *Bull. Ent. Res.*, XXII, pp. 132-133, (1931)[1]. [Host: *Sphenoptera gossypii*]. [This species is represented in a collection of Chalcid parasites of *Sphenoptera gossypii* received for identification from Dr. RAMAKRISHNA AYYAR, Coimbatore[2]].

[1]Sudan: Africa
Lyallpur: Punjab
[2]Coimbatore.

23. Pareuderus FERRIERE

Pareuderus, FERR., *Bull. Ent. Res.*, XXII, p. 287, (1931). [Genotype *P. torymoides* FERR.]

Torymoides FERRIERE

Pareuderus torymoides, FERR., *Bull. Ent. Res.*, XXII, pp. 287-288, figs. 3, *c, d, e*, (1931)[1], [Hosts : *Alcides leeuweni*. Also *Lixus iruncatulus*[2]].

[1]Buitenzorg: Java.
[2]Also Pusa.

FAMILY MYMARIDÆ

SUBFAMILY MYMARINÆ

TRIBE *ANAPHESARIÆ*

24. Paranagrus Perking

Paranagrus, PERK., *Bull. Hawaii. Sug. Plant. Assoc.,* I, p. 199, (1905). [Genotype *P. optabilis* Per.].

optabilis Perkins

Paranagrus optabilis, PERK., *Bull. Hawaii. Sug. Plant. Assoc.,* I, p. 199, (1905)[1]. — [1]Coimbatore.

The original descriptions of the following genera and species could not be traced in the literature available in this country :

Eridontomeroidella gibboni, Girault. RAMAKR., *Rep. Proc. Third Ent. Meet. Pusa,* p. 935, (1919)[1]. [Host : *Eurytoma indi* of *daincha* pods]. — [1]Coimbatore.

Chartocerys musciformis, MOTSCHULSKY, *Etud. Ent.* p. 171, (1859)[1]. [Genus not traced in the literature]. — [1]Ceylon.

Eurytoma samskowi, Vass. This species is represented in the Collections of the Imperial Agricultural Research Institute, New Delhi, by several examples taken at Hazipur, Hazara and N.W.F. Provinces, where it is a fruit pest.

FAMILY EURYTOMIDAE

TRIBE *EURYTOMARIAE*

25. Eurytoma ILLIGER

Samsonowi Vasiljev

Eurytoma samsonowi Vasiljev, *Petrograd Trd. B. po. entom.* 11 No. 7 (1915) pp. 1-15. Fletcher, *Rept. Imp. Ent.* (1928-29) p. 71[1], 1930-31 p. 90[2]. — [1]Haripur; Hazara. [2]Attack; Punjab.

PART XXIV – EVANIDÆ

Family EVANINÆ

LATREILLE, *Hist. Nat. Crust. Ins.*, III, p. 328, (1802). LEACH, *Edinb. Encyc.*, IX, p. 142, (1830). HALIDAY, *Ent. Month. Mag.*, V, p. 212, (1838). SHUCKARD, *Entomologist*, I, p. 120, (1839). WESTWOOD, *Intr. Mod. Class. Ins.*, II, p. 124, (1846). PROVANCHER, *Petite Faun. Ent. Hymen. Canada*, p. 244, (1883). SCHLETTERER, *Verh. zool.-bot. Ges. Wien*, XXXV, pp. 267-325, (1885); *Ann. naturh. Hofmus.*, IV, pp. 107-180, 289-338, 373-546, (1889). ASHMEAD, *U. S. Board Agric., Cat. Ins.*, p. 563, (1900); *Proc. U. S. N. Mus.*, XXIII, p. 6, (1900). DALLA TORRE, *Cat. Hymen.*, III, (2), pp. 1058-1086, (1902), [World List]. KIEFFER, *Gen. Ins.*, fas. 2, p. 13, (1902), [Keys and descriptions of genera]. LEFROY, *Ind. Ins. Life*, pp. 180-181, (1909), [General account]. KIEFFER, *Das Tierreich*, Lief. 30, pp. 432, (1912), [Monographic account]. HELLÉN, *Medd. Soc. Faun. Fenn.*, XLI, pp. 67-69, (1915), [Finlands spp.]. COCKERELL, *Proc. U. S. N. Mus.*, LI, p. 102, (1916), [Florissan Miocene spp.]. FRISON, *Trans. Amer. Ent. Soc.* XLVIII, pp. 1-33, (1922), [Neotropical forms]. BRUES, *Ann. S. Afric. Mus.*, XIX, p. 2, (1924), [Synonymy and notes on S. African spp.]. SEMENOV-TIANSHANSKIJ & KOSTYLEY, *Rev. russe Ent.*, XXII, pp. 86-91, figs. 7, (1928), [First additions to KIEFFER's monograph]. UCHIDA, *Trans. Sapporo Nat. Hist. Soc.*, XII, pp. 189-193, (1932), [Aulacidae of Japan'. WATANABE, *Trans. Sapporo Nat. Hist. Soc.*, XIII, pp. 280-281, 281-286, figs. 2, (1934), [Japanese Evanidae]. IMMS., *Text Book Ent.*, p. 571, (1934), [General account].

Subfamily I. EVANINÆ

Evaninæ CAMERON, *Biol. Centr. Amer. Hymen.*, p. 422, (1887). CRESSON, *Trans. Ent. Soc. Philadelphia, Suppl., (Synops. Hymen.)*, p. 36, (1887). ASHMEAD, *U. S. Board Agric., Cat. Ins.*, p. 563 (1900); *Proc. U. S. N. Mus.*, XXIII, p. 7, (1900), [Classification]. KIEFFER, *Gen. Ins., fas.* 2, p. 2, (1902), [Descriptions of genera]; *Das Tierreich*, Lief. 30, p. 6, (1912), [Monographic revision].

Evanidæ, HANDL., in Schroder's *Handb. Ent.*, III, p. 752, (1925), [General account].

1. **Hyptiogaster** Cockerell.

Hyptiogaster, COCKERELL, *Ann. Ent. Soc. Amer.,* X, p. 19, (1917), [Genotype *H. electrina* Cockl.].

electrina COCKERELL.

Hyptiogaster electrina, COCKERELL, *Ann. Ent. Soc. Amer.,* X, p. 20, (1917)[1], [In Burmese amber]. [1]Burma.

2. **Prosevania** Kieffer.

Evania, (subgen. *Prosevania*), KIEFFER, *Ann. Soc. Ent. France,* LXXX, p. 157, (1911).

Prosevania, KIEFFER, *Das Tierreich,* Lief. 30, p. 62, (1912), [Genotype: *P. afra* (Kieff.)].

subtangens (Kieffer).

Evania (*Prosevania*) *subtangens,* KIEFFER, *Ann. Soc. Ent. France,* LXXX, pp. 160-161 (1911).[2]

Prosevania subtangens, KIEFFER, *Das Tierreich,* Lief. 30, p. 62, (1912).[3] [2, 3]Singapore: Malaya.

3. **Evania** Fabricisu.

Ichneumon, REUAMER, (*partim*) *Mem. Hist. Ins.,* VI, p. 293, (1742). LINNAEUS, *Syst. Nat.,* (ed. 10), I, p. 560, (1758).

Sphex, LINNAEUS, *Syst. Nat.* (ed. 10), I, (2), p. 930, (1767).

Evania, FABRICIUS, *Syst. Ent.,* p. 345, (1775), [Genotype *E. appendigaster* (Linn.) Fab.]. SCHLETTERER, *Ann. naturh. Hofmus. Wien,* IV, p. 118, (1889). ASHMEAD, *Proc. U. S. N. Mus.,* XXIII, p. 7, (1900). KIEFFER, *Gen. Ins., fas.* 2, p. 2, (1902); *Das Tierreich,* Lief. 30, p. 65, (1912), [Keys & descriptions of spp.].

albitarsis Cameron.

Evania albitarsis, CAMERON, *Mem. Proc. Manchester Lit. Philos. Soc.,* XLIII, p. 53, (1889)[1]. LEFROY, *Ind. Ins., Life,* p. 181, (1909). KIEFFER, *Das Tierreich,* Lief. 30, p. 95, (1912)[2]. [1, 2]Khasia Hills.

antennalis Westwood.

Evania antennalis, WESTWOOD, *Trans. Ent. Soc. London*, III, p. 244, (1841)[3], SCHLETTERER, *Verh. zool.-bot. Ges. Wien*, XXXVI, p. 10, (1886); *Ann. naturh. Hofmus. Wien*, IV, p. 148, (1889). KIEFFER, *Gen. Ins.*, fas. 2, p. 3, (1902). LEFROY, *Ind. Ins. Life*, p. 181, (1909). KIEFFER., *Das Tierreich*, Lief. 30, p. 120, (1912). [3]Bombay, Ceylon.

Evania cariniformis, CAMERON, *Proc. Trans. Nat. Hist. Soc. Glasgow*, (2), p. 296, (1885-86).

appendigaster (Linnaeus) Fabricius.*

Ichneumon appendigaster, LINNAEUS, *Syst. Nat.*, (ed. 10), I, p. 566, (1758). DE GEER, *Mem. Hist. Ins.*, III, p. 594, (1773). RETZEUS, *Gen. Spec. Ins.*, p. 68, (1783).

Sphex appendigaster, LINNAEUS, *Syst. Nat.*, (ed. 12), I, p. 943, (1767).

Evania appendigaster, FABRICIUS, *Syst. Ent.*, p. 345, (1775); *Spec. Ins.*, I, p. 442, (1781); *Mant. Ins.*, I. p. 271, (1787). RÖMER, *Gen. Ins.*, p. 58, (1789). ROSSI, *Faun. Etrusca*, II, p. 59, (1790). OIVER, *Encyc. method Ins.*, VI, p. 453, (1791). FABRICIUS, *Syst. Ent.*, II, p. 192, (1793). LAMARK, *Syst. anim. vert.*, p. 267 (1801). FABRICIUS, *Syst. Piez.*, p. 178, (1804). LATREILLE, *Hist. nat. Crust. Ins.*, XIII, p. 193, (1805). NEES *Hymen. Ichneum. affin. Monogr.*, I, p. 311, (1834). BLANCHARD, *Hist. nat. Ins.*, III, p. 299, (1840). WESTWOOD, *Intr. Mod. Class. Ins.*, II, p. 134, (1840). SHUCKARD, *Entomologist*, p. 120, (1841). WESTWOOD, *Trans. Ent. Soc. London*, III, p. 241, (1843); (2), I, p. 213, (1851). SCHLETTERER, *Verh. zool.-bot. Ges. Wien*, XXXVI, p. 10, (1886). CAMERON, *Biol. Centr. Amer., Hymen.*, p. 427, (1887). SCHLETTERER, *Ann. naturh. Hofmus. Wien*, VII, p. 233, (1892). KIEFFER, *Ann. Soc. Ent. France*, p. 818, (1900). DALLA TORRE, *Cat. Hymen.*, III, (2), p. 1076, (1902). KIEFFER, *Gen. Ins.*, fas. 2, p. 3, (1902). LEFROY, *Ind. Ins. Life*, p. 181, (1909). KIEFFER, *Das Tierriech*, Lief. 30, p. 83, (1912), HABER, *Canad. Ent.*, LII, p. 248, (1920), [Oviposition]. [Hosts: *Periplaneta americana, P. orientalis, Panchlora madeiroe*].

---

* Found all over the world.

Evania loevigata, OLIVER, *Encyc. MethodIns.,* VI, p. 453, (1791). LATREILLE, *Gen. Crust. Ins.,* III, p. 251, (1801). LAMARK, *Hist. nat. anim. vert.,* IV, p. 147, (1817). LEACH, *Edinb. Encyc.* IX, p. 142, (1830). NEES, *Hymen. Ichneum. Monogr.,* I, p. 311, (1834), [Notes]. WESTWOOD, *Trans. Ent. Soc. London,* III, p. 241, (1843); (2), I, p. 213, (1851). SMITH, *Philos. Trans. Roy. Soc.,* CLXVIII, p. 538, (1879).

Evania flavicornis, OLIVER, *Encyc. Method Ins.,* VI, p. 453, (1791).

Evania fuscipes, ILLIGER, in Rossi's *Faun. Etrusca,* (ed. 2), II, p. 83, (1807); NEES, *Hymen. Ichneum. affin. Monogr.,* I, p. 310, (1834). WESTWOOD, *Trans. Ent. Soc. London,* III, p. 241, (1843); (2), I, p. 213, (1851).

Evania desjardini, BLANCHARD, *Hist. nat. Ins.,* III, p. 299, (1840). WESTWOOD, *Trans. Ent. Soc. London,* III, p. 242, (1843); (2), I, p. 214, (1851).

araticeps Kieffer.

Evania araticeps, KIEFFER, *Ann. Soc. Ent. France,* LXXX, p. 167, (1911);[1] *Das Tierreich,* Lief. 30, p. 121, (1912).[2] — [1, 2]Assam.

binghami Cameron.

Evania binghami, Cameron, *Ann. Mag. Nat. Hist.,* (7), XIX, p. 169, (1907).[1] KIEFFER, *Das, Tierreich,* Lief. 30, p. 94, (1912).[2] — [1, 2]India and Middle Tenasserim.

brachystylus Schletterer.

Evania brachystylus, SCHLETTERER, *Ann. naturh. Hofmus. Wien,* IV, p. 152, (1889).[1] KIEFFER, *Gen. Ins.,* fas. 2, p. 3, (1902); *Das Tierreich,* Lief. 30, p. 125, (1912). — [1]Ceylon.

curvicarinata Cameron.

Evania curvicarinata, CAMERON, *Mem. Proc. Manchester Lit. Philos. Soc.,* XLIII, p. 51, (1899).[1] LEFROY, *Ind. Ins. Life,* p. 181, (1909). KIEFFER, *Das Tierreich,* Lief. 30, p. 122, (1912). — [1]Khasia Hills.

deesænsis Cameron.

Evania deesoensis, CAMERON, *Journ. Bombay Nat. Hist. Soc.,* XVII, p. 583, (1907).[1] KIEFFER, *Das Tierreich,* Lief. 30, p. 124, (1912). — [1]Deesa: Bombay.

dolichopus Schletterer.

Evania dolichopus, SCHLETTERER, *Ann. naturh. Hofmus. Wien*, IV, p. 151, (1889).[1] KIEFFER, *Gen. Ins.*, fas. 2, p. 3, (1902); *Das Tierreich*, Lief. 30, p. 113, (1912). — [1]Ceylon.

erythrosoma Schletterer.

Evania erythrosoma, SCHLETTERER, *Verh. zool.-bot. Ges. Wien*, p. 18, (1886);[1] *Ann. naturh. Hofmus. Wien*, IV, p. 149, (1889). KIEFFER, *Gen. Ins.*, fas. 2, p. 3, (1902); *Das Tierreich*, Lief. 30, p. 112, (1912). — [1]Ceylon.

grisea Kieffer.

Evania grisea, KIEFFER, *Ann. Soc. Ent. France*, LXXX, p. 166, (1911); *Das Tierreich*, Lief. 30, p. 116, (1912).[1] — [1]Kandy: Ceylon.

interstitialis Cameron.

Evania interstitialis, CAMERON, *Spol, Zeyl.*, III, p. 74, (1905).[1] KIEFFER, *Das Tierreich*, Lief. 30, p. 98, (1912). — [1]Ceylon.

microstyla Kieffer.

Evania microstyla, KIEFFER, *Ann. Soc. Ent. France*, LXXX, p. 176 (1911);[1] *Das Tierreich*, Lief. 30, p. 108, (1912).[2] — [1, 2]India; Singapore: Malaya.

montana Kieffer.

Evania montana, KIEFFER, *Ann. Soc. Ent. France*, LXXX, p. 165, (1911);[1] *Das Tierreich*, Lief. 30, p. 99, (1912). — [1]Kangra.

nurseana Cameron.

Evania nurseana, CAMERON, *Journ. Bombay Nat. Hist. Soc.*, XVII, p. 99, (1906).[1] KIEFFER, *Das Tierreich*, Lief. 30, p. 110, (1912). — [1]Quetta.

origena Kieffer.

Evania origena, KIEFFER, *Ann. Soc. Ent. France*, LXXX, p. 164, (1911);[1] *Das Tierreich*, Lief. 30, p. 122, (1912). — [1]Kangra.

peradeniyae Cameron.

Evania peradeniyoe, CAMERON, *Spol. Zeyl.*, III, p. 75, (1905).[1] KIEFFER, *Das Tierreich*, Lief. 30, p. 98, (1912). — [1]Peradeniya: Ceylon.

rubrofasciata Brues.

Evania rubrofasciata, BRUES, *Bull. Amer. Mus.*, XXXV, p. 718, (1916).[1] — [1]India.

simlaensis Cameron.

Evania simlaensis, CAMERON, *Deuts. ent. Zeit.,* p. 660, (1909).[1] KIEFFER, *Das Tierreich,* Lief. 30, p. 110, (1912). [1]Simla.

solox Enderlein.

Evania solox, ENDERLEIN, *Zool, Anz.,* XXVIII, p. 704, (1905).[1] KIEFFER, *Das Tierreich,* Lief. 30, p. 114, (1912). [1]Burma, Tenasserim.

stenochela Kieffer.

Evania stenochela, KIEFFER, *Ann. Soc. Ent. France,* LXXX, pp. 171-172, (1911);[1] *Das Tierreich,* Lief. 30, p. 94, (1912).[2] [1, 2]Karachi.

szépligetii Bradley.

Evania parva, SZÉPLIGETT, *Ann. Mus. Hungar.,* I, p. 381, (1903).

Evania szépligetii, BRADLEY, *Trans. Amer. Ent. Soc.,* XXXIV, p. 167, (1908). KIEFFER, *Das Tierreich,* Lief. 30, p. 119, (1912).[1] [1]Singapore: Malaya.

Subfamily II, GASTERUPTIONINÆ

Evaninoe, (partim.), CAMERON, *Biol, Centr. Amer. Hymen.,* p. 422, (1887). ASHMEAD, *U. S. Board Agric., Cat. Ins.,* p. 563, (1900).

Aulacinoe, (partim), SCHLETTERER, *Ann. naturh. Hofmus. Wien,* IV, p. 118, (1889).

Gasteruptioninoe, ASHMEAD, *Proc. U. S. N. Mus.,* XIII, p. 7, (1900). SCHULZ, *Spol. Hymen.,* p. 133, (1906). KIEFFER, *Das Tierreich,* Lief. 30, p. 188, (1912).

Foeninoe, KIEFFER, *Gen. Ins., fas.* 2, p. 6, (1902).

Gasteruptionidœ, HANDL., in Schroder's *Handb. Ent.,* III, p. 751, (1925?).

4. Trichofoenus Kieffer.

Foenus, (partim), GUERIN-MÉNEVILLE, *Icongr. Régean.,* II, p. 406, (1829-38).

Gasteruption, (partim), SCHLETTERER, *Verh. zool.-bot. Ges. Wien,* XXXV, p. 283, (1885). SEMENOV; *Bull. Ac. St. Petersburg,* (N. S.), III, p. 13, (1892).

Trichofoenus, KIEFFER, *Ann. Soc. Ent. France*, LXXIX, p. 77, (1910), [Genotype *T. merceti* (Kieff.)]; *Das Tierreich*, Lief. 30, p. 213, (1912).

corynetes Kieffer.

Trichofoenus corynetes, KIEFFER, *Ann. Soc. Ent. France*, LXXX, p. 188, (1911);[1] *Das Tierreich*, Lief. 30, p. 217, (1912).

[1]Nasik: Bombay.

orientalis (Cameron).

Gasteruption orientalis, CAMERON, *Mem. Proc. Manchester Lit. Philos. Soc.*, (4), I, p. 19, (1888).[1] SCHLETTERER, *Ann. naturh. Hofmus. Wien*, IV, p. 435, (1889). KIEFFER, *Gen. Ins.*, fas. 2, p. 9, (1902). CAMERON, *Journ. Bombay Nat. Hist. Soc.*, XVII, p. 583, (1907). LEFROY, *Ind. Ins. Life*, p. 181, (1909).

[1]Barrackpore: Bengal.

Trichofoenus orientalis, KIEFFER, *Das Tierreich*, Lief. 30, p. 217, (1912).

rufithorax Kieffer.

Trichofoenis rufithorax, KIEFFER, *Ann. Soc. Ent. France*, LXXX, p. 189, (1911);[1] *Das Tierreich*, Lief. 30, p. 217, (1912).

[1]Calcutta.

5. Trigonofoenus Kieffer.

Gasteruption, (*partim*), CAMERON, *Journ. Bombay Nat. Hist. Soc.*, XVII, p. 582, (1907).

Trigonofoenus, KIEFFER, *Ann. Soc. Ent. France*, LXXX, p. 177, (1911); [Genotype *T. trianguliferum* (Kieff.) *Das Tierreich*, Lief. 30, p. 224, (1912)].

abuensis (Cameron).

Gasteruption abuensis, CAMERON, *Journ. Bombay Nat. Hist. Soc.*, XVII, p. 582, (1907).[1]

[1]Mt. Abu.

Trigonofoenus abuensis, KIEFFER, *Das Tierreich*, Lief. 30, p. 225, (1912).

6. Electrofoenus Cockerell.

Electrofoenus, COCKERELL, *Amer, Journ. Sc.*, XLIV, p. 364, (1917), [Genotype *E. gracilipes* Cockl.].

gracilipes Cockerell.

Electrofoenus gracilipes, COCKERELL, *Amer. Journ. Sc.,* XLIV, p. 365, (1917),[1] [From Burmese amber]. [1]Burma.

7. **Protofoenus** Cockerell.

Protofoenus, COCKERELL, *Ann. Ent. Soc. Amer.,* X, p. 19, (1917), [Genotype *P. swinhoei* Cockl.].

swinhoei Cockerell.

Protofoenus swinhoe, COCKERELL, *Ann. Ent. Soc. Amer.,* X, p. 19, (1917),[1] [From Burmese amber]. [1]Burma.

8. **Gasteruption** Latreille.

Ichneumon, (partim), LINNÆUS, *Syst. Nat.,* (ed. 10), I, p. 560, (1758).

Gasteruption, LATREILLE, *Precis caract. Ins.,* p. 113, (1796), [Genotype *G. assectator* (Fabr.)=*G. affector* (Linn)]. SCHLETTERER, *Verh. zool.-bot, Ges. Wien.* p. 268, (1885). ASHMEAD, *U. S. Board Agric., Cat. Ins.,* p. 563, (1900). KIEFFER, *Gen. Ins.,* fas. 2, p. 7, (1902). SCHULZ, *Spol. Hymen.,* p. 133, (1906). KIEFFER, *Das Tierreich,* Lief. 30, p. 226, (1912).

Foenus, FABRICIUS, *Ent. Syst., Suppl.,* p. 240, (1798). TOURNIER, *Ann. Soc. Ent. Belge,* XX, p. (1877). ABEILLE, *Bull. Soc. Hist. Nat. Toulouse,* XIII, p. 260, (1879).

baluchistanense Cameron.

Gasteruption baluchistanense, CAMERON, *Journ. Bombay Nat. Hist. Soc.,* XVII, p. 100, (1906).[1] KIEFFER, *Das Tierreich,* Lief. 30, p. 292, (1912). [1]Quetta.

brevicuspis Kieffer.

Gasteruption brevicuspis, KIEFFER, *Ann. Soc. Ent. France,* LXXX, p. 196, (1911);[1] *Das Tierreich,* Lief. 30, p. 293, (1912). [1]Assam.

ceylonicum Cameron.

Gasteruption ceylonicum, CAMERON, *Spol. Zeyl.,* III, p. 77, (1905).[1] KIEFF., *Das Tierreich,* Lief. 30, p. 294, (1912). [1]Punduloya: Ceylon.

kashmirense Cameron.

Gasteruption kashmirense, CAMERON, *Journ. Bombay Nat. Hist. Soc.,* XVII, p. 292, (1906).[1] KIEFFER, *Das Tierreich,* Lief. 30, p. 292, (1912). — [1]Kashmere.

mandibulare Cameron

Gasteruption mandibulare, CAMERON, *Mem. Proc. Manchester Lit. Philos. Soc.,* XLIII, p. 59, (1899).[1] LEFROY, *Ind. Ins. Life,* p. 181, (1909). KIEFFER, *Das Tierreich,* Lief. 30, p. 296, (1912). — [1]Khasi Hills.

nigrinerve Kieffer.

Gasteruption nigrinerve, KIEFFER, *Ann. Soc. Ent. France,* LXXX, p. 193, (1911);[1] *Das Tierreich,* Lief. 30, p. 281, (1912). — [1]Kandy: Ceylon.

quettaense Cameron.

Gasteruption quettaense, CAMERON, *Journ. Bombay Nat. Hist. Soc.,* XVII, p. 101, (1906).[1] KIEFFER, *Das Tierreich,* Lief. 30, p. 295, (1912). — [1]Quetta.

rotundicolle Cameron.

Gasteruption rotundicolle, CAMERON, *Journ. Bombay Nat. Hist. Soc.,* XVII, p. 582, (1906).[1] KIEFFER, *Das Tierreich,* Lief. 30, p. 293, (1912). — [1]Deesa.

sericeipes Kieffer.

Gasteruption sericeipes, KIEFFER, *Ann. Soc. Ent. France,* LXXX, p. 199, (1911).[1] *Das Tierreich,* Lief. 30, p. 298, (1912). — [1]Karachi, Peshawar.

tricoloratum Cameron.

Gasteruption tricoloratum, CAMERON, *Spol. Zeyl.,* III, p. 76, (1905).[1] KIEFFER, *Das Tierreich,* Lief. 30, p. 297, (1912). — [1]Peradeniya: Ceylon.

Subfamily III. AULACINAE

Aulacidoe, SHUCHARD, *Entomologist,* p. 12, (1840-2). HANDL., in Schroder's *Handb. Ent.,* III, p. 733, (1925).

Aulacinoe, CAMERON, *Biol. Centr. Amer. Hymen.,* p. 422, (1887). CRESSON, *Trans. Ent. Soc. Philadelphia,*

Suppl. (Synops. Hymen.), p. 36, (1887). ASHMEAD, *U. S. Board Agric., Cat. Ins.*, p. 563, (1900); *Proc. U. S. N. Mus.*, XXIII, p. 8, (1900). KIEFFER, *Gen. Ins.*, fas. 2, p. 10, (1902); *Das Tierreich*, Lief. 30, p. 342, (1912). UCHIDA, *Trans. Sapporo Nat. Hist. Soc.*, XII, pp. 189-193, figs. 2, (1932), [Japanese species].

9. **Aulacinus** Westwood.

Aulacinus, WESTWOOD, *Trans. Ent. Soc. London*, p. 331, (1868), [Genotype *Aulacinus* (*Aulacinus*) *moereus* Westw.]. KIEFFER, *Gen. Ins.*, fas. 2, p. 12, (1902); *Das Tierreich*, Lief. 30, p. 349, (1912).

Aulacus, (*partim*), SCHLETTERER, *Ann. naturh. Hofmus. Wien*, IV, p. 517, (1889).

bituberculatus (Cameron).

Aulacus (*Aulacinus*) *bituberculatus*, CAMERON, *Mem. Proc. Manchester Lit. Philos. Soc.*, XLIII, p. 55, (1899).[1] [1]Khasi Hills.

Aulacus bituberculatus, LEFROY, *Ind. Ins. Life*, p. 181, (1909).

Aulacinus bituberculatus, KIEFFER, *Das Tierreich*, Lief. 3, p. 350, (1912).

10. **Tetraulacinus** Kieffer.

Pristaulacus, (*partim*), CAMERON, *Zeit. Hymen. Dipt.*, VII, p. 222, (1907).

Tetraulacinus, KIEFFER, *Bull. Soc. Ent. France*, p. 350, (1910); [sans species]; *Das Tierreich*, Lief. 30, p. 363, (1912), [Genotype *T. rufobalteatus* (Kieff.)].

rufobalteatus leviceps Kieffer.

Tetraulacus rifobalteatus (*partim*), KIEFFER, *Ann. Soc. Ent. France*, LXXX, p. 225, (1911). [1]Darjeeling.

Tetraulacinus balteatus leviceps, KIEFFER, *Das Tierrerich*, Lief. 30, p. 364, (1912).[1]

11. **Aulacus** Jurine.

Bassus, FABRICIUS, *Syst. Piez.*, p. 93, (1804).

Aulacus, JURINE, *Nouv. Meth. Hymen. Dipt.*, p. 89, (1807), [Genotype *A. striatus* Jur.]. LATREILLE, *Gen. Crust. Ins.*, IV, p. 385, (1809). SCHLETTERER, *Ann. naturh. Hofmus. Wien*, IV, p. 489, (1889). KIEFFER, *Gen. Ins.*, fas. 2, p. 11, (1902); *Das Tierreich*, Lief. 30, p. 370, (1912).

Aulacostethus, PHILIPPI, *Ent. Zeit. Stett.*, XXXIV, p. 302, (1873).

Pammegischia,* PROVANCHER, *Petit, Faun. Ent. Canad. Hymen.*, p. 247, (1883).

flavipennis Cameron.

Aulacus flavipennis, CAMERON, *Mem. Proc. Manchester Lit. Philos. Soc.*, XXVI, p. 134, (1887).[1] KIEFFER, *Das Tierreich*, Lief. 30, p. 372, (1912). — [1]Dekaya: Ceylon.

signatus Schuckard.

Aulacus signatus, SHUCKARD, *Entomologist*, p. 124, (1841). KIEFFER, *Das Tierreich*, Lief. 30, p. 373, (1912).[1] — [1]Ceylon.

stigmaticus Westwood.

Aulacus stigmaticus, WESTWOOD, *Trans. Ent. Soc. London*, p. 327, (1868).[1] KIEFFER, *Das Tierreich*, Lief. 30, p. 372, (1912).[2] — [1, 2]Singapore: Malaya.

12. Pristaulacus Kieffer.

Bassus, FABRICIUS, *Syst. Piez.*, p. 93, (1804).

Aulacus, (*partim*), WESTWOOD, *Trans. Ent. Soc. London*, III, p. 262, (1843).

Pristaulacus, KIEFFER, *Ann. Soc. Ent. France*, LXVIII, p. 813, (1899), [Genotype *P. chlapowskii* Kieff.]. ASHMEAD, *Proc. U. S. N. Mus.*, XXIII, p. 8, (1900). KIEFFER, *Gen. Ins.*, fas. 2, p. 12, (1902); *Das Tierreich*, Lief. 30, p. 376, (1912).

Deraiodontus, BRADLY, *Trans. Amer. Ent. Soc.*, XXVII, p. 321, (1901).

Oleisoprister, BRADLY, *Trans. Amer. Ent. Soc.*, XXVII, p. 324, (1901).

---

* TOWNES *Canad. Ent.* LXX, pp. 254-255 (1938), [Synonymy, status, etc., discussed]. (Note added in second proof).

Neaulacus, BRADLY, *Trans. Amer. Ent. Soc.,* XXXIV, p. 121 (1908).

Psilaulacus, KIEFFER, *Bull. Soc. Ent. France,* p. 350, (1910).

beesoni Turner.

Pristaulacus beesoni, TURNER, *Ann. Mag. Nat. Hist.,* (9), X, p. 271, (1922).[1] — [1]India.

emarginaticeps Tonkin.

Pristaulacus emarginaticeps, TONKIN, *Ann. Mag. Nat. Hist.,* (9), X, p. 270, (1922).[1] — [1]India.

iridipennis (Cameron).

Aulacus iridipennis, CAMERON, *Mem. Proc. Manchester Lit. Philos. Soc.,* XLIV, p. 80, (1900).[1] — [1]Khasi Hills.

Pristaulacus iridipennis, KIEFFER, *Das Tierreich,* Lief. 30, p. 389, (1912).

magnificus (Schletterer).

Aulacus (?) congener, WESTWOOD, *Trans. Ent. Soc. London,* III, p. 267, (1843). — [1]Ceylon.

Aulacus magnificus, SCHLETTERER, *Ann. naturh. Hofmus., Wien,* IV, p. 509, (1889).[1]

Pristaulacus mangificus, KIEFFER, *Bull. Soc. Ent. France,* p. 338, (1900); *Gen. Ins.,* fas. 2, p. 13, (1902); *Das Tierreich,* Lief. 30, p. 388, (1912).

nigripes KIEFFER.

Pristaulacus nigripes, KIEFFER, *Ann. Soc. Ent. France,* LXXX, p. 229, (1911);[1] *Das Tierreich,* Lief. 30, p. 388, (1912). — [1]Sikkim.

PART XXV – THYSANOPTERA

Order THYSANOPTERA

Order THYSANOPTERA Haliday

Physapus DE GEER, *Kongl. Svensk. Vet.-Akad. Handl.*, V., p. 1, (1744). *Thrips* LINNE, *Syst. Nat.*, (*Ed.*), X, p. 457, (1758). *Vesitarses & Physapodes* DUMERIL, *Zool. anal. Methode nat. Class. animaux*, etc. pp. 261, 268, 269, (1806). *Thripsides* FALLEN, *Specimen novam Hemiptera desponendi methodum exhibens*, pp. 1-26, (1814). *Thrypsides*, LATREILLE, *Fam. nat. animal, etc.*, p. 429, (1825). *Thripidae* STEPHENS, *A system Catal, Brit. Ins., etc.*, Pt. II, p. 363, (1829). *Thripsites* NEWMAN, *Ent. Mag.*, II, pp. 380, 426, (1835). *Thysanoptera* HALIDAY, *Ent. Mag.*, III, pp. 439-451, (1836). [History and general account]. *Physopoda* BURMEISTER, *Handb. Ent.*, II Bd., p. 404, (1838). *Physapoda* HALIDAY-WALKER, *List Spec. Homop. Ins. Brit. Mus.*, Pt. IV, p. 1094, (1852). *Thripsina* NEWMAN, *Trans. ent. Soc. Lond.*, (New Ser.), III, p. 264, (1855?). *Thripsidae* FITCH, *First Rept. Nox. Benef. Ins. State New York*, p. 103, (1855). *Thripidae* FITCH, *Third Rept. Nox. Benef. Ins. State New York*, p. 305, (1856), [Synonymy]. JORDAN, *Zeit. wiss. Zool.*, XLVII, pp. 541-620, (1888), [Anatomy and bibliography]. GARMAN, *Bull. Essex Inst.*, XXII, pp. 24-27, (1890), [Mouth-parts]. UZEL, *Mon. Ordnung Thys.*, (1895), [General account]. HINDS, *Proc. U.S. Nat. Mus.*, XXVI, pp. 79-242, (1902), [Monograph]. MOULTON, *U.S. Dept. Agr. Bur. Ent. Techn. Ser.*, no. 21, pp. 1-56, (1911), [Classification]. BAGNALL, *Ann. Mag. Nat.* Hist., (8), X, pp. 220-222, (1912), [Classification]. KARNY & VAN LEEUWEN-REIJNVAAN, *Zeit. wiss. Ins.-Biol.*, (1914-16), [Gall-making forms]. BORDEN, *J. econ. Ent.*, VIII, pp. 354-359, (1915), [Mouth-parts]. HOOD, *Proc. Biol. Soc. Wash.*, XXVIII, pp. 53-60, (1915), [Classification]. PETERSON, *Ann. Ent. Soc. Amer.*, VIII, pp. 20-59, (1915), [Morphology of head and mouth-parts]. TAKAHASHI, *Zool. Mag., Tokyo*. XXXV, p. 85, (1921), [Metamorphosis]. KARNY, *Treubia*, I, pp. 211-261, (1921), [Classification]; *Treubia*, III, pp. 29-37, (1922), [Phylogeny]. WATSON, *Univ. Florida Agr. Exp. Stat., Bull.* 168, pp. 1-100, (1923) [Classification]. PRIESNER, *Sitz. Akad. wiss. Wien*, 132, pp. 1-18, (1923), [Larval forms]. DE GUYS & GERMAN, *Canad. Ent.*, LVI, pp. 177-182, (1924), [Genital apparatus]. KLOCKE,

Zeit. wiss. Zool., 128, I, pp. 1-36, (1926), [Anatomy and histology]. PRIESNER, *Die Thysanopteren Europas*, pp. 1-755, (1928), [Monograph]. KURTZ MULLER, *Zeit. wiss. Zool.*, 130., Bd., pp. 251-303, (1927), [Biology, anatomy, histology and metamorphosis]. BAGNALL, *Ann. Mag. Nat. Hist.*, (10), V, pp. 571-575, (1930), [Classification].

Indian forms: NEWMAN, *Trans. Ent. Soc. Land.*, p. 26, (1855). WALKER, *Ann. Mag. Nat. Hist.*, p. 224, (1859). UZEL, *Mon. Ordnung Thys.*, (1895). KIEFFER, *Marcellia*, VII, (1909). BAGNALL, *Rec. Ind. Mus.*, VII, p. 257, (1912); *Rec. Ind. Mus.*, VIII, p. 201, (1913); *Ann. Mag. Nat. Hist.* (1912 to recent years), *Bull. Ent. Res.* (1915-28). SCHMUTZ, *Sitz. Akad. Wiss. Wien.* CXXII, pp. 991-1091, (1913). WILLIAMS, *Bull. Ent. Res.*, VI, pp. 353-355, (1915). HOOD, *Insec. Inscit. Menstr.*, VII, pp. 99-103, (1919). RAMAKRISHNA, *Rept. Third Ent. Meeting, Pusa*, pp. 612-617, (1919); *J. Bombay Nat. Hist. Soc.*, XXX, pp. 861-871, (1925); *J. Bombay Nat. Hist. Soc.*, XXX, pp. 788-792 (1925); *Madras Agric. Journ.*, XIII, pp. 311-317, (1925); *Pusa Ent. Mem. Dept. Agr. Ind.*, X, (7), pp. 217-316; (1928); *Agri and Live-st. Ind.*, II, pp. 391-403, (1932); *Rec. Ind. Mus.*, XXXIV, pp. 279-281, (1932); *Ind. For. Rec.*, XX, pp. 1-12, (1934); *Rec. Ind. Mus.*, XXXVI, pp. 491-498, (1934). RAMAKRISHNA and MARGABANDHU, *J. Bombay Nat. Hist. Soc.*, XXXIV (4), pp. 1029-1040, (1931); *Rec. Ind. Mus.*, XLI, pp. 21-33, (1939), *Ind. J. Ent.*, I, (3), pp. 35-48, (1939); KARNY, *Pusa Ent. Mem. Dept. Agri. Ind.*, IX, (6), pp. 187-239, (1926). MOULTON, *Bull. Brook. Ent. Soc.*, XXII, no. 4, p. 184, (1927); *Ind. For. Rec.*, XIII, pp. 285-292, (1928); *J. Bombay Nat. Hist. Soc.*, XXXIV, (4), p. 1040, (1929); *Rec. Ind. Mus.* XXXI, pp. 93-100, (1929); PRIESNER, *Rec. Ind. Mus.*, XXXV, pp. 347-369, (1933).

SUBORDER I TEREBRANTIA Haliday.

Superfamily I AEOLOTHRIPOIDEA HOOD

Family I Orothripidae Bagnall.

Subfamily I Orothripinae Bagnall.

1. Orothrips Moulton.

Orothrips, MOULTON, *Tech. Bull.* 12, *U.S.A. Dept. Agr.*, p. 45, (1907), [Genotype *O. kelloggi* Moulton].

raoi Moulton.

Orothrips raoi, MOULTON, *Bull. Brook. Ent. Soc.* XXII, No. 4, p. 184, (1927)[1] RAMAKRISHNA, *Pusa Ent. Mem. Dept. Agr. Ind.*, X, (7), p. 247, (1928), [Unknown flowers][2]. [1, 2]Bangalore.

2. **Erythrothrips** Moulton.

Erythrothrips, MOULTON, *Tech. Bull. 21, U.S. Dept. Agr.*, p. 35, (1911), [Genotype *E. arizonae* Moulton].

asiaticus Ramakrishna & Margabandhu.

Erythrothrips asiaticus RAMAKRISHNA & MARGABANDHU, *J. Bombay Nat. Hist. Soc.*, XXXIV, pp. 1029-1030, (1931), [Distinguished from *arizonae* and *durango*, from grass sweepings.][1] RAMAKRISHNA, *Ind. For. Rec.*, XX, p. 3, (1934), [On shoots of sandal and on *Canthium didymum*.][2]

[1]Coimbatore.
[2]Bangalore, Salem.

Family II Aeolothripidae Uzel.

Subfamily I Aeolothripinae Bagnall.

3. **Aeolothrips** Haliday.

Aeolothrips, HALIDAY, *Ent. Mag.* III, p. 451, (1836), [Genotype *T. fasciata* (Linnaeus)]. MOULTON, *Tech. Bull. U.S. Dept. Agr.*, 12, p. 47, (1907), [Redescribed]. KARNY, *Treubia*, I, p. 235, (1921). [Characters].

fulvicollis Bagnall.

Aeolothrips fulvicollis, BAGNALL, *Ann. Mag. Nat. Hist.*, (9), IV, pp. 253-254, (1919), [In *Verbascum flowers*].[1] MOULTON, *Bull. Brook. Ent. Soc.*, XXII, No. 4, p. 184, (1927), [In blossoms][2]. RAMAKRISHNA, *Pusa Ent. Mem. Dept. Agr. Ind.*, X, (7), p. 247, (1928), [In mango flowers][3]. RAMAKRISHNA & MARGABANDHU, *J. Bombay Nat. Hist. Soc.*, XXXIV, No. 4, p. 1030, (1931), [Notes].[4]

[1]Cawnpore.
[2]Dehra Dun.
[3, 4]Pusa.

pandyani Ramakrishna & Margabandhu.

Aeolothrips pandiyani, RAMAKRISHNA & MARGABANDHU, *J. Bombay Nat. Hist. Soc.*, XXXIV, No. 4, pp. 1030-1031, (1931), [Distinguished from *A. fulvicollis*. On ears of *Pennisetum typhoideum*, and from grass sweepings][1].

[1]Koilpatti : (S. India).

Superfamily II Mymarothripoidæ Bagnall.

Family I Mymarothripidæ Bagnall.

4. Mymarothrips Bagnall.

Mymarothrips, BAGNALL, *Ann. Mag. Nat. Hist.,* (10), I, p. 306, (1928), [Genotype *m. ritchianus* Bagnal].

garuda Ramakrishna & Margabandhu.

Mymarothrips garuda, RAMAKRISHNA & MARGABANDHU, *J. Bombay Nat. Hist. Soc.,* XXXIV, No. 4, p. 1031, (1931), [On grape vine leaf with *Rhipiphorothrips cruentatus,* H. and on turmeric leaves with *Panchaetothrips indicus,* B][1]. — [1]Coimbatore, Bhavani.

Superfamily III Heterothripoidea Bagnall.

Family I Heterothripidæ Bagnall.

5. Adiheterothrips Ramakrishna.

Adiheterothrips, RAMAKRISHNA, *Pusa Ent. Mem. Dept. Agr. Ind.,* X, (7), p. 247, (1928), [Genotype *A. jambudvipae* Ramakrishna].

jambudvipae Ramakrishna.

Adiheterothrips jambudvipae, RAMAKRISHNA, *Pusa Ent. Mem. Dept. Agr. Ind.,* X, (7), p. 249, (1928), [From flowers of *Phoenix dactylifera.*][1] — [1]Nayudupeta (S. India).

Superfamily IV Thripoidea Hood.

Family I Thripidae Uzel.

Subfamily I Corynothripinae Karny.

6. Neocorynothrips Ramakrishna & Margabandhu.

Neocorynothrips, RAMAKRISHNA & MARGABANDHU, *Rec. Ind. Mus.,* XLI, pp. 21-22, (1939), [Genotype *N. asiaticus* Ram. & Marg.].

asiaticus Ramakrishna & Margabandhu.

Neocorynothrips asiaticus, RAMAKRISHNA & MARGABANDHU, *Rec. Ind. Mus.,* XLI, pp. 22-23, (1939), [On garden crotons].[1] — [1]Coimbatore.

Subfamily II Chirothripinae Karny.

7. **Chirothrips** Haliday.

Chirothrips, HALIDAY, *Ent. Mag.,* p. 444, (1836), [Genotype *C. manicatus* Haliday]. UZEL, *Mon. Ordnung Thys.,* p. 79, (1895). HINDS, *Proc. U.S. Nat. Mus.,* XXVI, p. 133, (1902). KARNY, *Treubia,* I, 4, p. 237, (1921). KNECHTEL, *Thys. Roman.,* p. 78, (1923). PRIESNER, *Die. Thys. Europas,* p. 136, (1928).

manicatus Haliday.

Chirothrips manicatus, HALIDAY, *Ent. Mag.,* III, p. 444, (1836). UZEL. *Mon. Ordnung Thys.,* p. 80, (1895). HINDS, *Proc. U.S. Nat. Mus.,* XXVI, p. 134, (1902)[1]. PRIESNER, *Die. Thys. Europas,* pp. 138-139, (1928), [Complete list of references and synonyms][2], RAMAKRISHNA & MARGABANDHU, *J. Bombay Nat. Hist. Soc.,* XXXIV, p. 1031, (1931), [From grass sweepings, tobacco and other plants][3].

[1]Coimbatore

[1]England, Germany, Finland, Russia, Bohemia. [2]United States of America.

8. **Limothrips** Haliday.

Limothrips, HALIDAY, *Ent. Mag.,* p. 445, (1836), [Genotype *L. carealium* Haliday]. UZEL, *Mon. Ordnung Thys.,* p. 85, (1895). HINDS, *Proc. U.S. Nat. Mus.,* XXVI, p. 138, (1902). KARNY, *Treubia,* I, 4, p. 237, (1921). KNECHTEL, *Thys. Roman.,* p. 82, (1923). PRIESNER, *Die. Thys. Europas,* p. 144, (1928).

cerealium Haliday.

Limothrips cerealium, HALIDAY, *Ent. Mag.,* III, p. 445, (1836). UZEL, *Mon. Ordnung, Thys.,* p. 89, (1895). HINDS, *Proc., U.S. Nat. Mus.,* XXVI, p. 139, (1902). KARNY, *Pusa, Ent. Mem. Dept. Agr. Ind.,* IX, (6), p. 188, (1926), [From *Mimusops elengi*]. BAGNALL, *Ann. Mag. Nat. Hist.,* (9), XVIII, p. 547, (1926), [Plentiful on wheat][2]. RAMAKRISHNA, *Pusa Ent. Mem. Dept. Agr. Ind.,* X, (7), p. 250, (1928), [In leaves of *Mimusops elengi* with *Arrhenothrips remukrishnai*][3]. PRIESNER, *Die. Thys. Europas,* pp. 150-154, (1928), [Complete list of references]. SHARGA, *Ann. Appl. Biol.,* XX, pp. 308-317, (1933), [Bionomics].

[1, 3]Coimbatore. [2]Australia, S. America, Europe, Egypt, Africa, Seychelles, Hawaii.

Subfamily III. Sericothripinae Karny.

9. **Hydatothrips** Karny.

Hydatothrips, Karny, *Engebn. Zentr. Afrik Exp.* IV, p. 281, (1912), [Genotype *H. fridericie* Karny].

ramaswamiahi Karny.

Hydatothrips ramaswamiahi, Karny, *Pusa Ent. Mem. Dept. Agr. Ind.*, IX, (6), p. 188, (1926), [From indigo leaves].[1] Ramakrishna, *Pusa Ent. Mem. Dept. Agr. Ind.*, X, (7), p. 251, (1928), [On shoots of sword beans][2]. Ramakrishna and Margabandhu, *J. Bombay Nat. Hist. Soc.*, XXXIV, No. 4, p. 1032, (1931), [On *Pongamia* leaves].[3]

[1, 2]Coimbatore. [3]Sholinghur : (N. Arcot Dt.), Koilpatti : (Tinnevelly Dt.)

10. **Scirtothrips** Shull.

Scirtothrips, SHULL, *Entom. News.* XX, No. 5, p. 222, (1909). [Genotype *S. ruthveni* Shull]. Hood, *Proc. Ent. Soc. Washington,* XVI, p. 40, (1914). Karny, Treubia, I, 4, p. 237 and p. 240, (1921).

dorsalis Hood.

Scirtothrips dorsalis, Hood, *Insec. Inscit. Menstr.*, VII, pp. 90-91, (1919), [On castor and chillies].[1] Ramakrishna, *Pusa Ent. Mem. Dept. Agr. Ind.*, X, (7) p. 251, (1928), [Notes][2]; *Agr. Livestk. Ind.*, II, pt. IV, pp. 393-394, (1932), [Bionomics]. Ramakrishna and Margabandhu, *J. Bombay Nat. Hist. Soc.*, XXXIV, No. 4, p. 1032, (1931). [On a variety of plants][3]; Indian J. Ent., I. (3), p. 36, (1939), [On chillies][4].

[1, 2]Coimbatore. [3]Travancore, Cochin, Southern Dts. [4]Coorg.

signipennis Bagnall.

Scirtothrips signipennis, Bagnall, *Ann. Mag. Nat. Hist.*, (8), XIII, pp. 22-23, (1924), [Under leaf sheaths of banana].[1]

[1]Peradeniya: Ceylon.

11. **Pseudodendrothrips** Schmutz.

Pseudodendrothrips, Schmutz, *Sitz. Akad. wiss. Wien.*, CXXII, p. 998, (1913), [Genotype *P. ornatissima* Schmutz].

ornatissmia Schmutz.

Pseudodendrothrips ornatissima, Schmutz, *Sitz. Akad. wiss. Wien.*, CXXII, pp. 999-1001, (1913), [On *Macrocarpus integrifolia*].[1] — [1]Peradeniya : (Ceylon).

12. **Dendrothrips** Uzel.

Dendrothrips, Uzel, *Mon. ordnung Thys.* p. 159, (1895), [Genotyp *D. tibaie* Uzel.] Karny, *Treubia,* I, 4, pp. 98 and 238, (1921), *Knechtel, Thys. Roman.*, P. 94, (1923).

Monochaetella, Karny, *Treubia,* I, 4, p. 238, (1921).

Dichaetella, Karny, *Treubia,* I, 4, p. 238, (1921).

bispinosus Bagnall.

Dendrothrips bispinosus, Bagnall, *Bull. Ent. Res.*, XIV, p. 455. (1924), [On tea][1]. — [1]Kotagiri : (Nilgiris).

dwivarna Ramakrishna and Margabandhu.

Dendrothrips dwivarna, Ramakrishna and Magrabandhu, *Journ. Bombay Nat. Hist. Soc.*, XXXIV, No. 4, pp. 1032-1033, (1931), [From jak shoots][1]. — [1]Kaladi : (Travancore State), Trichur: (Cochin State).

indicus Bagnall.

Dendrothrips indicus, Bagnall, *Ann. Mag. Nat. Hist.*, (9), IV, p. 261, (1919), [On arrow root leaves][1]. Ramakrishna, *Pusa. Ent. Mem. Dept. Agr. Ind.*, X, (7), pp. 251-252, (1928), [In tender shoots of Arrowroot plant in company with *Anaphotrips citricinctus* Bag., and *Panchaetothrips indicus* Bag.)].[2] — [1, 2]Taliparamba (Malabar).

sexmaculatus Bagnall.

Dendrothrips sexmaculatus, Bagnall, *Ann. Mag. Nat. Hist.*, (8), XVII, p. 401 (1916). — [1]Peradeniya: (Ceylon).

13. **Dendrothripiella** Bagnall.

Dendrothripiella, Bagnall, *Ann. Mag. Nat. Hist.*, (9), XIX, p. 567, (1927), [Genotype *D. Phyllireae* Bagnall]. Ramakrishna and Margabandhu, *Indian J. Ent.*, I, (3), (1939).

jasminum Ramakrishna and margabandhu.

Dendrothripiella jasminum., Ramakrishna and Margabandhu, *Indian J. Ent.* I, (3), pp. 36-37, (1935), [On jasmine][1]. — [1]Madura.

14. **Dendrothripoides** Bagnall.

Dendrothripoides, Bagnall, *Ann. Mag. Nat. Hist.,* (9), XII, pp. 624-625, (1923), [Genotype *D. ipomeae Bagnall*]. Ramkrishna, *Pusa Ent. Mem. Dept. Agr. Ind.,* X, (7), p. 271 (1928).

ipomeae Bagnall.

Dendrothripoides ipomeae, Bangall, *Ann. Mag. Nat. Hist.,* (9), XII, p. 625, (1923), [On *Ipomea staphylina*].[1] Ramakrishna, *Pusa Ent. Mem. Dept., Agr. Ind.* X, (7), p. 271, (1928).[2] — [1, 2]Maddur: (Mysore State).

15. **Anaphotrips** Uzel.

Anaphotrips, Uzel, *Mon. Ordnung Thys.,* p. 142, (1895), [Genotype *A. ferruginea,* UZel]. Hinds, *Proc. U.S. Nat. Mus.,* XXVI, p. 160, (1902), Karny, *Treubia,* I, 4, p. 242, (1921).KNECHTEl,*Thys. Roman.,* p. 113, (1923).

Euthrips, KARNY, *Zeit. wiss. Ins. Biol.,* P. 355, (1914).

citricinctus Bagnall.

Anaphotrips citricinctus, Bagnall. *Ann. mag. Nat. Hist.,* (9), IV, pp. 270-271, (1919).[1] Ramakrishna, *Pusa Ent. Mem. Dept. Agr. Ind.,* X, (7), p. 269, (1928). [On arrowroot leaves and from cholam shoots].[2] — [1, 2]Taliparamba: (Malabar). [2]Coimbatore.

flavicinctus (Karny).

Euthrips flavicinctus, Karny *Marecellia,* XI, p. 115, (1912).[1] — [1]Java.

Neophysopus medioflavus, Schmutz, *Sitz. Akad. wiss. Wien.* CXXII, p. 1017 (1913).[2] — [2]Peradeniya: (Ceylon).

Anaphotrips flavicinctus, Ramkrishna and Magabandhu, *J. Bombay Nat. Hist. Soc.,* XXXIV, No. 4 p. 1036, (1931), [Distinguishing features. Oncombu ears].[3] Yakhontov, *Rev. Ent. U. R. SS.* XXV, p. 174, (1933), [Morphology]. — [3]Koilpatti: (Tinnevelly Dt.)

fungivora Ramkrishna.

Anaphotrips fungivora, Ramakrishna, *Pusa Ent. Mem. Dept. Agr. Ind.,* X, (7), pp. 268-269, (1928), [Feeding on rust on wheat plants][1]. — [1]Coimbatore.

oligochaetus Karny.

Anaphotrips oligocahaetus, Karny, *Pusa Ent. Mem. Dept. Agr. Ind.,* IX, (6), p. 201, (1926), [In flowers and shoots of cotton, pomegranates and *Caesalpinia pulcherrima*].[1] Ramakrishna, *Pusa Ent. Mem. Dept. Agr. Ind.,* X, (7), p. 268, (1928).[2] Ramkrishna and Margabandu, *Indian J. Ent.,* I, (3), p. 37, (1939), [On mulberry shoots).[3] — [1, 2]Coimbatore. [3]Kollegal.

ramakrishnai Karny.

Anaphothrips ramakrishnai, Karny, *Pusa, Ent. Mem. Dept. Agr. Ind.,* IX, (6), p. 203, (1926), [In ears of *Pennisetum typhoideum*].[1] Ramkrishna, *Pusa Ent. Mem. Dept. Agr. Ind.,* X, (7), p. 268, (1928).[2] — [1, 2]Coimbatore.

16. **Hemianaphothrips** Priesner.

Anaphotrips (Hemianaphotrips) Preisner, *Die Thys. Europas,* p. 181, (1928), [Genotype *H. articulosus* Priesner].

palmae Ramakrishna.

Hemianaphotrips Palmae, Ramkrishna, *Rec. ind. Mus.,* XXXVI, pp. 491-492, (1934), [Both male and female recorded from flower heads of *Phoenix dactylifera* in company with *Adiheterothrips jambudvipae* Ram., on the banks of Tungabhadra river].[1] — [2]Bellary Dt.

17. **Oxythrips** Uzel.

Oxythrips, Uzel, *Mon. Ordnung Thys.,* P. 133, (partim), (1895), [Genotype *O. ajugae* Uzel] Karny, *Treubia,* I, 4, p. 243, (1921). Priesner, *Wiener Ent. Zeitg.,* XI, pp. 115-117, (1923). Knechtel, *Thys. Roman.,* p. 116, (1923). Priesner, *Die. Thys. Europas,* p. 210, (1928).

hemavarna Ramakrishna and Margabandhu.

oxythrips hemavara, Ramkrishna and Margabandhu, *J. Bombay Nat. Hist. Soc.,* XXXIV, No. 4, p. 1036, (1931), [From mango leaves][1]. — [1]Coimbatore.

Subfamily IV **Heliothripinae, Karny.**

18. **Tryphactothrips** Bagnall.

Tryphactothrips, Bagnall, *Ann. Mag. Nat. Hist.,* (9), VII, p. 264 (1921), Genotype *T. rutherfordi* Bagnall.

mundus Karny.

Tryphactothrips mundus, Karny, *Pusa Ent. Mem. Dept. Agr. Ind.,* IX, (6), p. 190, (1926), [on leaves of a wild plant causing small galls].[1] — [1]Taliparamba: (Malabar).

octarticulata (Schmutz.)

Parthenothrips octariticulata, Schumtz, *Sitz. Akad. wiss. Wien.* CXXII, p. 993, (1913).[1] — [1]Peradeniya: (Ceylon).

ruthefordi (Bagnall).

Dinurothrips rutherfordi, Bagnall, *Ann. Mag. Nat. Hist.,* (8), XV, pp. 319-320, (1915), [On leaves of *Allmands*].[1] — [1, 4]Ceylon.

Tryphactothrips rutherfordi, Bagnall, *Ann. Mag. Nat. Hist.,* (9), IV, p. 257, (1919)[2]; *Ann. Mag. Nat. Hist.,* (9), VII, p. 264, (1921), Ramakrishna, *Pusa, Ent. Mem. Dept. Agr. Ind.,* X, (7), p. 271, (1928), [In *Dhatura* flowers, in cotton and brinjal].[3] Ramakrishna and Margabandhu, *J. Bombay Nat. Hist. Soc.,* XXXIV, No. 4 pp. 1036-1037, (1931), [Further description][4]; *Indian J. Ent.,* I. (3), pp. 37-38, (1939)[5]. — [2]Coimbatore. [2]South Arcot. [5]Punjab.

19. **Rhipiphorothrips** Morgan.

Rhipihorothrips, Morgan, (Preocc.) *Proc. U.S. Nat. Mus.,* XLVI, p. 17, (1913), [Genotype *R. pulchellus* Morgan].

Retithrips (nec. Marchal), Bagnall, *Ann. Mag. Nat. Hist.,* (8), XII, p. 290, (1913).

bicolor Bagnall.

Retithripe bicolor, Bagnall, *Ann. Mag. Nat. Hist.,* (8), XII, pp. 290-291, (1913).[1]; *Ann. Mag. Nat. Hist.,* (8), XV, p. 320, (1915), [On vines. Probably identical with the genotpye from Philippines]. — [1]Ceylon.

cruentatus Hood.

rhipiphorothrips cruentatus, Hood, *Insec. Inscit. Menstr.,* VII, pp. 96-96, (1919), [On grape vine[1] leaves and *Careya* leaves].[1a] Ramakrishna, *Rep. Third Ent. Meeting, Pusa,* p. 327, (1920). Karny, *Pusa Ent. Mem. Dept. Agr. Ind.,* IX, (6), p. 192, (1926), [On leaves of *Odina odiar, Eugenia jambulana, Calotropis* flowers].[2] Ramakrishna, *Pusa Ent. Mem. Dept. Agr.* — [1]Coimbatore. [3]Bezwada, Bangalore. [4]Madura, Travancore. [2, 3, 5]Lyallpur. [1a]Peradeniya: (Ceylon).

Ind., X, (7), p. 252, (1928).[3] Ramakrishna & Margabandhu, *J. Bombay Nat. Hist. Soc.* XXXIV, No. 4, p. 1034, (1931), [On grape vine]; *Indian J. Ent.*, I, (3), p. 38, (1939), [On leaves of *Eugenia jambolona, Punica granatum, Terminalia arjuna,* and *Mangifera indica]*. Ramakrishna, *Agr. and Livestock in Ind.*, II. Pt. IV. p. 396, (1932), [*Bionomics*].[4] Rahman and Bhardwaj, *Ind. Journ. Agr. Sci.*, VII, Pt. IV, pp. 633-651, (1937), [Bionomics and control.][5]

20. **Heliothrips** Haliday.

Heliothrips, Haliday, *Ent. Mag.*, p. 443, (1836), [Genotype *H. Haemorrhoidalis* Bouche]. Uzel, *Mon. Ordnung Thys.*, p. 167, (1895), Hinds, *Proc. U.S. Nat. Mus.*, XXVI, p. 168, (1902), Karny, Trubia, I, 4, p. 239, (1921). Knechtel, Thys. Roman., p. 100 (1923).

brunneipennis Bagnall.

Heliothrips brunneipennis, Bagnall, *Ann. Mag. Nat.*, (8), XV, p. 318, (1915), [On leaves of *Litsea chinensis*].[1]

[1]Peradeniya: (Ceylon).

haemorrhoidalis Bouche.

Heliothrips haemorrhoidalis, Bouche, *Nat. schadl. Garten-Ins.*, p. 206, (1833). Hinds, *Proc. U.S. Nat. Mus.*, XXVI, pp. 168-169, (1902), [Synonymy and redescripton]. Uzel, *Mon. Ordnung Thys.*, 168, (1895). [Synonmymy]. Karny *Arkiv. for Zool., Band* 17-A, No. 2, p. 12, (1924)[1]; *Bull. Deli. Proefest.* No. 23, p. 4, (1925), [References][2]; *Pusa Ent. Mem. Dept. Agr. Ind.*, IX, (6), p. 192, (1926), [On leaves of garden erotons in shoots of *Melia azadirachi*][3]; Ramakrishna, *Pusa Ent. Mem. Dept. Agr. Ind.*, X, (7), p. 254, (1928), [On tender leaves of garden Croton, shoots of *melia azadirac* and a wild plant][4]; *Agr. and Livestk. Ind.*, II, p. 399, (1932), [Bionomics]. Priesner, *Die Thys. Europas*, pp. 126-128, (1928). [Complete list of references). Rivnay, *Bull. Ent. Res.*, XXVI, pp. 267-278, (1935), [Ecology]; *Bull. Soc. Ent. Egypte*, XIX, pp. 119-124, (1935).

Heliothrips adonidium, Haliday, *Ent. Mag.*, p. 443, (1836).

Heliothrips haemorrhoidalis var. *ceylonicus*, Schmutz, *Sitz. Akad. wiss. Wien.*, CXXII, p. 992, (1913).[5]

[3]Coimbatore.
[4]Bangalore, Taliparamba.
[5]Peradeniya.

---

[1]Queensland.
[2]Sumatra.

indicus Bagnall.

Heliothrips indicus, Bagnall, *Ann. Mag. Nat. Hist.,* (8), XII, p. 291, (1913), [On indigo, bersim, onion and brinjal][1]. Ramakrishna, *Rpt. Third Ent. Meeting, Pusa,* pp. 327, 619, (1920), [on onions][2]; *Agr. and Livestock in Ind.* II, Pt. IV, p. 397, (1932), [Bionomics]; *Rec. Ind. Mus.,* XXXVi, p. 493, (1934), [On *Cannabis sativa,* sannhemp flowers, date flowers]. Bedford, *Sudan Ent. Bull.,* XVIII, pp. 1-52 (1921), [On cotton, *Vicia, Vigna, Pisum, Medicago, Trifolium, Dolichos lablab, Sesbania, Lucerne, etc.*][3]. Karny, *Pusa Ent. Mem. Dept. Agr. Ind.,* IX, (6), p. 193, (1926), [On cabbage leaves][4]. Ramakrishna & Margabandhu, *Indian J. Ent.* 1, (3), p. 38, (1939), [On cauliflower].[5]

[1]Sirsiah: (Bihar), Surat (Bombay).
[2]Northern Circars.
[4]Bombay, Pusa, Coimbatore, Siruguppa: (Bellary Dist.).
[5]Lyallpur: Punjab

---

[3]Sudan.

kadaliphila Ramakrishna & Margabandhu.

Heliothrips kadaliphila, Ramakrishna & Margabandhu, *J. Bombay Nat. Hist. Soc.,* XXXIV, No. 4, pp. 1033-1034, (1931), [On banana leaves][1]; *Rec. Ind.* Mus., XLI, p. 23, (1939), [On banana].[2]

[1]Coimbatore.
[2]Ponnani: (Malabar Dt.).

minutissimus Bagnall.

Heliothrips minutissimus, Bagnall, *Ann. Mag. Nat. Hist.,* (9), IV, p. 260, (1919), [In violet flowers].[1]

[1]Surat: (Bombay).

21. Reticulothrips Faure.

Reticulothrips, Faure, *South African Journ. Nat. Hist.* V, pp. 144-145, (1925), [Genotype *Reticulothrips peringueyi* Faure].

peringueyi Faure.

Reticulothrips peringueyi, Faure. *South African J. Nat. Hist.,* V, pp. 145-150. (1925), [On base of stems of tall grasses].[1] Ramakrishna & Margabandhu, *J. Bombay Nat. Hist. Soc.,* XXXIV, No. 4, p. 1034, (1931), [On *Panicum* and other grass sweepings].[2]

[2]Coimbatore.

---

[1]Arcadia: (Pretoria).

Sub-family V THRIPINAE Karny

22. Selenothrps Karny.

Heliothrips (Selenothrips), Karny, *Ent. Rundsch.,* XVIII, p. 180, (1911), [Genotype *S. rubrocinctus (Giard)*] *Selenothrips,* Hood, *Insec. Inscit, Menstr.,* I, p. 150, (1913).

rubrocinctus Giard.

Physopus rubrocinctus, Giard, *Bull. Soc. Ent. Franc,e* p. 263, (1901). Green, *Tropic. Agric., Ceylon,* XXVII, p. 248, (1906).[1] Lefroy, *Ind. Ins. Life,* pp. 543-544, (1909). — [1,2,4]Ceylon.

Heliothrips rubrocinctus, Giard, *Urich Bull. Board of Agric. Trinidad,* (1914). Franklin, *Proc. U.S. Nat. Mus.,* XXXIII, pp. 719-723, (1908), [Redescription].

Selenothrips rubrocinctus, Hood. *Insec. inscit. Menstr.,* I, p. 150, (1913). Moulton, *Rec. Ind. Mus.,* XXXI, p. 98 (1928), [On *Aporosa* and *Cardenia* flowers][2]. Ramakrishna & Margabandhu, *J. Bombay Nat. Hist. Soc.,* XXXIV, p. 1034, (1931), [On Cashewnut leaves].[3] Ramakrishna, *Agr. and Livestk. Ind.,* II, pt. IV, pp. 398-399, (1932), [Bionomics].[4] — [2]Calcutta. [3]Trichur: (Malabar). [4]Cochin.

[1,4]Florida.

Selenothrips mendax, Schutz, *Sitz., Akad. wiss. Wien.* CXXII, p. 994, (1913), [On *Anacardium orientale].*[5] — [4,5]West Indies. [4]Uganda, Mauritius, Hawaii.

23. **Ayyaria** Karny.

Ayyaria, Karny, *Pusa, Ent. Mem. Dept. Agr. Ind.,* IX, (6), p. 193, (1926), [Genotype *A. chaetophora* Karny]. Ramakrishna, *Pusa Ent. Mem. Dept. Agr. Ind.,* X, (7), p. 255, (1928).

chaetophora Karny.

Ayyaria chaetophora, Karny, *Pusa Ent. Mem. Dept. Agr. Ind.,* IX, (6), p. 193, (1926), [Inside flowers of cow-pea—*Vigna catiang*].[1] Ramakrishna, *Pusa Ent. Mem.' Dept. Agr. Ind.,* X, (7), (1928).[2] Ramakrishna & Margabhandhu, *J. Bombay Nat. Hist. Soc.,* XXXIV, No. 4, p. 1034, (1931), [On castor shoots, on garden Croton an grass sweepings][3] *Rec. Ind. Mus.,* XLI, p. 23, (1939), [On *Dolichos lab-lab*].[4] — [1,2,3]Taliparamba. [3]Coimbatore, Samalkota. [4]Malabar.

24. **Monilothrips** Moulton.

Monilothrips, Moulton, *Rec. Ind. Mus.,* XXXI, p. 93, (1929), [Genotype *M. kempi* Moulton].

kempi Moulton.

Monilothrips kempi, Moulton, *Rec. Ind. Mus.,* XXXI, p. 94, (1929), [Unknown plant].[1] — [1]Mangphu: (Darjiling).

25. Scolothrips Hinds.

Scolothrips, Hinds, *Proc. U.S. Nat. Mus.,* XXVI, p. 157, (1902), [Genotype *S. sexmaculatus* (Pergande)].

asura Ramakrishna & Margbandhu.

Scolothrips asura, Ramakrishna and Margabandhu, *J. Bombay Nat. Hist. Soc.,* XXXIV, No. 4, p. 1035, (1931), [On banana leaves].[1] — [1]Coimbatore.

sexmaculatus (Pergande).

Thrips sexmaculatus, Pergande, *Trans. St. Louis Acad.,* V, p. 542, (1894).

Thrips pallida Beach, *Proc. Iowa Acad. Sciences,* III, pp. 226-227, (1895).

Scolothrips sexmaculatus, Hinds, *proc. U.S. Nat. Mus.,* XXVI, p. 157, (1902), [Redescription]. Karny, *Pusa Ent. Mem. Dept. Agr. Ind.,* IX, (6), (1926), [In pomegranate flowers].[1] Ramakrishna, *Pusa Ent. Mem. Dept. Agr. Ind.,* X, (7), p. 255, (1928), [In pogemgranate flowers and in cholam shoots, feeding on mites].[2] — [1, 2]Coimbatore.

26. Frankliniella Karny.

Frankliniella, Karny, *Mitt. Nat. Ver. Univ. wien.,* VIII, 2, p. 46, (1910), [Genotpye *F. intonsa* (Trybom)]; *Zool. Annal.,* IV, p. 334, (1912), [Tables]; *Treuia,* I, 4, p. 240, (1921), Hood, *Proc. Ent. Soc. Wash.,* XVI, p. 337, (1914), [Type fixed as *Thripsintonsa* Trybom]. Knechtel, *Thys. Roman.,* p. 151, (1923). Priesner, *Deut-sche Ent. Zeit.* pp. 14-19, (1925), [Synopsis of Known species].; *Die Thys. Europas* p. 246, (1928). [References].

sulphura Schmutz.

Franklinielaa sulphurea, Schmutz, *Sitz., Akad. wiss. Wien.,* CXXII, p. 1019, (1913), [In flowers of *Thrunbergia alata*].[1] Ramakrishna, *J. Bombay Nat. Hist. Soc.* XXX, p. 865, (1925). Karny, *Pusa Ent. Mem. Dept. Agr. Ind.,* IX, (6), pp. 195-196, (1926), [Inside *Chrysanthemum* flowers, *Lantana, Bignonia grandis, Lucerne, Datura, Hibiscus rosasinensis,* etc.].[2] Ramkrishna, *Pusa Ent. Mem. Dept. Agr. Ind.,* X, (7),

[1]Peradeniya.
[2, 3]Coimbatore.
[3]Kollegal, Kadiam: (Godaveri Dt.), Guntur.
[4]Barkuda: Ganjam.
[5]Adoni : (Bellary).
[6]Lyallpur: Punjab.

255, (1928), [On jasinine flowers, shoots of chillies, groundnut flowers, and cotton].[3] Moulton, *Rec. Ind. Mus.*, XXXI, p. 93, (1929) [On *Datura* flowers].[4] Ramakrishna & Margabandhu, *Rec. Ind. Mus.*, XLI, p. 23, (1939). [Prickly pear flowers][5], *Indian J. Ent.*, I, (3), pp. 38-39, (1939), [Food plants].[6]

27. **Taeniothrips** Amyot & Serville.

Taeniothrips, Amyot et Serville, *Ins., Heimpt.*, p. 644, (1843), [Genotype *T. primulae* Haliday] Karny, *Zool. Ann.*, IV, p. 340, (1912). Hood, *Proc. Ent. Soc. Wash.*, XVI, pp. 36-38, (1914); *Ent.*, LVIII. p. 132, (1925), [Notes]. priesner, *Mus. Linz. Anm.* 4, p. (1920); *Die Thys. Europas*, p. 269, (1928), [References]. Knechtel, *Thys. Roman.*, p. 120, (1923). Steinweden, *Trans. Ent. Soc. Amer*, LIX, pp. 269-293, (1933), [Key to species].

balsaminae Priesner.

Taeniothrips balsaminae, Prisener, Ramakrishna, *Ind. For. Rec.*, XX, Pt. IV, p. 4, (1934), [On Sandal. Some distinguishing characters from other South Indian forms of the genus].[1]

[1]Denkanikota: (Salem Dt.).

brunneicornis (Bagnall).

Physothrips brunneicornis, Bagnall, *Ann. May. Nat. Hist.*, (8), XVII, p. 218, (1916)[1]; *Ann. Mag. Nat. Hist.*, (9), XIX, p. 206, (1918), [On rose].[2] Ramakrishna, *Ent. Mem. Dept. Agr. ind.*, X, p. 256, (1928).[3]

Taeniothrips brunneicornis, Ramakrishna & Margabandhu, *Rec. Ind. Mus.*, XLI, P. 24, (1939), [On *Lucerne*].[4]

[2]Darjiling.
[4]Poona: Bombay.

[1, 3]Japan.

carcamomi Ramakrishna

Taeniothrips cardamomi, Ramakrishna *Bull. Ent. Res.*, XXVI, pp. 357-358, (1935), [On diseased cardamom spikes and blossoms].[1]

[1]Valparai: (Anamalai hills, South India).

chaetogastra Ramakrishna.

Taeniothrips chaetogastra, Ramakrishna, *Rec. Ind. Mus.*, XXXVi, Pt. IV, pp. 494-495, (1934), [On flowers of the Persiannim, in company with *Dolichothrips indicus* Hood-pr].[1]

[1]Coimbatore.

distalis Karny.

Taeniothrips distalis, Karny, *Arch. Naturg.,* 79A, p. 122, (1913)[1]; *Pusa Ent. Mem. Dept. Agr. Ind.,* IX, (6), p. 196, (1926), [On lab-lab flower].[2] Ramakrishna, *Pusa Ent. Mem. Dept. Agr. Ind.,* X, (7), p. 256, (1928), [On lab-lab flowers, sannhemp flowers and *Pongamia glabra*].[3] Ramakrishna & Margabandhu, *Rec. Ind. Mus.,* XLI, p. 25, (1939), [Flowers of *Nepelium litchi* and on Lucerne].[4]

[2,3]Coimbatore. [3]Bangalore, Anakapalli, Samalkot. [4]Pusa, Poona: (Bombay).

---

[1]Japan.

gladioli Moultan & Steinweden.

Taeniothrips gladioli, Moulton & Steinweden, *Canad. Ent.,* LXIII, p. 20 (1931).[1] Herr, *Ohio Agric. Expt. Sta., Bull.,* No. 537, (1934).[2] Ramakrishna & Margabandhu (*Indian J. Ent. I,* (3), p. 39 (1939), [On galdiolus and other hill plants].[3]

[3]Coonoor: Nilgiris.

[1,2]N. America.

longistylus Karny.

Taeniothrips longistylus, Karny, *J. Siam Soc.,* XVI, p. 99, (1933).[1]; *Bull. Ent. Res.,* XVI, p. 125, (1925).[2]; *Pusa Ent. Mem. Dept. Agr. Ind.,* IX, (6), p. 196, (1926), [On Cajanus indicus, *Melia azadirach, Crotalaria juncea,* inside flowers of cow-pea, and Lucerne][3] Ramakrishna *Pusa Ent. Mem. Dept. Agr. Ind.,* X, (7), p. 258, (1928). [On groundnut flower].[4]; *Ind., For. Rec.,* XX, Pt. IV, p. 4, (1934), [On sandal shoots]. Moulton, *Rec. Ind. Mus.,* XXXI, p. 95, (1925).[5]

[3]Coimbatore, Taliparamba. [4]Bangalore. [5]Buldana: (Berar).

---

[2]Sunda Island, Indo-China. [1]Siam.

major Bagnall.

Taeniothrips major, Bagnall, *Ann. Mag. Nat. Hist.,* (8), XVII, p. 216, (1916), [On *Rhododendron* flowers].[1] Ramakrishna, *Pusa Ent. Mem. Dept. Agr. Ind.,* X, (7), p. 258, (1928).[2]

[1,2]Garhwal: (N. India over, 11,000 ft.).

nsitatus (Bagnall).

Physothrips usitatus Bagnall, *Ann. Mag. Nat. Hist.,* (8), XII, pp. 293-294, (1913), [In flowers of *Butea frondosa*].[1] Karny, *Zeit. f. wiss. Ins.,* X, p. 365, (1914). Ramakrishna, *Pusa Ent. Mem. Dept. Agr. Ind.,* X, p. 261, (1928).[2]

Frankliniella obscuricornis, Schumutz, *Sitz. Akad. wiss. Wien.,* CXXII, p. 1022, (1913) [On Crotoloria].[3]

[1,2]Allahabad. [3]Paradeniya.

nigricornis (Schmutz).

Frankliniella nigricornis, Schumutz, *Sitz. Akad wiss. Wien.* CXXII, p. 1020, (1913), [On coffee].[1] Preiesner, *Deuts. ent. Zeit.* p. 16, (1925).

Physothrips vsitatus (nec Bagnall), Karny, *Zeit. f. wiss.* Ins., X, p. 365, (1914). Ramakrishna, *J. Bombay Nat. Hist. Soc.*, XXX, pp. 865-866, (1925), [List].

Taeniothrips nigricornis, Ramakrishna & Margabandhu, *Rec. Ind. Mus.*, XLI, p. 25, (1939), [On *Mimosa pudica* flower].[2]

[1]Peradeniya. [2]Wynaad Hills: Malabar.

niloticus Priesner.

Taeniothrips niloticus, Preisner, *Bull. Soc. Roy. Ent. Egypte*, p. 11, (1930). Ramakrishna & Margabandhu, *Rec. Ind. Mus.*, XLI, p. 25, (1939), [On *Acacia arabica*].[2]

[2]Kollegal, Guntur.

---

[1]Egypt.

28. Physothrips Karny.

Physothrips, Karny *Zool. Annal.*, IV, p. 336, (1912), [Genotype *P. ulmifoliorum* (Haliday)]; *Zeit. f. Wiss. Ins.*, X, pp. 364-369, (1914), [Key to species]. Bagnall, *Ann. Mag. Nat. Hist.*, (9). XVIII; p. 101, (1926), [Generic distinction]. Hood, *Proc. Biol. Soc. Wash.*, XVI, p. (1914).

andrewsi Bagnall.

Physothrips andrewsi, Bangall, *Ann. Mag. Nat Hist.*, (9), VIII, pp. 394-395, (1921), [On rose][1]. Ramakrishna, *Pusa Ent. Mem. Dept. Agr. Ind.*, X, (7), p. 256, (1928)[2].

[1, 2]Darjiling.

flavidulus Bagnall.

Physothrips flavidulus, Bagnall, *Ann. Mag. Nat. Hist.*, (9), XII, p. 628, (1923), [In *Eriobotrys* and *Clematis* flowers][1]. Ramakrishna, *Pusa Ent. Mem. Dept. Agr. Ind.*, X, (7), p. 256, (1928). Moulton, *Rec. Ind. Mus.*, XXXVI, p. 95, (1929), [Pine trees][2].

[1]Dehra Dun, Bhowali: (Kumaon). [2]Simla Hills.

immsi Bagnall.

Physothrips immsi, Bagnall, *Ann. Mag. Nat. His.*, (9), XVIII, p. 106 (1926), [On Clematis sp][1]. Ramakrishna, *Pusa Ent. Mem. Dept. Agr. Ind.*, X, (7), p. 256, (1928)[2].

[1, 2]Bhowali: (Kumaon).

lefroyi Bagnall.

Physothrips lefroyi, Bagnall, *Ann. Mag. Nat. Hist.,* (8), XII, p. 292, (1913), [On flowers of tea][1].; *Bull. Ent. Res.,* IX, pp. 63-64, (1918)[2]; Ramakrishna, *Pusa Ent. Mem. Dept. Agr. Ind.,* X, (7) p. 258, (1928). Moulton, *Rec. Ind. Mus.,* XXXI, p. 95, (1929)[3]. — [1, 2]Lebong: (Darjiling). [3]Sitong: (Darjiling).

longiceps Bagnall.

Physothrips longiceps, Bagnall, *Ann. Mag.* Nat. Hist. (8), XVII, pp. 220-221, (1916), [On flowers of *Rhododendron*][1]. Ramakrishna, *Pusa Ent. Mem. Dept. Agr. Ins.,* X, (7), p. 258, (1928)[2]. — [1, 2]Garhwal: (11700 feet).

minor Bagnall.

Pysothrips minor. Bagnall, *Ann. Mag. Nat. Hist.,* (9), VIII, pp. 393-394, (1921), [On *Ipomea staphylina*][1]. Karny, *Pusa Ent. Mem. Dept. Agr. Ind.,* IX, (6), p. 196 (1926). Ramakrishna, *Pusa Ent. Mem. Dept. Agr. Ind.,* X, (7), p. 258, (1928)[2]. — [1, 2]Coimbatore, Maddur: (Mysore).

peculiaris Bagnall.

Physothrips peculiaris, Bagnall., *Ann. mag. Nat. Hist.,* I, pp. 206-207, (1918), [on *Luecerne*][1]. Ramakrishna, *Pusa Ent. Mem. Dept. Agr. Ind.,* X, (7), p. 258, (1928)[2]. — [1, 2]Pusa.

pingala (Ramakrishna).

Taeniothrips pingala, Ramakrishna, *Pusa Ent. Mem. Dept. Agr. Ind.,* X, (7), pp. 260-261, (1928), [On *Pongamia glabra*][1]. — [1]Bangalore.

setiventris Bagnall.

Physothrips setiventris, Bagnall, *Bull. Ent. Res.,* IX, pp. 61-63, (1918), [On tea].[1] Ramakrishna, *Pusa Ent. Mem. Dept. Agr. Ind.,* X, (7), p. 261, (1928), [In Sun-flower and also injurious to tea][2]. And Rews, *Rept. Third Ent. Meeting Pusa,* p. 621, (1920). — [1]Darjiling. [2]Pusa.

vittata (Schmutz).

Taeniothrips vittata, Schmutz, *Sitz., Akad. wiss. Wien.* CXXII, p. 1023, (1923), [In *Thunbergia flowers*][1]. — [1]Peradeniya.

physothrips (Taeniothrips) vittata, Karny, *Zeit. f. wiss.,* X, p. 365, (1914).

Odontothrips vittata, Priesner, *Deuts. ent. Zeits.* p. 18, (1925).

jonnaphila (Ramakrishna).

Taeniothrips jonnaphila, Ramakrishna, *Pusa Ent. Mem. Dept. Agr. Ind.,* X, (7), pp. 256-258, (1928), [Tassels of maize, tender shoots of *Sorghum* and sugarcane leaves][1]. — [1]Guntur, Samalkot, Coimbatore.

29. Megaphysothrips Ramakrishna & Margabandhu.

Megaphysothrips, Ramakrishna & Margabandhu, *Rec. Ind. Mus.,* XLI, p. 25, (1939). [Genotype *M. subramanii* Ram. & Marg.].

subramanii Ramakrishna & Margabandhu.

Megaphysothrips subramanii, Ramakrishna & Margabandhu, *Rec. Ind. Mus.,* XLI, pp. 25-26, (1939), [Freeding on Cofee rust][1]. — [1]Mysore.

30. Euphysothrips Bagnall.

Euphysothrips, Bagnall, *Ann. Mag. Nat.* Hist., (9), XVIII, p. 646, (1926). [Genotype *E. menozzi* Bagnall].

menozzi Bagnall.

Euphysothrips menozzi, Bagnall, *Ann. Mag. Nat. Hist.,* (9), XVIII, p. 646, (1926), [From flowers of *Clematis vitalba*][1].; *Ann. Mag. Nat. Hist.,* (9), XIX, p. 570, (1927). Priesner, *Bull. Soc. Roy. Ent. Egypte,* 1-3, p. 4, (1933), [On grasses between the vines][2]. Ramakrishna & Margabandhu, *Rec. Ind. Mus.,* XLI. p. 27, (1939), [On ears of cumbu, *Pennisetum typhoideum* and on *Lucerne*][3]. — [3]Coimbatore. Koilpatti.

---

[1]France.
[1]Egypt.

31. Projectothrips Moulton.

Projectothrips, Moulton, *Rec. Ind. Mus.,* XXXI, p. 95, (1929), [Genotype *p. pruthi* Moulton].

pruthi Moulton.

Projectothrips pruthi, Moulton, *Rec. Ind. Mus.,* XXXI, p. 96, (1929), [On *Pandanus*][1]. Ramakrishna & Margabandhu, *Rec. Ind. Mus.,* XLI, p. 28, (1939), [On *Pandanus*][2]. — [1]Sibpore: (Calcutta) [2]Coimbatore.

32. **Isoneurothrips** Bagnall.

Isoneurothrips Bagnall, *Ann. Mag. Nat. Hist.,* (8), XV, pp. 15-16, (1924), [Notes on the genus]. [Genotype *I. australis* Bagnall]. Prisener, *Natuur. Tidschr. Nederlandsch-Indie,* XCIV, p. 258, (1934), [As a subgenus of *Thrips*].

orientalis Bagnall.

Isoneurothrips orientalis, Bagnall, *Ann. Mag. Nat. Hist.,* (8), XV, p. 593, (1915)[1]. Karny, *Pusa Ent. Mem. Dept. Agr Ind.,* IX, (6), p. 197, (1926), [In flowers of *Morinda tinctoria,* and Jasmine][2]. Ramakrishna. *Pusa. Ent. Mem. Dept. Agr. Ind.,* X, (7), p. 261, (1928)[3]. Ramakrishna & Margabandhu, *Rec. Ind. Mus.,* XLI, p. 27, (1939), [On jasmine][4].

[1]Sarawak.
[2, 3]Coimbatore.
[3]Kollegal.
[4]Poona.

33. **Thrips** Linnaeus.

Thrips, Linnaeus, *Syst. Natur. (ed). XX,* p. 457, (1758). [Genotype *T. physapus* Linnaeus (designated by Westwood 1840)] Linnaues, *Fauna Suecica, (Ed).* I, p. 220, (1776). Uzel, *Mon. Ordnung Thys.,* p. 173, (1895). Hinds, *Proc. U. S. Nat. Mus.,* XXVI, p. 178, (1902). Karny, *Treubia,* I, 4, p. 241, (1921), Knechtel, *Thys. Roman.,* p. 165, (1923). Priesner, *Die Thys. Europas,* p. 343, (1928). [References]. Speyer, *Ann. Appl. Biol.,* XXi, pp. 120-152, (1934), [Common species of the genus dealt with].

albipes Bagnall.

Thrips albipes, Bagnall, *Ann. Mag. Nat. Hist.,* (8), XIII, p. 26, (1914), [On *Nasturtius*][1].; *Ann. Mag. Nat. Hist.,* (9), XVIII, p. 110, (1926), [In flowers of *Eriobotrya japonica* and in flower of *Clematis* sp.][2]. Ramakrishna, *Pusa Ent. Mem. Dept. Agr. ind.,* X, (7), p. 261, (1928)[3].

[2]Dehra Dun.
[3]Bhowal: (Kumaon).

[1, 3]Japan.

apicatus Priesner.

Thrips apicatus, Priesner, *Natuur. Tidschr Nederlandsch-Indie,* XCIV, p. 264, (1934), [In *Acacia* flower][1].

[1]Guntur.

florum Schmuts.

Thrips florum, Schumtz, *Sitz. Akad wiss. Wien.* CXXII, p. 1003, (1913), [A colour variety][1]. Karny, *Arkiv. Zool.,*

[3, 10]Coimbatore, Maddur: (Mysore),

XVII, p. 15, (1924), [Synonymy][2].; *Pusa Ent. Mem. Dept., Agr. Ind.*, IX, (6), p. 198, (1926), [In flowers of *Crotalaria juncea, Lantana,* Pomegranate, *Bignonia grandis, Hibiscus rosa-sinensis,* Jasmine, *Caesalpinia pulcherrima,* Banana, Lab-lab, in ears of *Pennisetum typhoideum* and on leaves of Indigo and on *Eugenia jambolana*][3]. Ramakrishna, *Pusa Ent. Mem. Dept. Agr. Ind.*, X, (7), p. 261, (1928), [In orange flowers, *Thevetia flowers,* banana flower sheaths, on rose, Divi-divi and *Corropita guianensis*].; *Rec. Ind. Mus.*, XXXIV, p. 277, (1932).; *Rec. Ind. Mus.*, XXXVI, p. 493, (1934), [From rose flowers]. Karny, *J. Siam, Nat. Hist. Soc.*, XVI, p. 109, (1923), [*magnipes = rhodamniae*]. Moulton, *Rec. Ind. Mus.*, XXXI, p. 98, (1929), Priesner, *Natuur. Tidschr. Nederlandsch-Indie,* XCIV, p. 261, (1934). [Synonymy]. Ramakrishna & Margabandhu, *Rec. Ind. Mus.*, XLI, p. 28, (1939), [On tea and rose flowers]; Indian J. Ent., I, (3), pp. 40-41 (1939), [Food-plants][11].

Nilgiris, Kadiam, Bangalore. [10]Buldana: (Berar). [1, 2, 4, 6, 8, 9, 10, 11]Peradeniya Lyallpur: Punjab.

---

[1]Queensland.

Thrips rhodamniae, Schumtz, *Sitz. Akad. wiss. Wien.* CXXII, p. 1008, (1913)[4]. Ramakrishna & Margabandhu, *Indian J. Ent.*, I, (3), p. 41, (1939), [Sweet-pea, lupin etc.]

Thrips magnipes, Schmut, *Sitz. Akad. wiss. Wien.* CXXII, 1006, (1913)[5].

Thrips longalata, Schmutz, *Sitz. Akad. wiss. Wien.* CXXII, p. 1009, (1913), [In Cinnmon and *Clerodendron* flower][6].

Thrips pallida, Schmutz, *Sitz., Akad. wiss. Wien.* CXXII, p. 1015, (1913)[7].

Thrips nigriflava, Schmutz, *Sitz. Akad. wiss. Wien.* CXXII, p. 1012, (1913), [In Cinnamon flowers][8].

Thrips colorata, Schmutz, *Sitz. Akad.wiss. Wien.* CXXII, p. 1013, (1913), [A color variety][9].

Thrips sulphurea, Schmutz, *Sitz. Akad. wiss. Wien.* CXXII, p. 1011, (1913), [In Cinnamon flowers][10].

gossypicola Priesner.

Thrips gossypicola, in Ramakrishna & Margabandhu, *India J. Ent.* I, (3), p. 41, (1939), [On gingelly and a papilionaceous flower][1].

[1]Coimbatore, Sholinghur: N. Arcot Dt.

immsi Bagnall.

Thrips immsi, Bagnall, *Ann. Mag. Nat. Hist.,* (9), XVIII, p. 110, (1926), [In flowers of *Eriobotrys japonica* and *Clematis* sp][1]. Ramakrishna, *Pusa Ent. Mem. Dept. Agr. Ind.,* X, (7), p. 262, (1928)[2]. — [1, 2]Dehra Dun, Bhowal: (Kumaon).

melaneurus Bagnall.

Thrips melaneurus, Bagnall, *Ann. Mag. Nat. Hist.,* (9), XVIII, (1926), [On rose][1]. Ramakrishna, *Pusa Ent. Mem. Dept. Agr. Ind.,* X, (7), p. 262, (1928)[2]. — [1, 2]Darjiling.

nilgiriensis Ramakrishna.

Thrips nilgiriensis, Ramakrishna, *Pusa Ent. Mem. Dept. Agr. Ind.,* X, (7), pp. 262-263, (1928), [In orange flowers][1]. — [1]Runnymede: (Nilgiris).

oryzae Williams.

Thrips oryzae, Williams, *Bull. Ent. Res.,* VI, pp. 353-355, (1915)[1]. Fletcher, *Rept. Second. Ent. Meeting Pusa,* p. 154, (1917), [Injurious to young rice plants]. Karny, *J. Siam Nat. Hist. Soc.,* XVI, pt. 2, p. 109, (1921)[2]. Ramakrishna, *Pusa Ent. Mem. Dept. Agr. Ind.,* X, (7), p. 263, (1928)[3]. — [1]Chinglepet. [3]Tanjore, Malabar, Kollegal, Assam.

---

[2]Java.

pallidulus Bagnall.

Thrips palidulus, Bagnall, *Ann. Mag. Nat. Hist.,* (9), XIII, pp. 424-425, (1924)[1]. Ramakrishna, *Pusa Ent. Mem. Dept. Agr. Ind.,* X, (7), p. 263, (1928)[2]. — [1, 2]Pusa.

palmi Karny.

Thrips palmi, Karny, *Bull. Deli. Proef st.,* No. 23, p. 10, (1925), [On tobacco]; *Pusa Ent. Mem. Dept. Agr. Ind.,* IX, (6), p. 199, (1926), [On lab-lab flowers][2] Ramakrishna, *Pusa Ent. Mem. Dept. Agr. Ind.,* X, (7), p. 263, (1928), [In *Lantuna* flowers][3]. — [2, 3]Coimbatore. [3]Nilgiris.

---

[1]Sumatra.

pandu Ramakrishna.

Thrips, pandu, Ramakrishna, *Pusa Ent. Mem. Dept. Agr. Ind.,* X, (7), pp. 264-265, (1928), [From flowers of *Citrus*][1]. — [1]Coimbatore.

parvus Schmutz.

Thrips parvus, Schmutz, *Sitz., Akad. wiss. Wien.* CXXII, p. 1004, (1913), [In cinnamon flower][1]. Karny, *Pusa* — [1]Ceylon. [2, 3]Coimbatore.

Ent. Mem. Dept. Agr. Ind., IX, (6), p. 199, (1926), [On *Caesalpionia pulcherrima* flowers][2] Ramakrishna, *Pusa Ent. Mem. Dept. Agr. Ind.*, X, (7), p. 265, (1928)[3]. Priesner, *Natuur, Tidschr. Nederlandsch-Indie*, XCIV, pp. 268-269, (1934), [Brief description]. [From flowers of *Thunbergia grandiflora, Dendrobium crumenatum. Bauhinia tomentosa, Elaegnus latifolia* and *Beaumonita multiflora.*][4].

---

[4]Sumatra, Java.

setosus Moulton.

Thrips setosus, Moulton, *Rec. Ind. Mus.*, XXXI, p. 97, (1929).[1]

[1]Puri: (Orissa).

striatopennata Schmutz.

Thrips striatopennata, Schmutz, *Sitz. Akad. wiss. Wien.* CXXII, p. 1002, (1913)[1].

[1]Nuwara Eliya: (Ceylon).

subnudula Karny.

Thrips subnudula, Karny, *Ind. For. Rec.*, XX, pp. 4-5, (1934), [On sandal shoots][1]; Ramakrishna & Margabandhu, *Indian J. Ent.*, I, (3), p. 41, (1939), [On orange flowers][4].

Ramaswamiaheilla subnudula, Karny, *Pusa Ent. Mem. Dept. Agr. Ind.*, IX, (6), pp. 208-210, (1926), [From flowers of *Calotropis, Illecebrum lantanum, Caesalpinia pulcherrima,* mango, pomegranate and *Hibiscus rosa-sinensis*][2]. Ramakrishna, *Pusa Ent. Mem. Dept. Agr. Ind.*, X, (7), p. 266, (1928)[3].

[1]Bangalore.
[1, 2, 3]Coimbatore.
[4]Kollegal.

tabaci Lindeman.

Thrips, tabaci, Lindeman, *Die schadl. Ins. Tabak.*, p. 15, (1888)[1].; Hinds, *Proc. U.S. Nat. Mus.*, XXVI, pp. 179-184, (1902), [Synonymy and redescription]. Karny, *Pusa, Ent. Mem. Dept. Agr.*, IX, (6), p. 199, (1926), [In flowers and shoots of cotto and on cabbage levaes][2]. Ramakrishna, *Pusa Ent. Mem. Dept. Agr. Ind.*, X, (7), p. 265, (1928), [wild bush and onion][3]; *Agr. & Livestk. Ind.*, II, Pt. IV, pp. 394-395, (1932). [Bionomics]. Priesner, *Die. Thys. Europas*, pp. 433-434, (1928), [complete and exhaustive list of references]. Ramakrishna & Margabandhu, *Rec. Ind. Mus.*, XLI, p. 28, (1939), [On daisy flower][4]; *Indian J. Ent.*, I. (3), p. 41, (1939), [On tea].[5]

[2, 3]Coimbatore.
[4, 5]Ootacamund (Nilgiri Hills).

Bessarabia.

34. **Oxyrrhinothrips** Priesner.

Thrips (Oxyrrinothrips), Priesner, *Konowia,* XII, p. 312, (1933), [Genotype *O. scolopax* (Priesner)].

beharensis Ramakrishna & Margabandhu.

Oxyrrhinothrips beharensis, Ramakrishna & Margabandhu, *Rec. Ind. Mus.,* XLI, pp. 29-30, (1939), [On mango][1]. — [1]Pusa: Bihar.

rostrata Ramakrishna & Margabandhu.

Oxyrrhinothrips rostrata, Ramakrishna & Margabhandu *Rec. Ind. Mus.,* XLI, p. 30, (1939), [On cotton][1]. — [1]Coimbatore.

35. **Fulmekiola** Karny.

Fulmekiola, Karny, *Deli Proefsta.,* No. 23, p. 18, (1925). [Genotype F. *interrupta* Karny].

indica Ramakrishna & Margabandhu.

Fulmekiola indica, Ramakrishna & Margabandhu, J. *Bombay Nat. Hist. Soc.,* XXXIV, No. 4, pp. 1034-1035, (1931), [On sugarcane leaves][1]. — [1]Samalkota: (Godavery Dt.).

saccharicida Ramakrishna & Margabandhu. Fulmekiola saccharicida, Ramakrishna & Margabandhu *Rec. Ind., Mus.,* XLI, pp. 23-24, (1939), [On sugarcane leaves].[1] — [1]Coimbatore.

Sub-family VI APTINOTHRIPINAE KARNY.

36. **Aptinothrips** Haliday.

Aptinothrips, Haliday, *Ent. Mag.,* III, p. 445, (1836), [Genotype A. *rufa* Haliday]. Hinds, *Proc. U.S. Nat. Mus.,* XXVI, p. 166, (1902), [Generic characters]. Speyer, *Trans. R. Ent. Soc., Lond.,* 83, pp. 483-508 (1935), [Revision, habitat and distribution].

rufus var **connaticornis** Bagnall.

Aptinothrips rufus var *connaticornis,* Bagnall, *Ann. Mag. Nat. Hist. (9),* I, p. 205, (1918). [In tea flower][1]. Ramakrishna, *Pusa Ent. Mem. Dept. Agr. Ind.,* X, (7), p. 267, (1928)[2]. Speyer, *Proc. R. Ent. Lond.,* (A), XI, p. 120, (1936), [Larva]. — [1, 2]Darjeeling.

[1]Europe, N. America.

Subfamily VII MYCTEROTHRIPINAE Karny.

37. **Rhamphothrips** Karney.

Rhamphothrips, Karny, *Bull. Jard. Bot. Buit.* X, p. 123, (1913), [Genotype *R. tenuirostris* Karny]. *Zeit. wiss. Ins.* pp. 294-295, (1914). *[Key to species* and synonymy].

Rhynchothrips, Karny, *Zool Anz.,* p. 297, (1912), (*nec Rhynchothrips* Hood.)

Brachythrips, Schmutz, *Seitz. Akad. wiss. Wien.* CXXII, p. 997, (1913), (nec *Brachythrips* Reuter).

lineata (Schmutz).

Brachythrips lineata, Schmutz, *Sitz, Akad. wiss.* Wien. CXXII, p. 997, (1913). *Rhamphothrips lineata,* Karny, *Zeit. wiss* Ins. X, p. 295, (1914). — Peradeniya.

38. **Mycterothrips** Trybom.

Mycterothrips, Trybom, *Ent. Tidj.,* p. ?, (1912), [Genotype *M. laticauda* Trybom.] Karny, *Pusa Ent. Mem. Dept. Agr. Ind.,* IX, (6), p. 199, (1926), [Key to species].

Priesner, *Bull, de la Soc. Royale Ent. Egypte,* 6. p. (1932), [Key to species].

pseudosetiprivus Ramakrishna & Margabandhu.

Mycterothrips pseudosetiprivus, Ramakrishna & Margabandhu, *Indian J. Ent.,* I, (3), pp. 42-43, (1939), [On brinjal][1]. — [1]Coimbatore.

setiprivus Karny.

Mycterothrips setiprivus, Karny, *Pusa Ent. Mem. Dept. Agr. Ind.,* IX, (6), p. 199, (1926). [In shoots of a wild legume][1]. Ramakrishna, *Pusa Ent. Mem. Dept. Agr. Ind.,* X, (7), p. 267, (1928), [In *Acacia suma]*[2]. — [1, 2]Coimbatore.

Subfamily *VIII BELOTHRIPINAE* Karney.

39. **Bregmatothrips** Hood.

Bregmatothrips, Hood, *Proc. Biol. Soc. Wash.,* XXV, p. 66, (1912), [Genotype *B. venustus* Hood].

binervis Kobus.

Bregmatothrips, binervis, Kobus, Karny, *Pusa Ent. Mem. Dept. Agr. Ind.,* IX, (6), p. 205, (1926), [Notes][1]. — [2]Palur: (S. Arcot Dt.).

Ramakrishna, *Pusa Ent. Mem. Dept. Agr. Ind.*, X, (7), p. 267, (1928)[2]. [1]Java.

ramakrishnae Bagnall.

Bregmatothrips ramakrishnae, Bagnall, *Ann. Mag. Nat. Hist.*, (9), XII, p. 625, (1933), [On leaf tips of sugarane][1]. Ramakrishna, *Pusa Ent. Mem. Dept. Agr. Ind.*, X, (7), p. 267, (1928)[2]; *Agr. & Livestk. Ind.*, II, Pt. II, p. 397, (1932). [1, 2]Palur.

40. **Gnomonothrips** Ramakrishna & Margabandhu.

Gnomonothrips, Ramakrishna Marga-bandhu, *Rec. Ind. Mys.*, XLI, p. 28, (1939), [Genotype *G. coimbatorensis* Ram. & Marg.]

coimbatorensis Ramakrishna & Marga- bandhu. *Gnomonothrips coimba-torensis,* Ramakrishna & Marga- bandhu, *Rec. Ind. Mus.*, XLI, pp. 28-29, (1939), [On *Cyperus bulbosus*][1]. [1]Coimbatore.

41. **Perissothrips** Hood.

Perissothrips, Hood, *Insec. Inscit. Menstr.*, VII, p. 91, (1919), [Genotype *P. parviceps* Hood].

parviceps Hood.

Perissothrips parviceps, Hood, *Insec. Inscit. Menstr.*, VII, p. 92, (1919)[1]. Karny, *Pusa Ent. Mem. Dept. Agr. Ind.*, IX, (6) p. 204, (1926). [On *Ailanthus excelsa* and *Caesalpinia pulcherrima* flowers][2]. Ramakrishna, *Pusa Ent. Mem. Dept. Agr. Ind.*, X, (7), p. 269, (1928)[3]. [1, 2, 3]Coimbatore.

42. **Stylotrips Karny.**

Stylothrips, Karny, *Pusa, Ent., Mem. Dept. Agr. Ind.*, IX, (6), p. 205, (1926), [Genotype, *S. brevipalpis* Karny].

brevipalpis Karny.

Stylothrips, brevipalpis, Karny, *Pusa Ent. Mem. Dept. Agr. Ind.*, IX, (6), (1926), [In *Chrysanthemum flowers*][1]. Ramakrishna, *Pusa Ent. Mem. Dept. Agr. Ind.*, X, (7), p. 266, (1928), [Karny's '*Stylothrips*' sunk as a synonym of '*Microcephalothrips*' Bagnall][2]. [1]Coimbatore. [2]Coonoor.

43. **Brachyurothrips** Bagnall.

Brachyurothrips, Bagnall, *Ann. Mag. Nat. Hist.,* (9), VII, p. 265, (1921), [Genotype *B. anomalous* Bagnall].

indicus Bagnall.

Brachyurothrips indicus, Bagnall, *Ann. Mag. Nat. Hist.,* (9), XVIII, p. 98, (1926)/Ramakrishna, *Pusa Ent. Mem. Dept. Agr. Ind.,* X (7), p. 271, (1928)[2]. — [1, 2]Calcutta.

Family II PANCHAETOTHRIPIDAE Bagnall

44. **Panchaetothrips** Bagnall.

Panchaetothrips, Bagnall, *Rec. Ind. Mus.,* VII, p. 256, (1912), [Genotype *P. indicus* Bagnall].

indicus Bagnall.

Panchaetothrips, indicus, Bagnall, *Rec. Ind. Mus.,* VII, pp. 257-260, (1912), [On Turmeric.]/Ramakrishna, *Rept. Thrid Ent. Meeting,* p. 327; p. 619, (1920)[2]; *Pusa Ent. Mem. Dept. Agr. Ind.,* X, (7), p. 273, (1928), [In turmeric leaves, arrowroot leaves and banana][3]; *Agr. & Livestk, in Ind.* II, pt. IV, pp. 399-400, (1932), [Bionomics]; *Rec. Ind. Mus.,* XXXVI, pt. IV, p. 495, (1934), [One solitary female from cotton flower). — [1]Madras. [2]Coimbatore. [2, 3]Malabar. [3]Travancore, Pusa.

SUBORDER **II TUBULIFERA** Haliday.

Superfamily **I** PHLOEOTHRIPOIDEA Hood.

Family I ECACANTHOTHRIPDAE Bagnall.

45. **Ecacanthothrips** Bagnall.

Ecacanthothrips, Bagnall, *Trans. N.H.S. North,* III, p. 14, (1910) , [Genotype *A. sanguineus* Bagnall.] *Ann. Mag. Nat. Hist.,* (9), VII, p. 290, (1921), [Discussed]. Hood, *Stylops,* IV, Pt. 9, p. 195, (1935), [Systematic position and key to species.] ISHID, *Ins. Map.,* VIII. pp. 147-148, (1934), [Distribution and key.]

Acanthothrips, Bagnall, *Ann. Mag. Nat. Hist.,* (8), I, p. 361, (1908) Hood, *Bull. Ill, State Lab. Nat. Hist.,* VII, *Article II,* p. 374, (1908), [Key to species).

fletcheri Ramakrishna.

Ecacanthothrips Fletcheri, Ramakrishna, *Rec. Ind. Mus.,* XXXVI, Pt. IV, pp. 495-496, (1934), [At light.][1]. — [1]Pusa.

sanguineus Bagnall.

Acanthothrips sanguineus, Bagnall, *Ann. Mag. Nat. Hist.,* (8), I, p. 302, (1908)[1].

Ormothrips sanguineus, Buffa, *Redia,* V, p. 167, (1910)[2].

Ecacanthothrips sanguineus, Bagnall, *Trans. N. H. S.,* III, p. 14, (1910); *Rec. Ind. Mus.,* VIII, p. 201, (1913)[3]. Karny, *Treubia,* III, p. 277, (1923), [References][4]; *Bull. Deli. Profesta.* No. 23, p. 37, (1925)[5]. Ramakrishna, *Pusa Ent. Mem. Dept. Agr. Ind.,* X, (7), p. 274, (1928), [Under bark].[6] Priesner, *Treubia,* XI, pp. 357–362, (1930), [Relationship]. Hoop, *Stylops,* IV, Pt. 9, p. 195, (1935), [Distinguished from other species.]

[2, 4, 6]Ceylon.
[6]Dehra Dun, Central India, Bengal, Burma.

---

[1]Dorey (N. Guinea).
[4]N. Guinea, Saleyer, Java, Engano, Mantawei, Nias, Formosa.
[4, 5]Sumatra.
[6]Malayan Archipelago.

steinskyi (Schmutz).

Ormoth rips steinskyi, Schmutz, *Sitz. Akad. wiss. Wien.* CXXII, p. 1028, (1913)[1]

Ecacanthothrips steinskyi, Priesner, *Treubia,* XI, pp. 357–362, (1930), [Relationships.] Hood, *Stylops,* IV, Pt. 9, p. 195, (1935), [Distinguished from other species of the genus]. Ramakrishna and Margabandhu, *Rec. Ind. Mus.,* XLI, p. 30, (1939), [On cotton][2].

[1]Ceylon.
[2]Mysore.

Family II PHLOEOTHRIPIDAE Uzel.

Sub-family I PHLOEOTHRIPINAE Karny.

46. Phloeothrips Haliday.

Phloeothrips, Haliday, *Ent. Mag.,* p. 442, (1836), [*Genotype P. coriaceus* Haliday]. Uzel, *Mon. Ordnung Thys.,* p. 254. (1895). Hinds, *Proc. U.S. Nat. Mus.,* XXVI, p. 195, (1902), Karny , *Treubia,* I, 4, p. 246, (1921). Watson, *Florida Agri. Exp. Sta., Bull.* 168, p. 16, (1923). Knechtel., *Thys. Roman,* p. 234, (1923). Priesner, *Tijdschr. v. Ent.,* LXVI, pp. 96–104), (1923); *Die Thys. Europas,* p. 636, (1928), [For complete list of references.]

anacardii Newman.

Phloeothrips anacardii, Newman, *Trans. Ent. Soc. Lond.,* p. 266, (1855)[1]. Ramakrishna, *Pusa Ent. Mem. Dept. Agr. Ind.,* X, (7), p. 275, (1928.)[2]

[1, 2]Mysore.

indicus Ramakrishna & Margabandhu.

Phloeothrips indicus, Ramakrishna and Margabandhu, *Indian J. Ent.* I., (3), p.p. 43-44, (1939), [On cocoanut leaf][1]. [1]Coimbatore.

stenomelas Walker.

Phloeothrips stenomelas, Walker, *Ann. Mag. Nat. Hist.*, (), ? p. 224, (1859)[1]. [1]Ceylon.

47. **Thilakothrips** Ramakrishna.

Thilakothrips, Ramakrishnas, *Pusa Ent. Mem. Dept. Agr. Ind.*, X, (7), p. 275, (1928), [Genotype *T. babuli* Ramakrishna.]

babuli Ramakrishna.

Thilakothrips babuli, Ramakrishna, *Pusa Ent. Mem. Dept. Agr. Ind.*, X, (7), pp. 275-279, (1928), [From leaf-galls of *Acacia leucophloea*][1]. [1]Bangalore, Tanjore, Delhi

48. **Malacothrips** Binds.

Malacothrips, Hinds, *Proc. U.S. Nat. Mus.*, XXVI, p. 200, (1902), [Genotype *M. zonatus* Hinds]. Karny, *Treubia*, I, 4, p. 246, (1921). Watson, *Florida Agr. Exp. Sta., Bull.* 168, p. 16, (1923). Priesner, *Die Thys. Europas*, p. 670, (1928).

hispanicus Bagnall.

Malacothrips hispanicus, Bagnall, *Ann. Mag. Nat. Hist.*, (9), XIV, p. 638, (1924). [1]Garhwal (11,000 ft.)

Cephalothrips hispanicus, Bagnall, *Ann. Mag. Nat. Hist.*, (9), XII, p. 129, (1923), [From flowers of *Rhododendron*][1]. Ramakrishna, *Pusa Ent. Mem. Dept. Agr., Ind.*, X, (7), p. 279, (1928).

Subfamily II.–LIOTHRIPINAE Karny.

49. **Arrhenothrips** Hood.

Arrhenothrips, Hood, *Insec. Inscit, Menstr.*, VII, p. 99, (1919), [Genotype *A. ramakrishnae* Hood].

dhumrapaksha Ramakrishna.

Arrhenothrips dhumrapaksha, Rama-krishna, *Pusa Ent. Mem. Dept. Agr. Ind.*, X, (7), pp. 280-281, (1928), [In leaf-galls of *Ficus retusa*][1]. [1]Mysore.

ramakrishnae Hood.

Arrheno thrips ramakrishnae, Hood, *Insec. Inscit. Menstr.,* VII, pp. 99-101, (1919), [On *Mimusops*][1]. Ramakrishna, *Proc. Third Ent. Meeting,* p. 327, p. 619, (1920)[2]; *Pusa Ent. Mem. Dept. Agr. Ind.,* X, (7), p. 282, (1928)[3]. [1, 2, 3, 4, 5]Coimbatore.

Mesothrips ramakrishnae, Bagnall, *Ann. Mag. Nat. Hist.,* (9), XIV, p. 635, (1924)[4]. Karny, *Pusa Ent. Mem. Dept. Agr. Ind.,* IX, (6), p. 210, (1926), [Karny does not agree with Bagnall in identifying *Arrhenothrips* with *Mesothrips* giving reasons and retains the former][5].

50. **Eurhynchothrips** Bagnall.

Eurhynchothrips, Bagnall, *Ann. Mag. Nat. Hist.,* (9), I, p. 213, (1918), [Genotype *E. convergens* Bagnall].

ordinarius (Hood).

Liothrips ordinarius, Hood, *Insec. Inscit. Menstr.,* VII, pp. 101-103, (1919), [On *Sesbania*][1]. [1, 2, 3]Coimbatore.

Eurhynchathrips ordinarius, Karny, *Pusa Ent. Mem. Dept. Agr. Ind.,* IX, (6), p. 210, (1926), [In flowers of *Sesbania grandiflora*][2]. Ramakrishna, *Pusa Ent. Mem. Dept. Agr. Ind.,* X, (7), p. 282, (1928)[3].

51. **Rhynchothrips** Hood.

Rhynchothrips, Hood, *Proc. Ent. Soc. Wash.,* XIV, p. 141, (1912), [Genotype *R. pruni* Hood]. Karny, *Treubia,* I, 4, p. 246, (1921). Watson, *Florida Agr. Expt. Sta., Bull.* 168, p. 16 (1923). Priesner, *Die Thys. Europas,* p. 518, (1928).

Liophloeothrips, Priesner, *Sitz., Akad. Wiss Wien.* CXXVIII, p. 138, (1919); *Ent. Zeit. Frank* O. M. XXXVII, pp. 5-7, (1924), [Synonymy] Karny, *Treubia,* I, 4, p. 246, (1921).

champakae Ramakrishna & Marga-bandhu.

Rhynchothrips champakae , Ramakrishna and Margabandhu, *Indian J. Ent.,* I, (3), pp. 44-46, (1939), (On *Michelia Champaca*][1]. [1]Kalimpong, Bengal.

palasae Ramakrishna.

Liothrips palasae, Ramakrishna, *Ent. Mem. Dept. Agr. Ind.,* X, p. 287, (1928), [In steam-galls of *Butena frondosa.*][1] — [1]Kadiam, Godavari Dt. [2]Sholinghur, N. Arcot Dt.

Rhynchothrips palasae, Ramakrishna and Margabandhu, *Indian J. Ent.,* I, (3), p. 46, (1939), [On a papilionaceous flower][2].

pallipes Karny.

Rhynchothrips pallipes, Karny, *Pusa Ent. Mem. Dept. Agr. Ind.,* IX, (6), p. 211, (1926), [Inside psyllid galls on *Terminalia* leaves][1]. Ramakrishna, *Pusa Ent. Mem. Dept. Agr. Ind.,* X, (7), p. 282, (1928)[2]. — [1, 2]Travancore.

raoensis Ramakrishna.

Rhynchothrips raoensis, Ramakrishna, *Pusa Ent. Mem. Dept. Agr. Ind.,* X, (7), pp. 282-283, (1928), [From leaf-galls of *Mallotus philippinensis* and in flowers of *Anacardium occidentale*][1]. — [1]Taliparamba.

vichitravarna Ramakrishna.

Rhynchothrips vichitravarna, Ramakrishna, *Pusa Ent., Mem. Dept. Agr. Ind.,* X, (7), pp. 283-285, (1928), [On *Eugenia jambolana* and *Mimusops elengi.*][1]. — [1]Bangalore, Coimbatore.

52. Liothrips Uzel.

Liothrips Uzel, *Mon, Ordnung. Thys.,* p. 261, (1895). [Genotype *L. hradecensis* Uzel]. Karny, *Trans. Ent. Soc. Lond.,* p. 471, (1912); *Treubia,* I, 4, p. 246, (1921), Knechtel, *Thys. Roman.* p. 230, (1923). Watson, *Florida Agr. Exp. Sta., Bull.* 168, p. 16. (1923). Priesner, *Treubia,* VIII, *Supplem.,* p. 217 (1926) ; *Die Thys.* Europas, p. 500, (1928).

Hoodia, Karny, *Mitt. Nat. Ver. Univ. Wien.* p. 41, (1910); *Treubia,* I, 4, p. 255, (1921). Knechtel, *Thys. Roman,* p. 203, (1923). Watson, *Florida Agr. Exp. Sta.,* Bull. 168, p. 22, (1923).

bosei Moulton.

Liothrips bosei, Moulton, Ind. For. Rec., XIII, pp. 2-3, (1928), [From galls of *Mallotus philippinensis*][1]. — [1]Dehra Dun.

brevitubus Karny.

Liothrips brevitubus, Karny, *Marcellia,* XI, p. 157, (1912), [From galls of *Mallotus repandus*][1]. Moulton, *Bull. Brook. Ent. Soc.,* XXII, p. 196, (1927), [In the leaf-galls of *Mallotus repandus* and *M. philippinensis*][2].

[2]Dehra Dun.

[1]Java.

dampfi Karny.

Liothrips lampfi, Karney, *Verh, zool. bot. Ges. Wien* LXIV, p. 58, (1921), Ramakrishna, *Rec. Ind. Mus.* XXXVI, p. 496, (1934), [On *Tamarix gallica*][1].

[1]Cawnpore.

kiriti Ramakrishna.

Liothrips kirti, Ramakrishna, *Pusa Ent. Mem. Dept. Agr. Ind.,* X, (7), pp. 285-286, (1928), [On mango leaf][1].

[1]Coimbatore.

sp. Kieffer.

Liothrips sp. Kieffer, *Marcellia,* p. (1905), [In galls on *Quercus apicata*][1].

[1]N. India.

53. **Dolichothrips** Karny.

Dolichothrips, Karny, *Zool. Anz.,* XI, p. 289, (1912), [Genotype *D. longicollis* Karny]. Bangall, *Ann. Mag. Nat. Hist.,* (9), VII, p. 359, (1921). Karny *Treubia,* I, 4, p. 247, (1921).

ochripes Karny.

Dolichothrips ochripes, Karny, *Pusa Ent. Mem. Dept. Agr. Ind.,* IX, (6), p. 213, (1926), [In shoots of *Cassia tora*][1]. Ramakrishna, *Pusa Ent. Mem. Dept. Agr. Ind.,* X, (7), p. 288, (1928). [In tender shoots of *Cassia auriculata*][2].

[1]Coimbatore.
[2]Guntur.

gracilipes Ramakrishna & Margabandhu.

Dolichothrips (Dolicholepta) gracilipes, Ramakrishna and Margabandhu, *Indian J. Ent.,* I, (3), pp. 46-47, (1939), [*Abutilon indicum*][1].

[1]Pattupur, Ganjam, Dt.

Subfamily III. HAPLOTHRIPNAE Karny

54. **Neoheegeria** Schmutz.

Neoheegeria, Schmutz, *Ann. Nat. Mus. Wien,* p. 344, (1909), [Genotype *N. dalmatica* Schmutz]. Karny, *Treubia,* I, 4, pp. 247-249 (1909). Knechtel, *Thys. Roman,* p. 229, (1923). Watson, *Florida Agr. Exp. Sta., Bul.*

168, pp. 17-18, (1923). Priesner, *Die Thys. Europas,* p. 628, (1928).

citripes Bagnall.

Neoheegeria citripes, Bagnall, *Ann. Mag. Nat. Hist.,* (9), VII, pp. 360-361, (1921), [On *Abutilon indicum*][1], Ramakrishna, *Pusa Ent. Mem. Dept., Agr. Ind.,* X, (7) p. 288, (1928)[2]. — [1]Pusa. [2]S. India.

fumipennis Bagnall.

Neoheegeria fumipennis, Bagnall, *Ann. Mag. Nat. Hist.,* (9), VII, p. 360, (1921)[1]. Ramakrishna, *Pusa Ent. Mem. Dept. Agr. Ind.,* X, (7), p. 288, (1928)[2]. — [1, 2]Kurseong.

hrasvamukha Ramakrishna.

Neoheegeria hrasvamukha, Ramakrishna, *Pusa Ent. Mem. Dept. Agr. Ind.,* X, (7), pp. 288-289, (1928), [On *Mallotus philippinensis*][1]. — [1]Malabar.

indica Hood.

Neoheegeria indica, Hood, *Insec. Inscit. Menstr.* VII, pp. 96-98, (1919), [On *Ailanthus*][1]. Karny, *Pusa Ent, Mem. Dept. Agr. Ind.,* IX, (6), p. 215, (1926), [On *Ailanthus,* in shoots of Persian nim and on *Cassia emarginata*][2].

Dolichothrips varipes, Bagnall, *Ann. Mag. Nat. Hist.,* (9), VII, 359, (1921)[3] Karny, *Pusa Ent, Mem. Dept. Agr. Ind.,* IX, (6), p. 215, (1926)[4], Ramakrishna, *Pusa Ent. Mem. Dept. Agr. Ind.,* X, (7), p. 289, (1928), [On a wild plant and in schoots of *Cochlospermum gossypium*][5]. — [1, 2, 3, 4]Coimbatore. [5]Malabar, Bangalore.

zizyphi Bagnall.

Neoheegeria zizyphi, Bagnall, *Ann. Mag. Nat. Hist.,* (9), XII, p. 629, (1923)[1]. Ramakrishna, *Pusa Ent, Mem. Dept. Agr. Ind.,* X, (7), p. 290, (1928)[2]. — [1, 2]Paresnath (Bengal) 4,400 feet.

55. **Hindsiana** Karny.

Hindsiana Karny, *Mitt. Nat. Ver. Univ. Wien.* VIII, p. 51, (1910), [Genotype *H. flavicinctus* Karny]; *Treubia,* I, 4, p. 248, (1921). Priesner, *Die Thys. Europas,* p. 622, *Haplothrips* (*Hindsiana*), (1928).

apicalis Bagnall.

Hindsiana apicalis, Bagnall, *Ann. Mag. Nat. Hist.,* (8), XV, p. 323, (1915), [Swept from a jungle plant][1]. Ramakrishna, *Pusa Ent., Mem. Dept. Agr. Ind.,* X, (7), p. 290, (1928)[2]. — [1, 2]Almora, 5,500 ft.

Haplothrips var *veroniae,* Priesner, Ramakrishna, *Pusa, Ent. Mem. Dept. Agr. Ind.,* X, (7), p. 291, (1928), [In some wild flowers][3]. — [3]Coimbatore.

56. **Zygothrips** Uzel.

Zygothrips, Uzel, *Mon Ordnung Thys.,* p. 243, (1895), [Genotype Z. *minuta* Uzel]. *Haplothrips (Zygothrips),* Priesner, *Ent. Zeitschr. Frankfurt,* XXXVII, Nr. 21/22, (1924); *Die Thys Europas,* p. 625, (1928).

andhra Ramakrishna.

Zygothrips andhra, Ramakrishna, *Pusa, Ent. Mem. Dept. Agr. Ind.,* X, (7), pp. 290-291. (1928). [On cholam shoots]. — Guntur.

57. **Haplothrips** Amyot & Serville.

Haplothrips, Serville, *Hist. Ins. Hemipt.,* p. 640, (1843), [Genotype *H. albipennis* (Burm.)]. Karny, *Zool. Annal.,* IV, p. 324, (1912); *Treubia,* I, 4, p. 248, (1921); *Treubia,* II, 1, pp. 22-36, (1921). Hood, *Mem. Queens, Mus.,* VI, p. 126, (1918), [Synonymy and notes]. Priesner, *Treubia,* II, 1, pp. 1-20, (1921) [Brief account of the genus]; *Rec. Ind. Mus.,* XXXV, pp. 347-369 (1933), [Indomalayan species]. Knechtel, *Thys. Roman.,* p. 209, (1923). Watson, *Florida Agr. Exp. Sta., Bull.* 168, p. 15, 18, (1923). Priesner, *Die Thys. Europas,* p. 564, (1928).

Anthothrips, Uzel, *Mon. Ordnung, Thys.,* p. 237, (1895). Hinds, *Proc. U.S. Nat. Mus.,* XXVI, p. 188, (1902).

Zygothrips, Uzel, *Mon. Ordnung. Thys.,* p. 243, (1895).

Hindsiana, Karny, *Mitt. Nat. Ver, Univ., Wien,* VIII, p. 51, (1921).

ceylonicus Schmutz.

Haplothrips ceylonicus, Schmutz, *Sitz. Akad. wiss. Wien.* CXXII, p. 1038, (1913), [*Crotalaria* flowers][1]. Priesner, *Treubia,* — [1]Peradeniya. [2]Java.

II, 1, p. 7 (1921); *Rec. Ind. Mus.*, XXV, 358, (1933), [References and notes]. Karny, *Deli Priefsta. Medan*, No. 23, p. 44, (1925)[2]; Karny, *Pusa Ent. Mem. Dept. Agr. Ind.*, IX, (6) p. 217, (1926)[3]. Ramakrishna, *J. Bomb. Nat. Hist. Soc.*, XXX, p. 868, (1925); *Pusa Ent. Mem. Dept. Agr. Ind.*, X, (7), p. 291, (1928)[4]. Moulton, *Rec. Ind. Mus.*, XXXI, p. 98, (1929)[5].

[3]Coimbatore.
[4]Kadiam, Anakapalli.
[4]Bangalore.
[5]Buldana (Berar).

inquilinus Priesner.

Haplothrips inquilinus, Priesner, *Treubia*, II, 1, pp. 4, 6, (1921)[1]; *Treubia*, VIII, *Supplem.*, pp. 95, 96, 245, 252, (1926); *Rec. Ind. Mus.*, XXV, p. 349, (1933), [References]. Karny, *Treubia*, III, p. 309, (1923); *J. Siam, Soc.*, XVI, p. 112, (1923)[2]; Pusa Ent. Mem. Dept. Agr. Ind., IX, (6), p. 216, (1926)[3]; Van Leeuwen *Zoocecid. Nederl. East Indies*, pp. 212, 1431, (1926). Ramakrishna, *Pusa Ent. Mem. Dept. Agr. Ind.*, X, (7), p. 292, (1928)[4]; *Rec. Ind. Mus.*, XXXVI, p. 496, (1934), [On *Mallotus philippinensis* with a *Liothrips*].

[4]Nellore, Coimbatore.
[1, 2, 4]Java.

---

[2]Celebes, Southern Siam.

Haplothrips aculeatus, Karny, *Marcellia*, XI, p. 121, (1912). Karny & Van Leeuwen, *Bull. Jard. Bot. Buitenzorg*, (2), X, pp. 7, 19, 65, (1913); *Zeitschr. wiss. Ins. Biol.*, XI, p. 87, (1915).

ganglbaueri Schmutz.

Haplothrips ganglbaueri, Schmutz, *Sitz. Akad, wiss. Wien*, p. 1034, (1913), Preisner, *Treubia*, II, 1, p. 2, 7, (1921); *Rec. Ind. Mus.*, XXV, p. 355, (1933), [References][2]. Karney, *Treubia*, II, pp. 2-4, (1921); *Pusa Ent. Mem. Dept. Agr. Ind.*, IX (6), p. 217, (1926), [On a wild plant][3]. Ramakrishna, *J. Bombay Nat. Hist. Soc.*, XXX, p. 868, (1925); *Pusa Ent. Mem. Dept. Agr. Ind.*, X, (7), p. 292, (1928), [In Jasmine flowers][4].

[2]Coimbatore.
[2]Kollegal
[1]Peradeniya,

---

[2]Java.

Haplothrips aculeatus, Karny, *Bull. Jard. Bot. Butitenzorg*, X, p. 13, 65, (1913) (partim); *Zeit, wiss, Ins. Biol.*, XI, p. 87, (1915) (partim).

gowdeyi (Franklin).

Haplothrips gowdeyi, Ramakrishna and Margabandhu *Indian J. Ent.*, I, (3), pp. 47-48, (1939), [On *leaves of Sola. num]*[1].

[1]Kallar: Foot of Nilgiris.

pictipes Bagnall.

Haplothrips pictipes, Bagnall, *Ann. Mag. Nat. Hist.,* (9), IV, p. 273, (1919)[1]. Ramakrishna, *J. Bombay Nat. Hist. Soc.,* XXX, p. 868, (1925); *Pusa Ent. Mem. Dept. Agr. Ind.,* X, (7) p. 292, (1928), [On diseased berries of black pepper][2]. Priesner, *Rec. Ind. Mus.,* XXV, Pt. III, p. 251, (1933), [References].

[1, 2]Taliparamba.

ramakrishnai Karny.

Haplothrips ramakrishnai, Karny, *Pusa Ent. Mem. Dept. Agr. Ind.,* IX, (6) p. 218, (1926), [Inside chrysanthemum flowers][1]. Ramakrishna, *Pusa Ent. Mem. Dept. Agr. Ind.,* X, (7), p. 292, (1928), [Notes][2]. Ramakrishna and Margabandhu, *J. Bombay Nat. Hist. Soc.,* XXXIV, 4, p. 1038, (1931). Priesner, *Rec. Ind. Mus.,* XXV, p. 361, (1933), [References].

[1, 2]Coimbatore.

soror Schmutz.

Haplothrips soror, Schmutz, *Sitz. Akad. wiss. Wien,* CXXII, p. 1039, (1913)[1]. Karny, *Bull, Ent. Res.,* XVI, p. 125, (1925)[2]; *Pusa Ent. Mem. Dept. Agr. Ind.,* IX, (6), p. 125, (1926), [Inside chrysanthemum and *Lantana* flowers][3]. Ramakrishna, *Pusa Ent. Mem. Dept. Agr. Ind., X, (7),* p. 292, (1928)[4].

[1]Peradeniya.

[3, 4]Coimbatore.
[2]Fiji, Indo-China.

sororcula Schmutz.

Haplothrips sorocula, Schmutz, *Sitz. Akad. wiss. Wien,* p. 1036, (1913)[1]. Karny, *Denkschr. Akad. wiss Wien.,* p. 16 (1921). Ramakrishna, *J. Bombay Nat. Hist. Soc.,* XXX, p. 868, (1925). Priesner, *Treubia,* II, 1, p. 7, (1921); *Rec. Ind. Mus.,* XXV, p. 351, (1933).

[1]Ceylon.

reuteri (Karny).

Anthemothrips reuteri, Karny, *Berl. ent. Zeit.,* LII, p. 51, (1907).

Haplothrips tritici, Preisner, *Sitz. Akad. wiss. Wien,* CXXVIII, p. 133, (1919).

Haplothrips reuteri, Priesner, *Treubia,* II, 1, p. 8, 14, (1921); *Kanowia,* IV, p. 152, (1925); *Teubia,* VIII, *Supplem,* p. 95, (1926); *Die Thys. Europas,* p. 582, (1928), [References]. Knechtel, *Thys, Roman.,* p. 223, (1923), John, *Ent. Mitt.,* XIV, p. 22, (1925). (*partim* ?) Moulton, *Ind. For. Rec.,* XIII, Pt. VI, p. 4, (1928), [Notes][1].

[1]Dehra Dun.

tenuipennis Bagnall.

Haplothrips tenuipennis, Bagnall, *Ann. Mag. Nat. Hist.*, (9), I, pp. 210-211, (1918), [In tea bushes and on rose][1]. Andrews, *Rept. Third. Ent. Meeting*, p. 621, (1920) . Ramakrishna, *J. Bombay Nat. Hist., Soc.*, XXX, p. 868, (1925); *Pusa Ent. Mem. Dept. Agr. Ind.*, X, (7), p. 292, (1928), [Pest on tea with *Physothrips setiventris*][2]. Blunck, *Handb. Phlanzenkrkht.*, p. 268, (1925), Priesner, *Die Thys, Europas*, p. 355, (1928), [References]. — [1,2]Darjeeling.

Adraneothrips tenuipennis, Bagnall, *Ann. Mag. Nat. Hist.*, (10), III, p. 606, (1929). Andrews, *Rept. Second. Ent. Meeting*, p. 27, (1917), ["Black Thrips of Tea"].

tirumalraoi Ramakrishna & Margabandhu.

Haplothrips tirumalraoi, Ramakrishna & Margabandhu, *J. Bombay Nat. Hist. Soc.*, XXXIV, 4, p. 1038, (1931), [On *Colocasia*][1]. — [1]Northern Circars.

58. **Podothrips** Hood.

Podothrips, Hood, *Insec. Inscit. Menstr.*, I, pp. 67-68, (1913), [Genotype *P. semiflavus* Hood]; *Psyche*, XXIII, pp. 8-9, (1916). Priesner, *Bull. Soc. Roy. Ent. Egypte*, p. 61, (1929). Ramakrishna and Margabandhu, *Indian J. Ent.*, I, (3), p. 48, (1939).

javanus Priesner.

Podothrips javanus, in Ramakrishna and Margabandhu, *Indian J. Ent.*, I, (3), p. 48, (1939). [On mango inflorescence and carried by wind][1]. — [1]Coimbatore.

59. **Gnophothrips** Hood.

Gnophothrips, Hood, *Journ. New York Ent. Soc.*, XXIII, p. 133, (1915), [Genotype *G. megaceps* Hood].

kabandha Ramakrishna.

Gnophothrips kabandha, Ramakrishna, *Pusa Ent. Mem. Dept. Agr. Ind.*, X, (7), p. 293, (1928), [Under bark of *Pithecolobium saman*][1]. — [1]Coimbatore.

60. **Euryaplothrips** Ramakrishna and Margabandhu.

Euryaplothrips, Ramakrishna and Margabandhu, *J. Bombay Nat. Hist. Soc.,* XXXIV, 4, p. 1037, (1931), [Genotype *E. crassus* Ramakrishna & Margabandhu].

crassus Ramakrishna & Margabandhu.

Euryaplothrips crassus, Ramakrishna & Margabandhu, *J. Bombay Nat. Hist. Soc.,* XXXIV, 4, p. 1038, (1931), [On *Amaranthus* Flower heads][1]. Priesner, *Bull, Soc. Roy. Ent. Egypte,* Fasc., 1-2, pp. 19-23, (1932), [Larvae and pupae described, on *Areva tomentosa*][2].

[1]Coimbatore.
[2]Egypt.

Sub-family IV.–TRICHOTHRIPNAE Karny.

61. **Brachythrips** Reuter.

Brachythrips, Reuter, *Act,* Soc., Faun. Flor, Fenn., XVII, 2, p. 17, 28, (1899), [Genotype *B. flavicornis* Reuter]. Karny, *Treubia,* I, 4, p. 249, (1921). Watson, *Flor-Agr. Exp. Sta., Bull.* 168, p. 19, (1923). Priesner, *Die. Thys. Europas,* p. 524, (1928).

dantahasta Ramakrishna.

Brachythrips dantahasta, Ramakrishna, *Pusa, Ent. Mem. Dept. Agr. Ind.,* X, (7), p. 294, (1928), [In leaf-galls of *Memexylon*][1].

[1]Mangalore.

dirghavadana Ramakrishna.

Brachythrips dirghavadana, Ramakrishna, *Pusa, Ent. Mem. Dept. Agr. Ind.,* X, (7), pp. 295-296, (1928), [From leaf-galls of *Ficus,* and on *Sesbania aegyptica*][1].

[1]Mysore, Gudiyattam.

perandaphaga Ramakrishna.

Brachythrips perandaphaga, Ramakrishna, *Pusa, Ent. Mem. Dept. Agr. Ind.,* X, (7), pp. 296-297, (1928), [On *Vitis quadrangularis*][1].

[1]Coimbatore.

62. **Austrothrips** Brethes.

Austrothrips, Brethers, *Ann. Mus. Buenos. Aires.,* 89, (1915), [Genotype *A. verae* Brethes].

cochinchinensis Karny.

Austrothrips cochinchinensis, Karny, *J. Siam. Soc.,* XXI, p. 113, (1923), [In leaf-galls of *Hymenodictyon*

[2, 3]Taliparamba.
[3]Tenmalai.

parviflorum][1]. Ramachandra Rao, *Agr. J. India*, p. 436, (1924), [Galls of *Calycopteris floribunda*][2]. Karny, *Pusa Ent. Mem. Dept. Agr. Ind.*, IX, (6), p. 219, (1926)[3]. Ramakrishna, *Pusa, Ent. Mem. Dept. Agr. India*, X, (7), p. 297, (1928), [In galls of *Cordia myxa*][4]. (Travancore) [4]Dehra Dun.

---

[1]Siam.

63. **Cecidothrips** Kieffer.

Cecidothrips, Kieffer, *Marcellia*, VII, p. 165, (1909), [Genotype *C. bursarum* Kieffer].

bursarun, Kieffer.

Cecidothrips bursarum, Kieffer, *Marcellia*, VII, p. 165, (1908), (Making galls][1].

64. **Trichothrips** Uzel.

Trichothrips, Uzel, *Mon. Ordnung Thys.*, p. 246, (1895), [Genotype *T. pedicularia* Haliday)]. [1]Trichinopoly.

hadrocerus Karny.

Trichothrips hadrocerus, Karny, *Pusa Ent. Mem. Dept. Agr. Ind.*, IX, (6), p. 220, (1926), [On a wild plant]. Ramakrishna *Pusa Ent. Mem. Dept. Agr. Ind.*, X, (7), p. 298, (1928)[2]. [1, 2]Sangleteri (Tinnevelly).

65. **Eothrips** Hood.

Eothrips, Hood, *Eng.*, XLVIII, p. 106, (1915), [Genotype *E.* crassicornis (Karny)].

aswamukha Ramakrishna.

Eothrips aswamukha, Ramakrishna, *Pusa Ent. Mem. Dept. Agr. Ind.*, X, (7), pp. 299-300, (1928), [On *Vitis quadrangularis* with *Brachythrips* perandaphaga][1]. [1]Coimbatore.

coimbatorensis Ramakrishna.

Eothrips coimbatorensis, Ramakrishna, *Pusa Ent. Mem. Dept. Agr. Ind.*, X, (7), pp. 298-299, (1928), [On a wild plant][1]. [1]Coimbatore.

foliiperda Karny.

Eothrips foliiperda, Karny, *Pusa Ent. Mem*[1]*., Dept. Agr. Ind.*, IX, (6), p. 222, (1926), [On leaves of a wild creeper][1]. Ramakrishna, *Pusa, Ent. Mem. Dept. Agr. Ind.*, X, (7), p. 298, (1928). [In leaf-galls of Pothos scandens)[2]. [1, 2]Taliparamba.

jambuvasi Ramakrishna.

Eothrips jambuvasi, Ramakrishna, *Pusa Ent. Mem. Dept. Agr. Ind.,* X, (7), pp. 300-301, (1928), [From leaf-galls on *Eugenia* sp][1]. — [1]Marudamalai (Coimbatore).

66. Eumorphothrips Schmutz.

Eumorphothrips, Schmutz, *Sitz. Akad. wiss. Wien.* CXXII, p. 1049, (1913), [Genotype *E. albicornis* Schmutz].

albicornis Schmutz.

Eumorphothrips, albicornis, Schmutz, *Sitz. Akad. wiss. Wien.* CXXII, p. 1050, (1913), [Cinnamon flowers][1]. Karny, Zeit. *Insektenbiol.,* XI, p. 145, 147, (1915). — [1]Peradeniya.

67. Oedemothrips Bagnall.

Oedemothrips, Bagnall, *Ann. Mag. Nat. Nist.,* (9), I, p. 20, (1918), [Genotype *Oe. porpinquus* Bagnall].

ceylonicus Karny.

Oedemothrips ceylonicus, Karny, Bull. *Ent. Res.,* XVI, p. 137, (1925)'. — [1]Peradeniya.

Subfamily V. CRYPTOTHERIPINAE Karny.

68. Androthrips Karny.

Androthrips, Karny, *Marcellia,* XI, p. 152, (1912) [Genotype *A. melastomae* (Zimmermamn]; *Zeitz, Wiss. Insektenbiol.,* X, p. 90, (1914) [Key to species].

coimbatorenis Ramakrishna.

Androthrip coimbatorensis, Ramakrishna, *Rec. Ind. Mus.,* XXXVI, pp. 497-498, (1934), [Collected on bamboon][1]. — [1]Coimbatore.

flavipes Schmutz.

Androthrips flavipes, Schmutz, *Sitz. Akad, wiss. Wien.* CXXII, p. 1031, (1913)[1]. Bagnall, *Ann. Mag. Nat. Hist.,* (8), XIII, pp. 27-28, (1918), [On Memexylon *umbellatum*].; *Ann. Mag. Nat. Hist.,* (8) XV. p. 324, (1925), [Synonymy], Karny, *Zeits, wiss. Insektenbiol.,* XI, p. 90, (1915); *Pusa Ent. Mem. Dept. Agr. Ind.,* IX, (6) p. 224, (1926), [Damaging leaves of *Ficus retusa* and *Mimusops, elengi,* and inside Psyllid gall, — [1]Peradeniya. [2,3]Kollegal, Coimbatore, Travancore.

and in *Terminalia* leaves][2]. Ramakrishna, *Pusa Ent. Mem. Dept., Agr. Ind.*, X, (7), p. 301, (1928)[2].

melastomae Zimmermann.

Mesothrips melastamae, Zimmermann, *Bull, Inst. Bot., Buitenzorg*, VII, p. 17, (1900).; van Leeuwen-Relinvaan, *Marcellia*, VIII, p. 108, (1909). — [1]Nilgiri Hills.

Androthrips melstomae, van Leeuwen-Reijnvaan, *Marcellia*, IX, p., 175, (without name); *Bull, Jard. Bol. Buitenzorg*, (2), XV, p. 6, 29, 30, 47, 48, (1914).; *Bull. Jard. Bot. Buitenzorg*, (2), XXI, p. 28, (? 1916). [without name]. Karny, *Centralbl. Bakt. Parasitenk.*, II, Abt., XXX, p. 560, (1911).; *Marcellia* , XI, p. 153, (1912); *Treubia*, III, 3-4, p. 314, (1923), [References]. Karny & van Leeuwen-Reijnvaan, *Zeitschr, wiss, Insektenbiol.*, X, .p. 202, 203, 205, 206, 207, 289, 290; XI, p. 90, 138, (1914-1915).; *Bull. Jard. Bot. Buitenzorg*, (2), XXI, . 28, (? 1916). Ramakrishna & Margabandhu, *Rec. Ind. Mus.* XLI, p. 31, (1939)[1].

ramachandrai Karny.

Androthrips ramachandrai, Karny, *Pusa Ent. Mme. Dept. Agr. Ind.* IX, (6) p. 226, (1926), [From *Calycoteris floribunda*][1]. Ramakrishna, *Pusa Ent. Mem. Dept. Agr. Ind.* X, (7), p. 301 (1928)[1]. — [1]Taliparamba, Tenmalai.

69. **Aleurodothrips** Franaklin.

Aleurodothrips, Franklin, *Ent. News*, XX, p. 228, (1909), [Genotype *A. fasciapennis* (Franklin).

fasciapennis (Franklin).

Cryptothrips fasciapennis, Franklin, *Proc., U.S. Nat. Mus.*, XXXIII, p. 727, (1908)[1]. — [1]Peradeniya.

Aleurodothrips fasciapennis, Franklin, *Ent. News*, XX, p. 228, (1909), *Moulton, U.S. Dept. Agr. Tech. Bull.*, 21, p. 29, (1911)[2]. Bagnall, *Ann. Mag. Nat. Hist.*, (8), XV, p. 324, (1915), [On the insect *Aspidiotus lataniae*][3]. — [2, 3]Florida., [3]Barbados.

70. **Dichaetothrips** Hood.

Dichaetothrips, Hood, *Proc. Biol. Soc. Wash.*, XXVII, pp. 164-165, (1914), [Genotype *D. brevicollis* Hood].

beesoni Moulton.

Dichotetothrips beesoni, Moulton, *Ind. For. Rec.,* XIII, Pt. VI, pp. 5-6, (1928), [From twigs of *Ficus religiosa* attacked by Scolytiae][1].

[1]Jhajra: (Dehra Dun).

71. Chromatothrips Schmutz.

Chromatothrips, Schmutz, *Sitz. Akad. wiss. Wien.* CXXII, p. 1043, (1913), [Genotype *C. fasciata* Schmutz].

annulicornis Schmutz.

Chromatothrips annulicornis, Schmutz, *Sitz. Akad. wiss. Wien.* CXXII, p. 1045, (1913).

[1]Peradeniya.

fasciata Schmutz.

Chromatothrips fasciata, Schmutz, *Sitz., Akad. wiss. Wien.,* CXXII, p. 1044, (1913)[1].

[1]Peradeniya.

plantaginis Schmutz.

Chromatothrips plantaginis, Schmutz, *Sitz. Akad. wiss. Wien.* CXXII, p. 1047, (1913)[1].

[1]Nuwara Eliya (Ceylon).

72. Gynaikothrips Zimmermann.

Gynaikothrips, Zimmermann, *Bull, Inst. Bot. Buit.,* VII, p. 12, (1900), [Genotype *G. uzeli* Zimmermann] Karny, *Zeit. f. wiss. Ins.,* X, p. 324, (1915), [Key to species].; *Pusa Ent. Mem. Dept. Agr. Ind.,* IX, (6), pp. 234-237, (1926), [Key to species]. Priesner, *Rev. Zool. Bot. Afr.,* XXII, Fasc., 2, p. 196, (1932), [Key to species].

chavicae Zimmermann.

Gynaikothrips chavicae, Zimmermann, *Bull. Inst. Bot. Buit.,* VII, p. 14, (1900). Karny & van Leeuwen-Reijnvaan, *Bull. Inst. Bot. Buit.,* X, p. 109, (1913). Karny, J. Siam. Soc., XVI, p. 137, (1923)[1].; *Pusa Ent. Mem. Dept. Agr. Ind.,* IX, (6), p. 234, (1926), [On Pepper leaf][2]. Ramakrishna, *Pusa Ent. Mem. Dept. Agr. Ind.,* X, (7), p. 302, (1928). [In leaf-galls of *Piper nigrum*][3].

[2]India.
[3]Wynaad: (S. India).

---

[1]Sumatra, Philippine Islands, Celebes Siam.

flaviantennatus Moulton.

Gynaikothrips flaviantennatus, Moulton, *Rec. Ind. Mus.,* XXXI, p. 98, (1929)[1].

[1]Balighai: (Orissa).

flavitibia Moulton.

Gynaikothrips flavitibia, Moulton, *Rec. Ind. Mus.,* XXXI, p. 99, (1929)[1]. — [1]Calcutta.

interlocatus Karny.

Gynaikothrips interlocatus, Karny, *Pusa Ent. Mem. Dept. Agr. Ind.,* IX, (6), p. 237, (1926), (Inside psyllid galls in *Terminalia* leaves][1]. Ramakrishna, *Pusa Ent. Mem. Dept. Agr. Ind.* X, (7), p. 302, (1928)[1]. — [1]Travancore.

kannani Moulton.

Gyncikothrips Kannani Moulton, J. Bombay Nat. Hist. Soc. XXXIII, 3, p. 667, (1929), [On *Eugenia sp.*][1]. ; Ramakrishna & Margabandhu, *J. Bombay Nat. Hist. Soc.,* XXXIV, p. 1040, (1931), [From leaf-galls on *Ficus refusa.* Key to species of the genus][3]. — [1]Bangalore. [2]Maddur: (Mysore).

karnyi Bagnall.

Gynaikothrips karnyi, Bagnall, *Ann. Mag. Nat. Hist.,* (8), XIII, pp. 28-29, (1914), [On marginal leaf-galls of *Piper nigraum*][1]. — [1]Peradeniya.

malabarica Ramakrishna.

Gynaikthrips malabarica, Ramakrishna *Pusa Ent. Mem. Dept. Agr. Ind.,* X, (7), pp. 302-303, (1928). [From leaves of *Ficus bengalensis* is company with *Gigantothrips elegans*][1]. — [1]North Malabar.

mirabilis Schmutz.

Gynaikothrips mirabilis, Schmutz. *Sitz. Akad. wiss. Wein.* CXXII, p. 1041, (1913), [On *pavetta hispida*][1]. — [1]Ceylon.

moultoni Ramakrishna.

Gynaikothrips moultoni, Ramakrishna, *Pusa Ent. Mem. Dept. Agr. Ind.,* X, (7), pp. 303-304, (1928), [On a wild shrub][1]. — [1]Marudamalai: (Coimbatore).

pallicrus Karny.

Gynaikothrips pallicrus, Karny, *Treubia,* III, p. 315, (1923). Ramakrishna & Margabandhu, *Rec. Ind. Mus.,* XLI, (1939), [On some wild plant glass][1]. — [1]Nilgiri Hills.

pallipes Karny.

Gynaikothrips pallipes, Karny, *Bull. Bot. Buit.,* X, p. 110 (1913).; *Treubia,* III, p. 314, (1923), [References][1] *Pusa Ent. Mem. Dept. Agr. Ind.,* IX, (6), p. 238, (1926), [In tender leaf rolls of pepper][2]. Ramakrishna, *Pusa Ent. Mem. Dept. Agr. Ind.,* X, (7), p. 304, (1928)[3].

[2, 3]Wynaad.

[1]Java.

uzeli Zemmermann.

Gynaikothrips uzeli, Zimmermann, *Bull. Inst. Bot. Buit.,* VII, p. 12, (1900), Hood *Insec. Inscit. Menstr., p. 153 (1913),* [Synonymy] , Bagnall, *Ann. Mag. Nat. Hist.,* (8) , XVII, p. 409, (1916), [Synonymy]. Karny, *Treubia,* III, p. 325, (1923) [References and synonymy][1]. *Pusa Ent. Mem. Dept. Agr. Ind.* IX, (6), p. 234, (1926), [Damaging the leaves of *Ficus retusa*][2]. Ramakrishna, *Pusa Ent. Mem. Dept. Agr. Ind.,* X, (7), p. 304, (1928), [On a wild shrub][3]. Moulton, *Rec. Ind. Mus.* XXXI, p. 100, (1929), [From *Ficus obtusa*][4].

[2]Kollegal.
[2, 3]Marudamalai: (Coimbatore).
[4]Barkuda: (Ganjam).

[1]Java, Algiers, Madeira, Canary Isle's Cuba Porto Rico, Florida.

watsoni Karny.

Gynaikothrips watsoni, Karny, *Ent. Mitt.,* IX, p. 104, (1920)[1].

[1]Bentota: (Ceylon).

73. Chiridothrips Ramakrishna & Margabandhu.

Chiridothrips, Ramakrishna & Marga-bandhu, *Rec. Ind. Mys.,* XLI, p. 32, (1939), [Genotype *C. indicus* Ram. & Marg.].

indicus Ramakrishna & Margabandhu.

Chiridothrips indicus, Ramakrishna & Margabandhu, *Rec. Ind. Mus.,* XLI, pp. 32-33, (1919), [From *Dolichos lab lab*][1].

[1]Coimbatore.

74. Ramakrishniella Karny

Ramakrishniella, Karny, *Pusa Ent. Mem. Dept. Agr. Ind.,* IX, (6), p. 232 (1926), [Genotype *R. unispina* Karny].

nirmalapkasha Ramakrishna.

Ramakrishniella nirmalapaksha, Rama-krishna, *Pusa Ent. Mem. Dept. Agr. Ind.,* X, (7), pp. 304-305, (1928), [From *Eugenia* sp and from a wild shrub.][1].

[1]Marudamalai (Coimbatore).

unispina Karny.

Ramakrishniella unispina, Karny, *Pusa Ent. Mem. Dept. Agr. Ind.,* IX, (6), p. 232 (1926), [On leaves of a

[1]Taliparamba.

wild plant][1]. Ramakrishna, *Pusa, Ent. Mem. Dept. Agr. Ind.*, X, (7), p. 304 (1928)[1].

75. **Cryptothrips** Uzel.

Cryptothrips, Uzel, *Mon. Ordnung. Thyps.*, p. 228 (part), (1895), [Genotype *C. latus* Uzel]. Hinds, *Proc. U.S. Nat. Mus.*, XXVI, p. 205, (part), (1902). Karny, *Zeit. wiss. Ins. Biol.*, p. 90, (1916); *Treubia* I, 4, p. 253, 255; (part), (1921). Hood, *Mem. Queens. Mus.*, VI, p. 142, (1918), [Notes]. Knechtel, *Thys, Roman.*, p. 200, (part), (1923). Watson, *Florida Agr. Exp. Sta., Bull.* 168, p. 21, 69, (part), (1923). Priesner *Treubia,* VI, *Suppl.*, p. 90, (1926).; *Die Thys, Europas,* pp. 484-485, (1928), [Notes on the genus].

longus Schmutz.

Cryptothrips (Mesothrips) longus, Schmutz, *Sitz. Akad. wiss. Wien.* CXXII, p. 1054, (1913). [On *Pavetta*][1]. — [1]Peradeniya.

Cryptothrips pavettae, Schmutz, *Sitz. Akad. wiss. Wien.* CXXII, p. 1055, (1913).

Cryptothrips sp. Kieffer, *Marcellia,* (1905).

76. **Neosmerinthothrips** Schmutz.

Neosmerinthothrips, Schmutz, *Sitz. Akad. wiss. Wien.* CXXII, p. 1052, (1013). Priesner, *Philip. J. Sic.*, LVII, pp. 369-370, (1935), [Characters and key]. [Genotype *N. fructum* Schmutz].

fructum Schmutz.

Neosmerinthothrips frudctum, Schmutz, *Sitz. Akad. Wiss. Wien.* CXXII, p. 1052, (1913)[1]. Ramakrishna, *Ind. For. Rec.*, XX. p. 12, (1934)[2]. — [1]Ceylon. [2]Coorg.

gloveri Ramakrishna & Margabandhu.

Necsmerinthothrips gloveri, Ramakrishna & Margabandhu, *Rec. Ind. Mus.*, XLI, pp. 31-32, (1939), [On *Zizyphus jujuba*][1]. — [1]Namkum: (Ranchi).

77. **Mesothrips** Zimmommann.

Mesothrips, Zimmeomann, *Bull. Inst. Bot. Buit.*, VII, p. 16, (1900). Karny, *Marcellia,* XI, p. 147, (1912);

Karny, *Zeit. f. wiss. Ins. Biol.*, p. 129, (1916), [Key to species]; *Treubia*, I, 4, p. 255, (1921). [Genotype *M. jordani* Zimmer-mann].

angusticornis Bagnall.

Mesothrips angusticornis, Bagnall, *Ann. Mag., Nat. Hist.*, (9), XIV, pp. 638-939, (1924)[1]. — [1]Peradeniya.

apatelus Karny.

Mesothrips apatelus, Karny, *Pusa Ent. Mem. Dept. Agr. Ind.* IX, (6), p. 229, (1926), [Damaging the leaves of *Ficus retusa*][1]. Ramakrishna, *Pusa Ent. Mem. Dept. Agr. Ind.*, X, (7), p. 305 (1928)[1]. — [1]Kollegal.

bhimabahu Ramakrishna.

Mesothrips bhimabahu, Ramakrishna, *Pusa Ent. Mem. Dept. Agr. Ind.*, X, (7), p., 306, (1928), [In leaf-galls of *Ficus retusa*][1]. — [1]Kollegal.

brevis Bagnall.

Mesothrips brevicollis, Bagnall, *Ann. Mag. Nat. Hist.*, (9) XIV, pp. 636-637, (1924), [Making galls on pepper][1]. — [1]Ceylon.

indicus Bagnall.

Mesothrips indicus, Bagnall, *Ann. Mag. Nat. Hist.*, (9) VII, pp. 365-366, (1921), [Bark of sal][1]. — [1]Dehra Dun.

karnyi Bagnall.

Mesothrips karnyi, Bagnall, *Ann. Mag. Nat. Hist.*, (9) IV, pp. 639-640, (1924)[1], Ramakrishna, *Pusa Ent. Mem. Dept. Agr. Ind.*, X, (7), p. 307, (1928), [Inside dead pods of *Albizzia lebbek*][2]. — [1, 2]Dehra Dun.

malloti Moulton

Mesothrips malloti, Moulton, *Ind. For. Rec.*, XIII, pp. 3-4, (1928), [In galls of *Mallotus philippinensis* along with *Liothrips brevitubus* and *L. bosei]*[1]. — [1]Dehra Dun.

melinocnemis Karny.

Mesothrips melinocnemis, Karny, *Pusa, Ent. Mem. Dept. Agr. Ind.*, IX, (6), p. 229, (1926), [On leaves of a wild creeper][1]. Ramakrishna, *Pusa Ent. Mem. Dept. Agr. Ind.*, X, (7), p. 307, (1928), [On galled leaves of *Pothos scandens*][2]. — [1, 2]Taliparamba.

78. **Karnyothrips** Watson.

Karnyia, Watson, *Fla. Ent.,* VI, No. 1, p. 6, (1922), [Genotype *K. weigeli* Watson].

Karnyothrips, Watson, *Florida Agr. Exp. Sta. Bull.* 168, p. 23, 70, (1923), [New name as the previous name is preoccupied]. Priesner, *Konowia,* XII, 3-4, p. 305, (1934), [Critical notes].

nigriflaves Ramakrishna.

Karnyothrips nigriflavus, Ramakrishna, *Rec. Ind. Mus.,* XXXVI, Pt. IV, pp. 496-497, (1934), [On bamboo with *Veerabahu thrips bambusae* Ram. *and Androthrips coimbatorensis* Ram][1]. — [1]Coimbatore.

79. **Euoplothrips** Hood.

Euoplothrips, Hood, *Mem. Queens land Mus.,* VI. p. 143, (1918), [Genotype *E. bagnalli* Hood].

malabarica Ramakrishna & Marga-bandhu.

Euoplothrips malabarica, Ramakrishna & Margabandhu, *J. Bombay Nat. Hist. Soc.,* XXXIV, 4, pp. 1039-1040, (1931), [On *Piper nigrum*][1]. — [1]Taliparamba and North Arcot. Dt.

80. **Eugynothrips** Preisner.

Eugynothrips, Priesner, *Treubia,* VIII, *Suppl.,* p. 157, (1926), [Genotype *E. conocephali* Karny].

priesneri, Ramakrishna.

Eugynothrips priesneri, Ramakrishna, *Pusa Ent. Mem. Dept. Agr. Ind.,* X, (7), pp. 307-308, (1928), [On a garden shrub][1]. — [1]Coonoor: (Niligiri Hills).

81. **Mallothrips** Ramakrishna.

Mallothrips, Ramakrishna, *Pusa Ent. Mem. Dept. Agr. Ind.* X, (7), p. 308, (1928), [Genotype *M. indica* Ram.].

indica Ramakrishna.

Mallothrips indica, Ramakrishna, *Pusa Ent. Mem. Dept. Agr. Ind.,* X, (7), pp. 308-310, (1928), [In leaf-galls of *Eugenia jambolana*][1]; *Rec. Ind. Mus.,* XXXVI, p. 498, (1934), [In *Eugenia jambolana fruits*][2]. — [1]Coimbatore. [2]Cawnpore.

Subfamily VI Macrothripinae Karny.

82. **Dinothrips** Bagnall.

Dinothrips, Bagnall, *Trans. Nat. Hist. Soc., Northumber.,* 1., N.S., III, p. 11, (1908), [Genotype *D. sumatrensis* Bagnall]. Karny, *Treubia,* I, 4, p. 282, (1921), [Notes and key to species].

sumatrensis Bagnall.

Dinothrips sumatrensis, Bagnall, *Trans. Nat. Hist. Soc. Northumber.,* III, p. 11, (1908). Karny, *Treubia,* III, p. 293, (1922)[1]; *J. Siam. Soc.,* XVI, p. 151, (1923)[2]; *Pusa Ent. Mem. Dept., Agr. Ind.,* IX, (6), p. 239, (1926), [Under bark of a wooden post in a shed][3]; *Philip. J. Sci.,* XVII, p. 203, (1920)[4]. Moulton, *Rec. Ind. Mus.,* XXXI, p. 100, (1929)[5].

Dinothrips furcifer, Schmutz, *Sitz. Akad. Wiss. Wien.* CXXII, p. 1026, (1913)[7]. Ramakrishna, *Ind. For. Rec.,* XX, pp. 6-7, (1934) [References. On sandal][6]. Karny, *Treubia,* III, p. 293, (1923).

[1]Ceylon, Burma, Bengal,[5] [3]Taliparamba. [4,7]Peradeniya., [6]Coorg. [1]Tonkin. [1, 2]Penang, Singapore, Sumatra, Nias, Mentawei, Engano, Java, Borneo Buru, Neuguinea. [2]Siam, Benghalis, Sab roang, Sarawak. [2, 4]Philippine Isles.

gardneri, Moulton.

Dinothrips gardneri, Moulton, *Ind. For. Rec.,* XIII, pp. 6-8, (1928), [Bred from logs of *Shorea robusta* attacked by Scoly-tidae (*Sphaerotypes*)][1]. [1]Dehra Dun.

83. **Ischyrothrips** Schmutz.

Ischyrothrips, Schmutz, *Sitz. Akad. wiss. Wien.* CXXII, p. 1074, (1913), [Genotype *I. crassus* Schmutz].

crassus Schmutz.

Ischyrothrips crassus, Schmutz, *Sitz. Akad. wiss. Wien.* CXXII, p. 1075, (1913), [On *Bauhinia triandra* shoots][1]. [1]Ceylon.

niger Schmutz.

Ischyrothrips niger, Schmutz, *Sitz. Akad. wiss. Wien.* CXXII, p. 1080, (1913), [Probably a *Liothrips*][1]. [1]Peradeniya.

obscurus Schmutz.

Ischyrothrips obscurus, Schmutz, *Sitz. Akad. wiss. Wien.* CXXII, p. 1074, (1913), [Karny thinks it a *Mesothrips*][1]. [1]Ceylon.

spinosus Schmutz.

Ischyrothrips spinosus, Schmutz, *Sitz. Akad. wiss. Wien.* CXXII, p. 1078, (1913).

84. Diaphorothrips Karny.

Diaphorothrips, Karny, *Ent. Mitt.,* IX, p. 186, (1920), [Genotype *D. unguipes* Karny].

unguipes Karny.

Diaphorothrips unguipes, Karny, *Ent. Mitt.,* IV, p. 186, (1920)[1]; *Treubia,* III, p. 299, (1923).

[1]Bentota: (Ceylon).

Family III IDOLOTHRIPIDAE Bagnall.

85. Dicaiothrips Buffa.

Dicaiothrips, Buffa, *Redia,* V, p. 169, (1908), [Genotype *D. scotti* (Heeger); *J. Linn. Soc.,* XXX, p. (1910), [Table with male characters of the species given]. Schmutz, *Sitz. Akad. Wiss. Wien.* CXXII, p. 1063, (1913), [Key to species]. Karny, *Philip. J. Sci.,* XVII, 2, p. 209, (1920), [Species of the genus distinguished]; *Treubia,* I, p. 258, (1921), [Characters].

bouvieri Vuillet.

Dicaiothrips bouvieri, Vuillet, *Bull. Soc. Ent. France,* p. 276, (1914)[1]. Ramakrishna, *Pusa, Ent. Mem. Dept. Agr. Ind.,* X, (7), p. 311, (1928)[2].

[1, 2]Coonoor: (Nilgiri Hills).

brunneitarsis Schmutz.

Dicaiothrips brunneitarsis, Schmutz, *Sitz. Akad. wiss. Wien,* CXXII p. 1070, (1913)[1].

Dicaiothrips brunneitarsis var. *levis,* Schmutz, *Sitz. Akad. wiss. Wien,* CXXII, 3, 1072, (1913)[2]. Karny, *Ent. Mitt.,* IX, 189, (1920)[3].

[1, 2]Peradeniya.
[3]Puttalam.

crassiceps Bagnall.

Dicaiothrips crassiceps, Bagnall, *Ann. Mag. Nat. Hist.,* (9), VII, p. 399, 1921)[1].

[1]Myawadi: (Burma-Siam frontier).

dallatorensis Schmutz.

Dicaiothrips dallatorensis, Schmutz, *Sitz. Akad. wiss. Wien.* CXXII, p. 1067, (1913)[1]. Karny, *Treubia,* II, p. 65, (1921)[2].

[1]Peradeniya.

---

[2]Java.

greeni Bagnall.

Dicaiothrips greeni, Bagnall, *Ann. Mag. Nat. Hist.* (8), XIII, pp. 289-290, (1914), [From decayed pods of *Phaseolus* sp.][1]. — [1]Peradeniya.

novus Schmutz.

Dicaiothrips novus, Schmutz, *Sitz. Akad. wiss. Wien.* CXXII, p. 1066, (1913)[1]. — [1]Ceylon.

procer Schmutz.

Dicaiothrips procer, Schmutz, *Sitz. Akad. wiss. Wien.* CXXII, p. 1063, (1913)[1]. — [1]Ceylon

proximus Bagnall.

Dicaiothrips proximus, Bagnall, *Ann. Mag. Nat. Hist.,* (8), XIII, p. 289, (1914), [Pods of *Crotalaria* sp.][1]. — [1]Peradeniya.

86. Cercothrips Hood.

Cercothrips, Hood, *Insec. Inscit. Menstr.,* VII, pp. 73-74, (1919), [Genotype *G. nigrodentatus* (Karny)].

tibialis (Bagnall).

Gigantothrips tibialis, Bagnall, *Ann. Mag. Nat. Hist.,* (9), VII, pp. 364-365, (1921), [On Careya arborea][1]. Karny, *Pusa, Ent. Mem. Dept. Agr. Ind.,* IX, p. 239, (1926), [On Careya arborea][2]. Ramakrishna, *Pusa Ent. Mem. Dept. Agr. Ind.* X, p. 311, (1928)[3]. — [1]Ceylon. [2, 3]Taliparamba. [4]Siruguppa: Bellary Dt.

Cercothrips tibialis, Ramakrishna, *Rec. Ind. Mus.,* XXXVI, p. 498, (1934), [In numbers on a wild shrub near the Tungabhadra river][1].

87. Gigantothrips Zimmermann.

Gigantothrips, Zimmermann, *Bull. Inst. Bot. Buit.* VII, p. 18, (1900), [Genotype *G. elegans* Zimmermann]. Bagnall, *Ann. Mag. Nat. Hist.,* (8), XVII, p. 403, (1916), [Synonymy]; *Ann. Mag. Nat. Hist.,* (9), p. 364, (1921), [Key to species]. Karny, *Treubia,* I, p. 258, (1921), [Characters].

Panurothrips, Bagnall, *Trans. N. H. Soc. North, Durham,* (N.S.), III, p. 208, (1908).

elegans Zimmermann.

Gigantothrips elegans, Zimmermann, *Bull. Inst. Bot. Buit.,* VII, p. 18, (1900)[1]. Van Leeuven-Reijnvaa, *Marcellia,* X, p. 75, (without name), (1911). Karny, *Centralbl. Backteriol.,* II, Abt. XXX, pp. 559-562, (1911); *Marcellia,* XI, p. 158, (1912); *J. Siam. Soc.,* XVI, p. 152, (1923), [Synonymy and notes][2]. *Pusa Ent. Mem. Dept. Agr. Ind.,* IX, (6), p. 239, (1926), [Damaging leaves of *Ficus retusa*][3]. Karny & Von Leeuwen-Reijnvaan, *Bull, Jard. Bot. Buit.,* (2), K, p. 4, 8, 117, (1913); *Zeits. wiss, Ins. Biol.,* XII, p. 189, (1916). Bagnall, *Ann. Mag. Nat. Hist.,* (8), p. 403, (1916), [*G. elegans* and *G. gracilis*). Uichanco, *Philip. J. Sci.,* XIV, (5). p. 547, (1919). Ramakrishna, *Pusa, Ent. Mem. Dept. Agr. Ind.,* X, (7). p. 360, (1928), [On different species of *Ficus*][4]. Moulton, *Rec. Ind.* Mus., XXXI, p. 100, (1929), [On *Ficus obtusa* and *F. glomerata*][5].

Panurothrips gracilis, Bagnall, *Trans. Nat. Hist. Soc. Northumberl.,* (N.S.), III, 1, p. 28, (1908). *Trans. Nat. Hist. Soc. Northumberl.,* (N.S.), III, 2, p. 3, (1910).

[3]Perur: (S. India).
[4]Coimbatore, Kollegal, Bangalore, Godaveri.
[5]Barkuda: (Ganjam Dt.).
[2]Pusa

---

[1]Siam, Java.
[2]Philipine Isles.

88. Elaphrothrips Buffa.

Elaphrothrips, Buffa, Priesner, *Rev. Zool., Bot. Afr.,* XXII, Fasc., 3, pp. 323-325, (1932), [Key to species], [Genotype not fixed].

achaetus Bagnall.

Elaphrothrips achaetus, Bagnall, *Ann. Mag. Nat. Hist.,* (10), XIII, p. 633, (1934)[1]. — [1]Ceylon.

agasthya Ramakrishna.

Elaphrothrips agasthya, Ramakrishna, *Ind. For. Rec.,* XX, Pt. IV, pp. 10-12, (1934), [On sandal][1]. — [1]Coorg, North Salem.

approximatus Bagnall.

Elaphrothrips approximatus, Bagnall, *Ann. Mag. Nat. Hist.,* (10), XIII, p. 635, (1934)[1]. — [1]Ceylon.

beesoni Ramakrishna.

Elaphrothrips beesoni, Ramakrishna, *Ind. For. Rec.,* XX, Pt. IV, pp. 7-9, (1934), [On sandal][1]. — [1]Coorg.

chandana Ramakrishna.

Elaphrothrips chandana, Ramakrishna, *Ind. For. Rec.,* XX, Pt. IV, pp. 9-10, (1934), [On sandal][1]. — [1]Coorg, North Salem.

coronatus Bagnall.

Elaphrothrips coronatus, Bagnall, *Ann. Mag. Nat. Hist.,* (10), XIII, p. 631, (1934)[1]. — [1]Ceylon.

graveleyi Bagnall.

Elaphrothrips graveleyi, Bagnall, *Ann. Mag. Nat. Hist.,* (10), XIII, p. 628, (1934)[1]. — [1]Hiamalayas: (India).

89. Phoxothrips Karny.

Phoxothrips, Karny, *Suppl. Ent. Berl.,* p. 132, (1913), [Genotype *P. pugilator* Karny].

breviceps Bagnall.

Phoxothrips breviceps, Bagnall, *Ann. Mag. Nat. Hist.,* (8), XIV, pp. 380-381, (1914)[1]. Ramakrishna, *Pusa Ent. Mem. Dept. Agr. Ind.,* X, (7), p 311. (1928)[2]. — [1, 2]Simla.

90. Kleothrips Schmutz.

Kleothrips, Schmutz. *Sitz. Akad. Wiss. Wien.* CXXII, p. 1057, (1913), [Genotype, *K. gigans* Schmutz].

Dracothrips, Bagnall, *Ann. Mag. Nat. Hist.,* (8), XIV, p. 289, (1914), [Characters]; *Ann. Mag. Nat. Hist.,* (9), I, p. 219, (1918).

ceylonicus Bagnall.

Dracothrips ceylonicus, Bagnall, *Ann. Mag. Nat. Hist.,* (8), XIII, p. 290, (1914), [Swept from bushes][1]; *Ann. Mag. Nat. Hist.,* (9), I, p. 219, (1918). — [1]Peradeniya.

gigans Schmutz.

Kleothrips gigans, Schmutz, *Sitz. Akad. Wiss. Wien.* CXXII, p. 1058, (1913), [On banana fruit][1]. — [1]Peradeniya.

subramanii Ramakrishna.

Kleothrips subramanii, Ramakrishna, *J. Bombay Nat. Hist. Soc.,* XXX, p. 788, (1925), [In leaves of palmyra — [1, 2]Chinglepet.

palm][1]; *Pusa, Ent. Mem. Dept. Agr. Ind.*, X, (7), p. 311, (1928)[2].

91. **Idoldothrips** Haliday.

Idolothrips, Haliday, *Hom. Ins. Mus.*, p. 1096, (1852), [Genotype *I. marginata* Haliday]. Froggatt, *Proc. Soc. N. S. W.*, Pt. 1, p. (1904). Bagnall, *Ann. Mag. Nat. Hist.*, XVII, p. 404, (1916), [Synonymy]. Karny, *Treubia*, I. p. 259, (1921), [Characters].

Acanthinothrips, Bagnall, *Trans. Nat. Hist. Soc. Northumberland & Durham*, (N. S.), III, p. 207, (1908), [Discusses the genus].

halidayi Newman.

Idolothrips halidayi, Newman, *Trans. Ent. Soc.*, p. 265, (1856)[1]. Ramakrishna, *Pusa Ent. Mem. Dept. Agr. Ind.*, X, (7), p. 311, (1928), [On *Anacardium* sp.][2]. — [1,2]Mysore.

schenklingi Karny.

Idolothrips schenklingi, Karny, *Ent. Mitt.*, IX, p. 189, (1920)[1]. — [1]Puttalam: (Ceylon).

Family IV MEGATHRIPIDAE Karny.

Subfamily I Bactrothripinae Karny.

92. **Bactrothrips Karny.**

Bactrothrips, Karny, *Zeit. f. wiss. Ins. Biol.,p. 116, (1918).* Bagnall, *Ann. Mag. Nat. Hist.*, (9), XVIII, pp. 557-568, (1926), [Key to species]. [Genotype *B. idolomorphus* Karny].

serraticornis Bagnall.

Bactrothrips serraticornis, Bagnall, *Ann. Mag. Nat. Hist.*, (9), VII, p. 397, (1921)[1]. — [1]Pundaluoya: (Ceylon).

Subfamily II Megatheripinae Karny.

93. **Veerabahuthrips** Ramakrishna.

Veerabahuthrips, Ramakrishna, *Rec. Ind. Mus.*, XXXIV, pp. 277-278, (1932), [Genotype *V. bambusae* Ramakrishna].

bambusae Ramakrishna.

Veerabahuthrips bambusae, Ramakrishna, *Rec. Ind. Mus.,* XXXIV, pp. 278-279, (1932), [In company with the insect *Asterolecanium* which is usually found on bamboo][1]. — [1]Coimbatore.

Family V HYSTRICOTHRIPIADE Karny.

94. **Leeuwenia** Karny.

Leeuwenia, Karny, *Marcellia,* XI, p. 161, (1912). *Treubia,* I, p. 260, (1921), [Genotype *L. gladiatrix* Karny].

coriaceus (Bagnall).

Panurothrips coriaceus, Bagnall, *Ann. Mag. Nat. Hist.,* (8), IX, pp. 216-217, (1912), [On leaves of *Eugenia operculata*][1]. — [1, 2]Dehra Dun.

Leeuwenia coriaceus, Bagnall, *Ann. Mag. Nat. Hist.,* (9), XIV, p. 640, (1924). Ramakrishna, *Pusa Ent. Mem. Dept. Agr. Ind.,* X, (7), p. 313, (1928)[2].

eugeniae Bagnall.

Leeuwenia eugeniae, Bagnall, *Ann. Mag. Nat. Hist.,* (9), XIV, p. 640, (1924)[1]. Ramakrishna, *Pusa Ent. Mem. Dept. Agr. Ind.,* X, (7), p. 313, (1928), [On *Eugenia jambolana*][2]. — [1, 2]Kodaikanal: (Palni Hills).

indicus Bagnall.

Leeuwenia indicus, Bagnall, *Ann. Mag. Nat. Hist.,* (8), XIV, pp. 277-378, (1914)[1]. Ramakrishna, *Pusa Ent. Mem. Dept. Agr. Ind.,* X, (7), p. 313, (1928)[2]. — [1, 2]Moulmein: (Burma Lower).

karnyi Ramakrishna.

Leeuwenia karnyi, Ramakrishna, *J. Bombay Nat. Hist. Soc.,* XXX, p. 791, (1925). [On *Eugenia jambolana*][1]. Ramakrishna, *Pusa Ent. Mem. Dept. Agr. Ind.,* X, (7), p. 313, (1928)[2]. — [1, 2]Bangalore.

www.ingramcontent.com/pod-product-compliance
Lightning Source LLC
LaVergne TN
LVHW020836091026
849796LV00003B/34
* 9 7 8 9 3 8 6 0 7 1 0 9 5 *